20 SIMPLE SOLAR PROJECTS

20 SIMPLE SOLAR PROJECTS

Fun, Practical, Attractive, Easy-to-Build Projects,
All Powered by the Sun

Edited by Ray Wolf

Written by Elizabeth Calhoun

Rodale Press, Emmaus, Pennsylvania

Printed in the United States of America on recycled paper containing a high percentage of de-inked fiber.

Library of Congress Cataloging in Publication Data

Calhoun, Elizabeth.
20 simple solar projects.

1. Solar energy—Amateurs' manuals. 2. Building—Amateurs' manuals. I. Wolf, Ray. II. Title.
III. Title: Twenty simple solar projects.
TJ810.C33 1983 621.47 83-9673
ISBN 0–87857–476–X hardcover

2 4 6 8 10 9 7 5 3 1 hardcover

The Rodale Press Design Center staff, shown above, developed all of the projects for this book. The designers are front row, left to right: *Dennis Kline, John Kline, and Jim Eldon;* back row, left to right: *Fred Matlack, Alan Tenenbaum, Philip Gehret, and Ed Wachter.*

Editor: Ray Wolf
Writer: Elizabeth Calhoun
Assistant Editor: Roger A. Moyer
Technical Illustrator: Frank Rohrbach
Secretary: Kim Greenawalt
Photography: Mitchell T. Mandel
Pat Seip
Ray Wolf
Cover Design: Barbara Field
Book Design: Jerry O'Brien

Copy Editor: Felicia D. Knerr
Project Designers: Jim Eldon
Philip Gehret
Dennis Kline
John Kline
Fred Matlack
Alan Tenenbaum
Ed Wachter
Thermal Engineer: Robert Flower
Project Testing and Evaluation: Harry E. Wohlbach

Contents

Foreword

Simple solar . . . to many that may sound like a contradiction. Many people feel solar energy is complex, expensive, and the exclusive province of those with special scientific training. Nothing could be further from the truth. There are practical, inexpensive, and simple ways to use solar energy, and you will find 20 of them in this book.

The projects we've selected, designed, and built for this book run the gamut of solar possibilities. Our goal was to find a mix of projects that would enable everyone to find at least one project that interested them. Every project in this book was conceived, designed, built, rebuilt, and in some cases rebuilt again by the staff of designers and carpenters at the Rodale Press Design Center. We know that every project works, and we believe that every project can be built from the directions in this book.

The projects were designed not only to appeal to you, but to teach you something about solar energy as well. Thus, the principles involved in the various projects are all illustrative of more complex uses of solar energy.

Not only are the operating principles behind the projects simple, but the construction of the projects is equally as simple. I think that illustrates an important point: that the use and construction of solar installations can be quite simple. A lot of the solar installations you read about in the news are quite complex and large-scale. However, there is a great deal of work being done on small-scale, simple solar applications. It is that kind of application we want to expose you to in this book.

Our goal throughout this book was to assemble a wide variety of projects that would demonstrate the entire spectrum of potential solar applications. To that end, the projects range from a simple solar clock (the slate sundial) to a solar electric battery charger. But, in every project the operating principles remain simple. There are no complex collectors and heat storage systems. In most cases, the sun shines on the target area, and the heat is used to accomplish a chore, be it generating electricity or telling time.

However, that does not mean that solar energy must be relegated to the world of the sundial. Not by a long shot.

The projects presented here are only the tip of the iceberg. The promise of what lies ahead for solar energy is truly mind-boggling. However, every journey must begin with a single step, and the projects in this book can start you on your journey to a more rewarding lifestyle—a lifestyle based on solar energy.

The sun offers us the promise of a clean, safe, and unlimited, but unique, source of energy. No one can overpower or control the sun. No amount of manhandling or technology will force the sun to do as we wish.

Of the 20 projects, one stands out in my mind as representative of the philosophy behind all of the projects. The project is the solar barbecue. When the designer, Dennis Kline, first suggested it to me, I quickly dismissed it, saying it wasn't feasible and, even if it were, it could never be made simple. Dennis made me eat my words in more ways than one. The final project has a set of reflectors, a griddle, and focusing linkage. You can cook a steak or bake a potato on it on a sunny day. I know; I've done it.

To me, a steak cooked on a solar barbecue tastes better than a steak cooked on a conventional stove. I know there is no technical difference, but to me there is still a very real perceived difference. Skeptics will cry that it isn't cost-effective to build a solar barbecue, and I agree, if all you look at is dollars. But I find this unique invention to be utterly fascinating, and I have a hard time putting a price on that. In fact, I get a lot of enjoyment out of its uniqueness. The Wright brothers' first airplane was not very cost-effective either, but look at what it has done to today's lifestyle. In a small way, I like to think of the solar barbecue in the same light. It is the beginning of a technological revolution.

When my children are my age, I hope they will be able to look back at some of the projects in this book, as I look back at equipment of 30 years ago, and laugh at how primitive our ideas, materials, and approaches were.

I believe the use of solar energy will continue to grow both in practicality and in application. I would like this book and its projects to be viewed as a starting point, as one of the early efforts to demonstrate that the use of solar energy is not only feasible, but is also attainable by average individuals.

Introduction

This is a book of projects. It is designed to be used—to be taken into your shop and to be your helper when building a project. But, before you go off to your shop to start your first solar project, there are a few things you should know about using the book.

Before you cut your first board, pound your first nail, or tighten your first screw, we recommend that you read through an entire project write-up to become familiar with what you will be doing. It is important that you understand how early steps may influence later steps. If you decide to change a procedure, you should know exactly what the ramifications of that change will be further along in the project. Don't start working until you know what you will be doing and why.

Every project has a materials list. These lists are designed to act as guides, not as exact shopping lists. Many of the projects are designed to be custom-sized to your house, window, roof, or what-have-you. With the more involved projects, there are separate sizing charts to help you calculate the size of each piece. On other, simpler, projects, the sizing chart is included in the materials list. Before buying any materials, go through the materials list and the text, and mentally build the project to get an idea of how much you will need of each material.

We have tried to use commonly available materials wherever possible. You should be able to buy the majority of the items you will need from a lumberyard and a good hardware store. When shopping for materials, don't overlook quality in quest of a bargain. The projects have been designed to last a long time, if you use quality materials.

Although we tried to use readily available materials in the projects, in some cases, obviously, we had to use specialized solar materials. These materials are generally available. Consult the list of solar suppliers below, and get catalogs from several. All of the companies listed specialize in mail-order sales and have good reputations. In addition, in some cases we have listed the source we used for a particular material at the end of the materials lists. For those of you on the East Coast, this may be helpful, but readers in the rest of the country should consult the solar suppliers list and do their shopping by mail. If you are lucky enough to have a retail solar store near you, shop there. Give them the parameters of what the project requires, and they should be able to help you out.

Every project also has a tools list. This includes only the special tools you will need. We assume you have a basic complement of hand tools, so items such as a hammer, tape measure, pencil, screwdrivers, and so on are not listed, nor are finishing tools.

We urge you to be careful when building a project. In some of our photographs, safety guards have been removed for clarity, but you should always use all guards and safety devices. Nothing can ruin a building experience like an injury, so please pay attention to safety. We have tried to simplify procedures and use the safest techniques we could find, but if you are not familiar with a recommended procedure, or do not feel you can do it safely, don't try it. Either get a friend to help out, or find another method. You will be working with power tools and should exercise proper respect for the tools when you use them.

Some of the projects will require a helper, either to hold material or to help lift and position items. Don't overextend yourself. Read through the instructions before starting and decide if you will need help, and, if so, make plans in advance.

You should now be ready to build your first solar project. Have fun.

Peoples' Solar Sourcebook
Solar Usage Now, Inc.
P.O. Box 306
420 East Tiffin Street
Bascom, OH 44809

Solar Age Catalog
P.O. Box 4934
Manchester, NH 03108

Solar Components Corporation
P.O. Box 237
Manchester, NH 03106

Solar Engineering and Contracting Master Catalog and Buyer's Guide
P.O. Box 3600
Troy, MI 48099

The Solar Wonderbook
28 Wade Court
Gaithersburg, MD 20878

Zomeworks Corporation
P.O. Box 25805
Albuquerque, NM 87125

1 Slate Sundial

A sundial can be an elegant addition to your garden or patio. Not only does it tell time by marking the sun's path through the sky, it also reminds us of the constancy of the sun's energy. Our first project is a horizontal, or polar, sundial, a duplicate of a centuries-old style of sundial.

The dial plate, the level surface on which the hours are scribed, characterizes the horizontal sundial. The gnomon, a triangular fin pointing north, casts a shadow onto the dial plate. The sundial's gnomon is angled to show solar time at one particular latitude; the spacing of the hour lines also varies with the latitude. The angle of the gnomon and

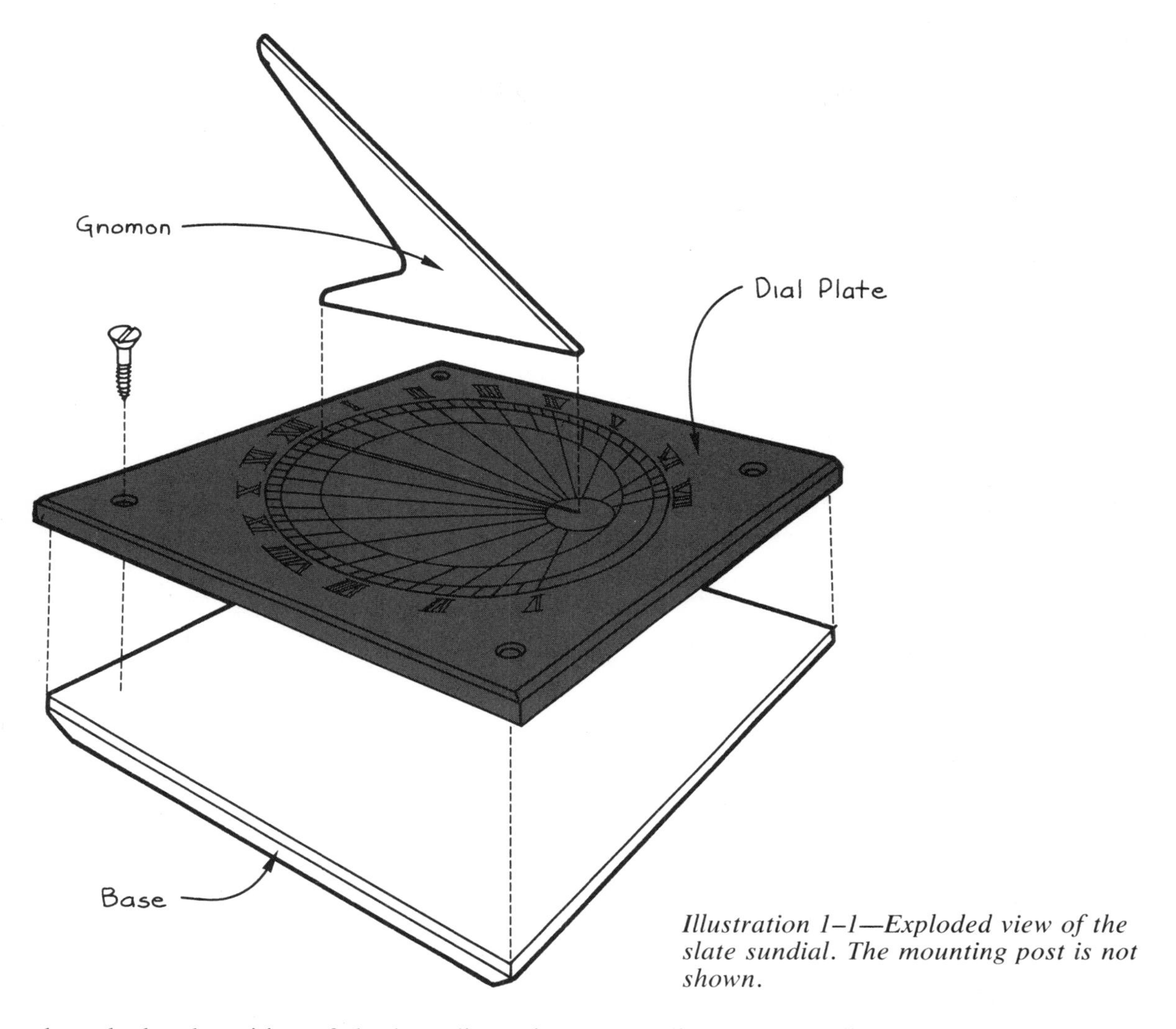

Illustration 1–1—Exploded view of the slate sundial. The mounting post is not shown.

the calculated position of the hour lines give the sundial accuracy all year long.

The beauty of this sundial lies in the contrasting materials—bright brass against dark slate—and the simplicity of the design. The shadow cast by the brass gnomon shows up clearly against the hour lines scribed into the slate.

The sundial is made from a piece of ordinary slate and a piece of brass plate. It is mounted on a plywood base that is screwed to a post set into the ground. The materials are durable and allow for personalization of the design. Constructing this project will reward you with an unusually good-looking addition to your landscape.

Chart 1–1 lists the materials for the sundial. The tools you will use are listed in chart 1–2. An exploded view of the slate sundial is shown in illustration 1–1.

The first part of the horizontal sundial to make is the dial plate. We used slate for this piece because it has a durable surface and is easy to shape, scribe, and polish. Start the sundial's construction by cutting a piece of ⅜-inch-thick slate to 9 × 9 inches, as shown in photo 1–1, using almost any saw that will cut slate, including a hacksaw. If your slate is rough, use either a belt sander or hand sanding block to smooth

CHART 1–1—
Materials

DESCRIPTION	SIZE	AMOUNT
	Lumber	
Redwood, Cedar, or Pressure-Treated Pine	4 × 4 × 60″	1
A-C Exterior Plywood	¾″ × 8¾″ × 8¾″	1
	Hardware	
#12 Flathead Wood Screws	2½″	4
#10 Brass Flathead Wood Screws	1″	4
	Miscellaneous	
Slate	⅜″ × 9″ × 9″	1
14-Gauge Brass	6″ × 6″	1
Epoxy	...	1 oz.
Terrazzo and Slate Sealer	...	1 qt.
Primer	...	1 pt.
Flat Black Enamel	...	1 pt.

CHART 1–2—
Tools

Tools
Hacksaw
Saber Saw
Drill and Bits
Plane
Level
Scriber
Files
Protractor
Compass
Dividers or Trammel Points
Buffing Wheel or Fine Steel Wool

Photo 1–1—Cut the slate to 9 × 9 inches with a hacksaw or any other saw that will cut slate.

the top and edges. Start with a coarse grit, and work your way through the medium and fine grits to achieve a smooth surface. Finish the plate, including the edges, with 400- and 600-grit wet/dry paper. Use sandpaper wrapped around a small block of wood or a file to bevel the corners of the top edge to about 45 degrees.

Lay out the design shown in illustration 1–2 on the slate with a pencil. Find the center of the plate by drawing two diagonal lines from opposing corners. Their intersection is the center point. Draw a squared line, which we designate the north/south line, through the center point from one side to the other.

Use the center point and compass to draw three concentric circles on the dial plate. The outer circle has a radius of 3¾ inches, the middle circle has a radius of 3¼ inches, and the inner circle's radius is 2½ inches.

Designate one end of the centerline *south* and the other end *north*. Measure from the center point along the centerline to the south 1¾ inches, and mark a second point. This point will be the focal point. Illustration 1–3 shows how to find this point, using the center point. Using this second mark as a center point, draw a circle with a radius of ¾ inch. This circle should just touch the inner circle.

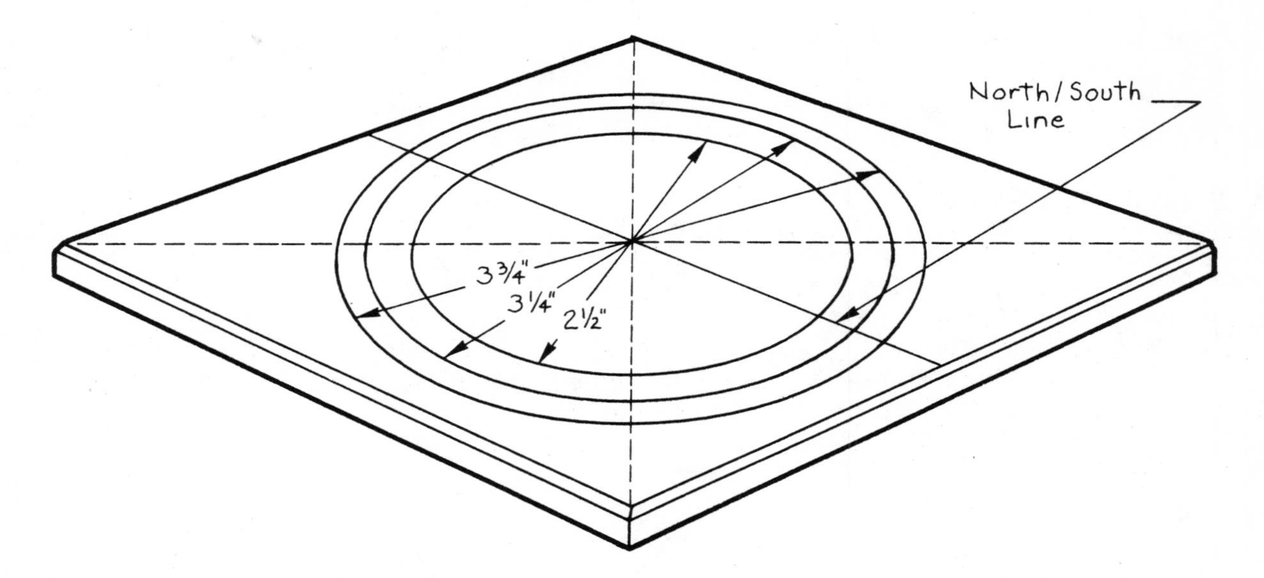

Illustration 1–2—Lay out the three concentric hour-line circles and the north/south line, as shown.

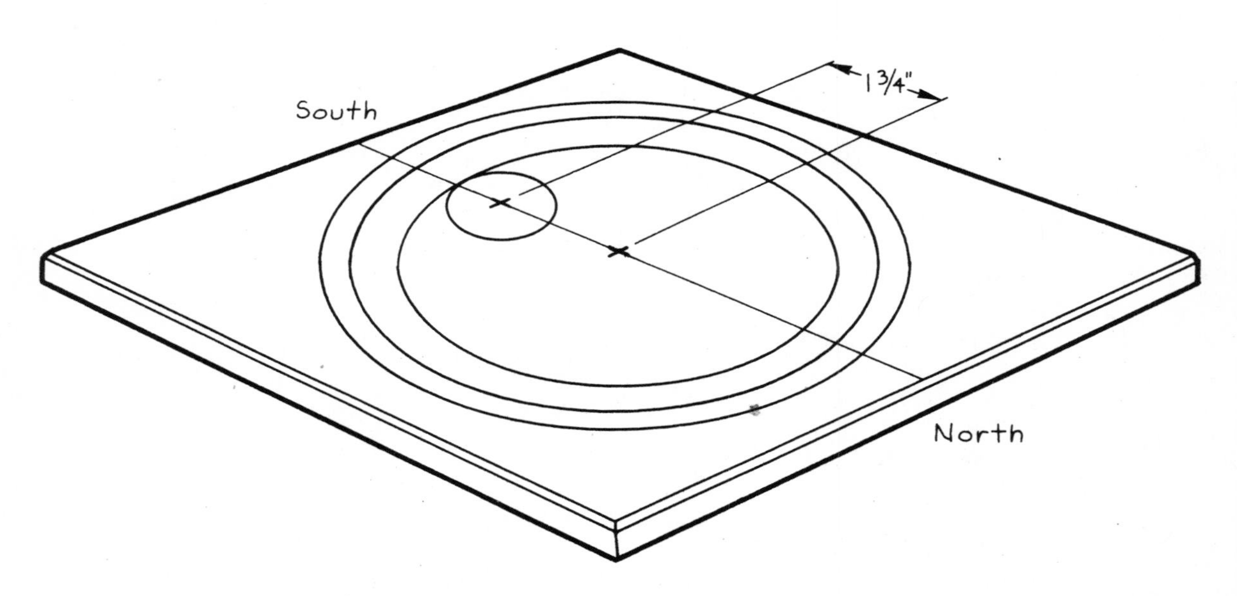

Illustration 1–3—The focal point is 1¾ inches from the center point.

Each sundial should be designed specifically for its own latitude. You can find your latitude by looking on a road map or asking a local surveyor. Both the angle of the gnomon and the layout of the hour marks are determined by the latitude at which the sundial will operate.

Find the latitude closest to your location on chart 1–3, and plot the lines marking the hours and half hours, using a protractor and the angles given. The angles at which to draw the time lines are listed under the latitudes.

The hours are given from 6:00 A.M. standard time until noon in half-hour increments. From noon until 6:00 P.M., the divisions mirror exactly the morning angles, and the same angles are used, but in reverse. The hour lines before 6:00 A.M. and after 6:00 P.M. are located by continuing the appropriate lines from the opposite side of the focal circle.

To draw the dial-plate lines, set your protractor with the center point on the focal point (not the center point) of the dial and the 90-degree line on the north/south line, as shown in photo 1–2. Start by marking 6:00 A.M. to your left and 6:00 P.M. to your right at 0 degrees. Use the hour-line angles from chart 1–3 to mark each half-hour and hour

CHART 1–3—

Hour-Line Angles for Differing Latitudes

		Latitudes					
A.M.	P.M.	30°	35°	40°	45°	50°	55°
6:00		0°	0°	0°	0°	0°	0°
6:30	5:30	15°	13°	12°	10½°	10°	9°
7:00	5:00	30°	25°	23°	21°	19½°	18°
7:30	4:30	40°	36°	34°	30°	28½°	27°
8:00	4:00	49½°	44½°	42°	39°	37°	35½°
8:30	3:30	57°	53°	50°	47°	45°	43°
9:00	3:00	63½°	60°	57°	54½°	52°	51°
9:30	2:30	69°	66°	64°	61½°	59½°	58°
10:00	2:00	74°	72°	69½°	68°	66°	65°
10:30	1:30	78½°	77°	75°	73½°	72½°	71°
11:00	1:00	82°	81½°	80°	79°	78°	77½°
11:30	12:30	86°	86°	85°	84½°	84°	84°
Noon		90°	90°	90°	90°	90°	90°

Photo 1–2—Mark the hour lines on the dial plate, using a protractor and the hour-line angles for your location from chart 1–3.

Photo 1–3—Scribe the curved lines with trammel points, as shown, or a strong compass. Use the center point of the dial plate, not the focal point.

line until you reach noon at 90 degrees. Then, mark the quarter hours by accurately measuring halfway between the hour and half-hour lines.

Draw a pencil line extending from the outer circle to the focal point circle, using a straightedge with one end on the focal point and the other end on the appropriate degree mark for each degree mark that corresponds to an hour line. The half-hour lines connect the outer circle with the inner circle, with the straightedge still on the focal point. The quarter-hour lines, halfway between, connect the outer circle with the middle circle, always directing the lines toward the focal point. Your completed pattern should closely resemble the one shown in illustration 1–4.

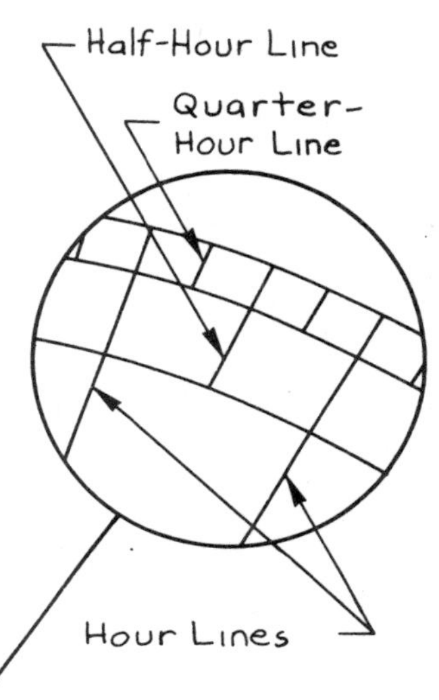

When the design has been laid out, you can scribe the lines into the slate. Use a scriber or any other sharp scratching tool and a straightedge for the straight lines. Scribe the curved lines with dividers or trammel points (two metal pins connected to a thin board), as shown in photo 1–3. You will find the slate scratches easily, so be careful to set your tools correctly at the start. Go over each line several times until it is clear; you will find that a line of a width and depth of 1⁄32 inch is easily visible.

The Roman numerals are a handsome decorative element to add to the dial plate. We show them full size in illustration 1–5. To give you guidelines for positioning the numerals, draw, but do not scribe, two additional circles in pencil on the dial plate. The radius of one circle is 3⅞ inches and of the other, 4⅜ inches, measured from the center point, not the focal point. Again—do not scribe these circles into the slate.

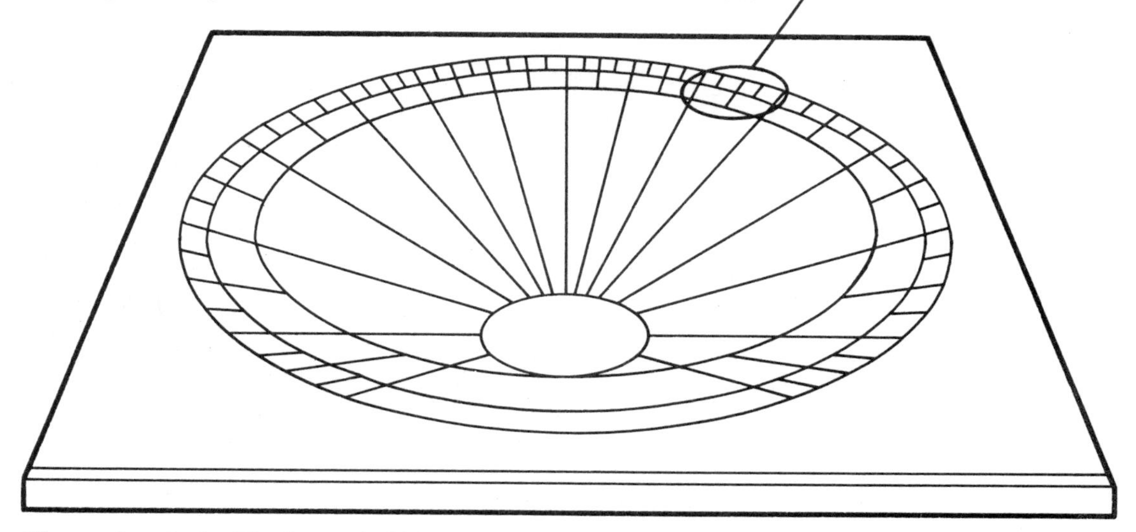

Illustration 1–4—The hour, half-hour, and quarter-hour lines are laid out on the dial plate, as shown.

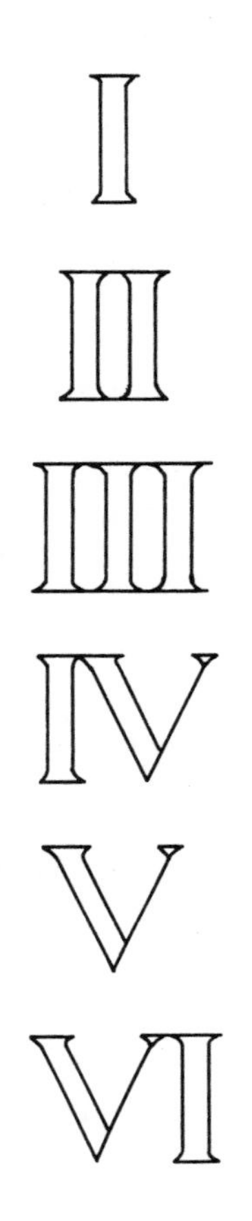

Illustration 1–5— Above and on the page opposite. *Roman numeral patterns.*

Trace the Roman numerals, and transfer them to the dial plate with carbon paper. However, if you want the numbers to continue the fan design of the hour lines, follow a more challenging procedure. Draw each Roman numeral individually, with a straightedge and pencil, having the straightedge intersect the focal point for all vertical lines.

You are marking the sundial in standard time, from 5:00 A.M. to 7:00 P.M., and you will read it clockwise around the dial plate with the morning numbers on the left, as you look from the south side. Remember that 6:00 A.M. and 6:00 P.M. are directly opposite each other and perpendicular to the north/south line. Using a straightedge, scribe the Roman numerals onto the dial plate.

To mount the dial plate on its base, you need four 7/32-inch-diameter holes. Lay out, drill, and countersink these holes, one at each corner of the plate, 5 inches from the center point.

The gnomon is shaped from a 6 × 6-inch piece of 14-gauge brass plate. Check to be certain that at least one edge is perfectly straight. It is not necessary to finish the edges at this time, provided they are square.

Draw a line 1/8 inch from the straight edge of the piece, as shown in illustration 1–6. This compensates for the depth that the gnomon will be mounted into the dial plate.

Position the plate with the edge that will become the base toward you, and mark the angle of your latitude, using a protractor with the center point at the lower left-hand corner of the plate and the 0-degree

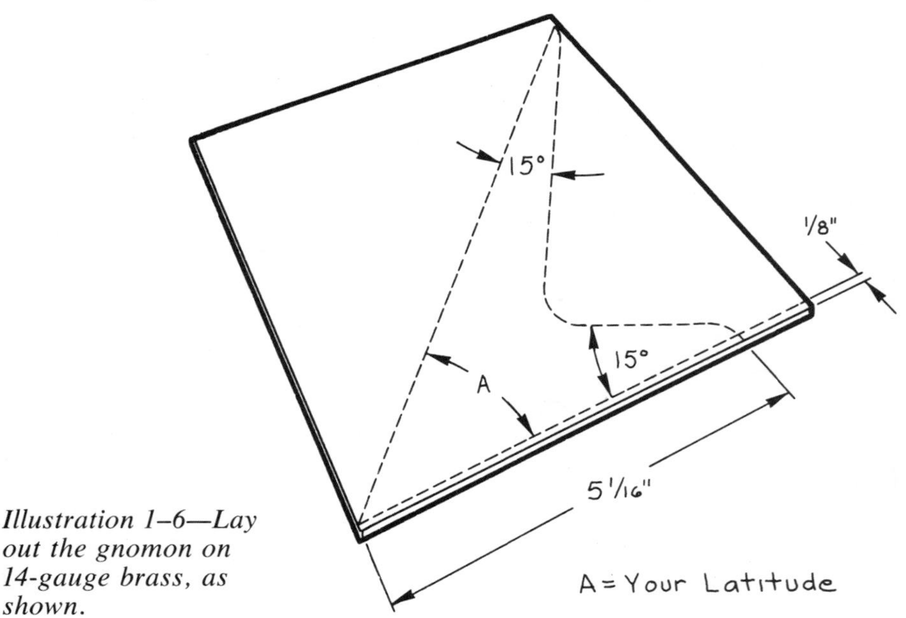

Illustration 1–6—Lay out the gnomon on 14-gauge brass, as shown.

line on the depth line, 1/8 inch from the bottom. Draw a latitude line from the depth line to the latitude mark you made.

Measure 5 1/16 inches from the lower left-hand corner along the depth line to mark the length of the gnomon base. The southern point of the gnomon will be set in the center of the focal circle, while the northern end of the gnomon will touch the middle circle line.

The shape of the back edge of the gnomon is optional and can be any design you wish, provided the base is the correct length and the latitude angle is maintained.

For the design shown, draw straight lines angled at 15 degrees from the depth line at the 5 1/16-inch mark and at the top of the latitude line. This shape results in a general guide, which can be altered to suit your taste. We have thickened the points to 1/8 inch at the top and 1/4 inch at the base. The intersection point is rounded with a 3/4-inch radius. Finish the shape by slightly rounding off the latitude line point and also the base where it meets the depth line. Shape the northern end of the base until it matches the distance between the center of the focal circle and the edge of the inner circle.

Cut the gnomon to shape with a hacksaw. Smooth the edges with a flat file, and shape the transitions from straight to curved lines with a round file. Finally, polish the gnomon with a buffing wheel or fine steel wool.

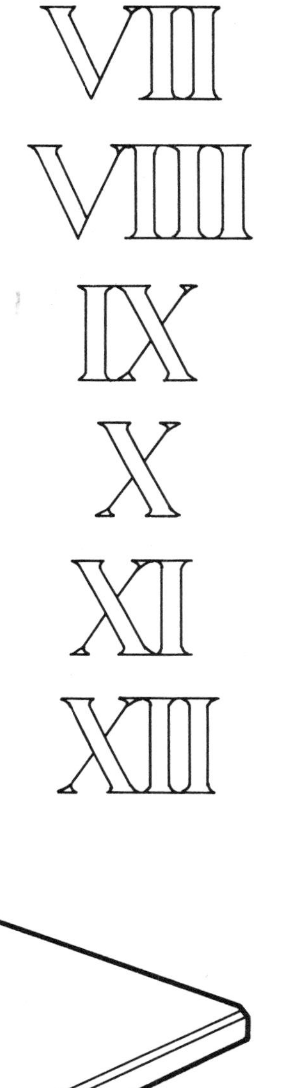

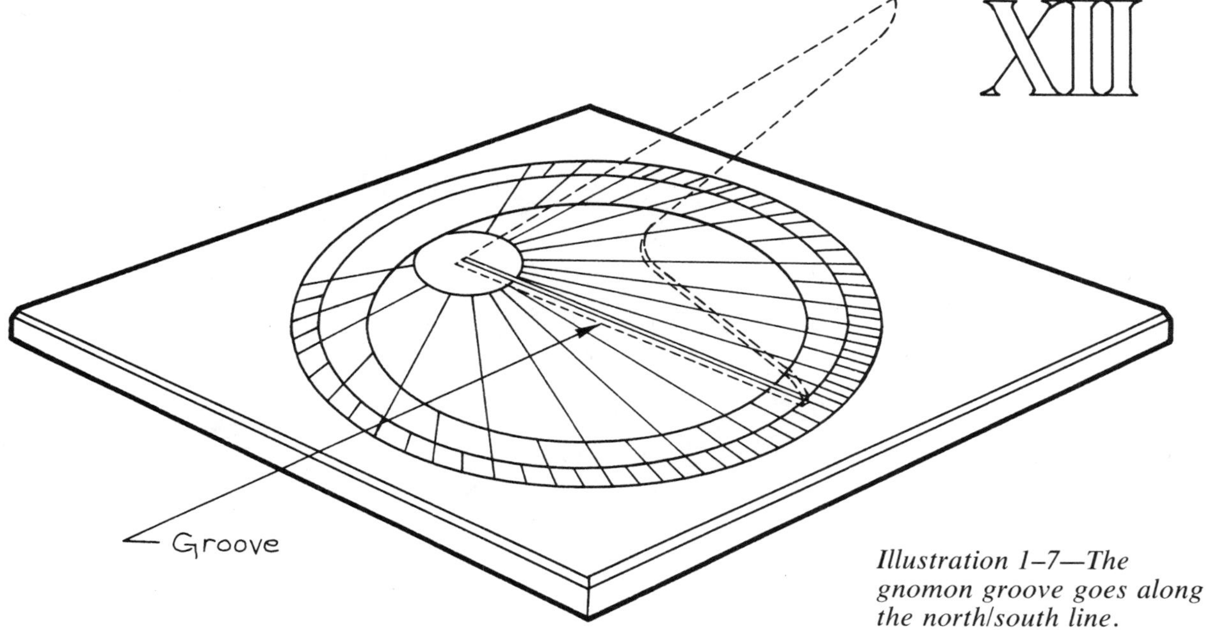

Illustration 1–7—The gnomon groove goes along the north/south line.

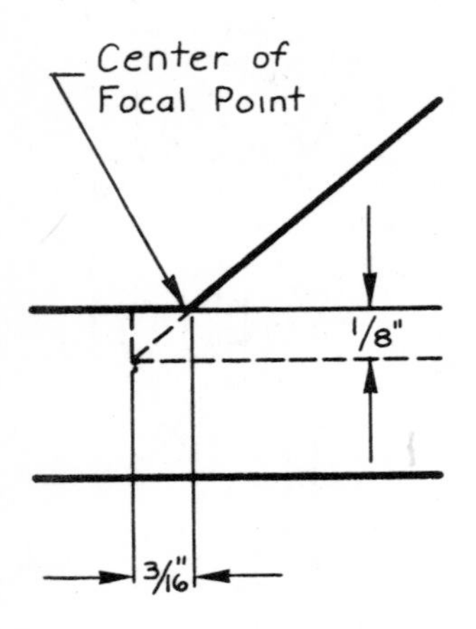

*Illustration 1–8—
Gnomon point details.*

Scribe a 5¹⁄₁₆-inch groove along the north/south line to hold the gnomon on the dial plate, as shown in illustration 1–7. Use a straightedge and scriber to cut out the groove for the gnomon. Gradually make the groove longer, wider, and deeper, trial-fitting the gnomon frequently, until it fits to a depth of ⅛ inch. The gnomon is positioned so that it is vertical and the latitude edge intersects the surface of the dial at precisely the center of the focal circle. See illustration 1–8. Fasten the gnomon to the dial plate, using epoxy. Support it at precisely 90 degrees to the dial plate while the epoxy hardens. Use a combination square to be sure the angle is correct, as we show in photo 1–4.

Seal and finish the dial plate with several coats of sealer for terrazzo and slate, such as tung oil.

Now you are ready to construct the base and mount for your sundial. The base is slightly smaller than the dial plate and chamfered, or angled, on the lower edge to set off the dial plate. From ¾-inch exterior plywood, cut a square 8¾ × 8¾ inches. Mark a line on one side ½ inch from each edge. Chamfer the base to the ½-inch line with a hand plane or a saw set for an angle cut.

The base will be screwed onto a mounting post, as shown in illustration 1–9. To position the screw holes, mark the center point of the base on the unchamfered side. Measure out 1⅜ inches from this point

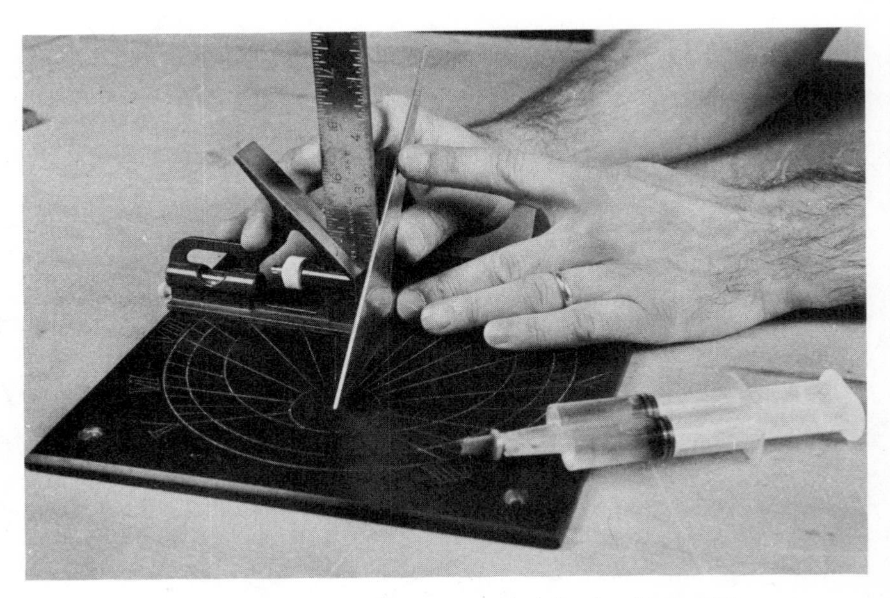

Photo 1–4—Apply epoxy to the gnomon and, checking with a square, set it precisely at 90 degrees to the dial plate.

toward each corner. Drill four ¼-inch-diameter holes at these points, and then countersink the holes.

To join the dial plate to the base, set the dial plate on the unchamfered side of the base, being sure to align the pieces evenly on all sides. Mark through the dial plate pilot holes at each corner. With the dial plate removed, drill four ⅛-inch-diameter holes in the unchamfered side of the base. Prepare and paint the base with primer followed by two coats of flat black paint. Mount the plate on the base with four 1-inch #10 brass flathead wood screws.

Sundials are usually mounted on a post so they are more visible, and they should be set in a sunny, open spot. A redwood, cedar, or pressure-treated 4 × 4, or any other material that is durable and rot-resistant, may be used for the post. Paint wood posts with creosote or some other wood preservative on the section that will be in the ground. Cut the post to a length of 32 inches plus the depth of the frost line for your area. You may want to round the corners with a hand plane to give the post a more finished look. Then set the dial base onto the post, and mark the screw holes. Drill four ⅛-inch-diameter pilot holes in the top of the post, then mount the base on the post with four 2½-inch #12 flathead wood screws.

Choose a location that receives sun all day long, if possible. Plan to direct the sundial toward solar south. To find solar south, find the exact times of sunrise and sunset. Contact your local weather service, or check your local newspaper. Note the exact times, and find the midpoint between the two. For instance, if the time of sunrise is 6:00 A.M. and sunset is 7:30 P.M., then solar noon is 12:45 P.M. Solar noon will change slightly every day.

Dig a hole to the depth of your frost line, and place the post in the hole. Level the base in both directions so that the gnomon will have the correct angle when it is in place. Assemble the sundial, base, and post, as shown in illustration 1–9, and then orient the sundial to solar south by lining up the gnomon-cast shadow to fall exactly on the 12 o'clock line at solar noon. Brace the post in two directions with temporary boards tacked to the post and pushed into the ground. Make sure they are long enough not to interfere with filling the hole. Fill the posthole with small rocks and dirt, tamping it firmly every few inches.

Your sundial is now set to solar time, which is to say that the sundial time will not necessarily correspond to your watch. We now have very technologically advanced methods of keeping time; we've divided the surface of the earth into time zones; and we've implemented daylight saving time and standard time. The sundial is a nice, simple reminder of times when life was slower and a day was measured by the movement of the sun, not the digital readout of a microchip.

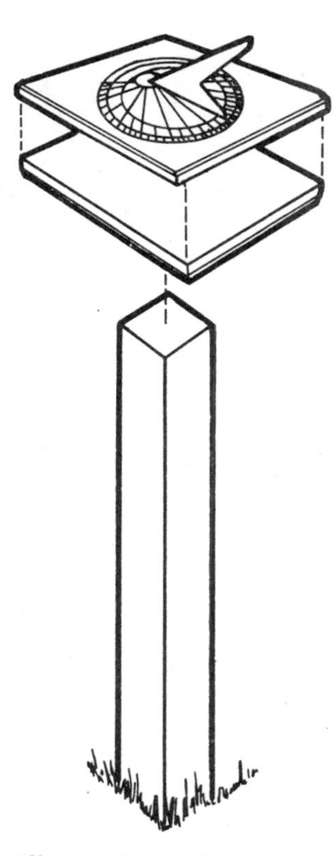

*Illustration 1–9—
Mounting post details.*

2 Equatorial Sundial

Sundials have served for centuries as both decorative and functional ornaments. The design presented here is known as an equatorial sundial. It is a striking combination of gleaming brass rods and bands with a base of heavy steel plate painted flat black. It should be permanently mounted outdoors in a sunny location.

The equatorial sundial is so named because the equator band, carrying the numbers that mark the hours, is positioned parallel to the

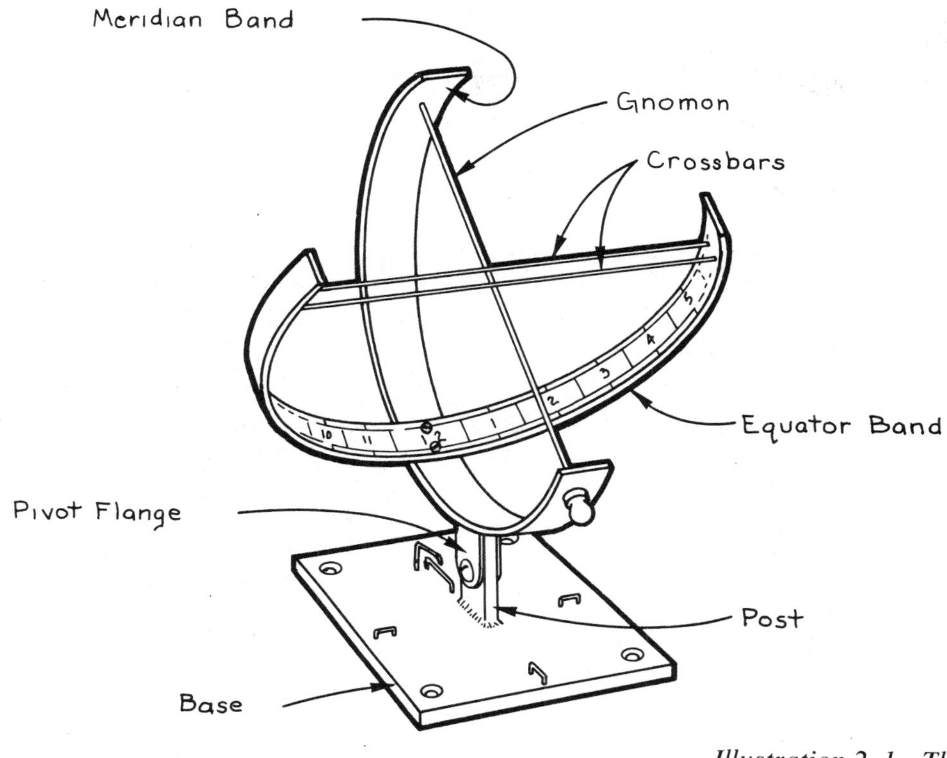

Illustration 2–1—The complete equatorial sundial. The mounting post is not shown.

earth's equator; the second band, the meridian band, supports the shadow-casting device called the gnomon.

The sundial indicates standard time, not daylight saving time. During the months when daylight saving time is in effect, correct by adding one hour. It's important to realize that the sundial registers solar time, which will differ from the time shown on your watch, depending on your location within your time zone.

Most sundials are made for one location and are only accurate at that latitude. However, we've designed this unit to be adjustable, so you may take it with you if you move.

This sundial is a project that uses basic metal-working skills but is not difficult for a person with some soldering or welding experience. It would make a handsome gift and could also be used as a learning project in a metal shop or for the hobbyist at home.

The parts of the equatorial sundial are shown in illustration 2–1. A materials list, chart 2–1, and a tools list, chart 2–2, follow.

Begin making your sundial by cutting the equator band from 14-gauge brass, using a hacksaw. You need a strip 1¼ inches wide and 20 inches long. It is best to mark the piece with a scratching instrument, or scriber, and a straightedge before cutting. Then, cut slightly to the outside of the line with the hacksaw, leaving some room for stray cuts. Finish the band by filing to the cut line to give the band a finished edge. When buying your brass, see if you can get the strip cut to width.

CHART 2–1—
Materials

DESCRIPTION	SIZE	AMOUNT
Hardware		
#10 Brass Flathead Wood Screws	1¼″	4
6–32 Flathead Brass Machine Screws with Washers and Acorn Nuts	¼″	2
¼–20 Brass Roundhead Bolt with Washer and Acorn Nut	¾″	1
Brass Finials	Baldwin #1090	2
Miscellaneous		
14-Gauge Brass	1¼″ × 4′	1
Brass Rod	3/32″ × 3′	1
Brass Rod	⅛″ × 1′	1
Steel Plate	¼″ × 8″ × 8″	1
Epoxy	...	2 oz.
Flat Black Spray Paint	...	1 can
Silver Solder	...	½ oz.
Flux	...	½ oz.
Steel or Brass Filler Rod	...	1

CHART 2–2—
Tools

- Hacksaw
- Drill and Bits
- Scriber
- Cold Chisel: ¼-inch
- Compass
- Center Punch
- Tap-and-Die Set: No. 6–32
- Grinding Wheel or File
- Arc Welder (optional)
- Brazing Equipment
- Steel Numeral Punch Set: ¼-inch (optional)
- 400- and 600-Grit Silicon Carbide Cloth

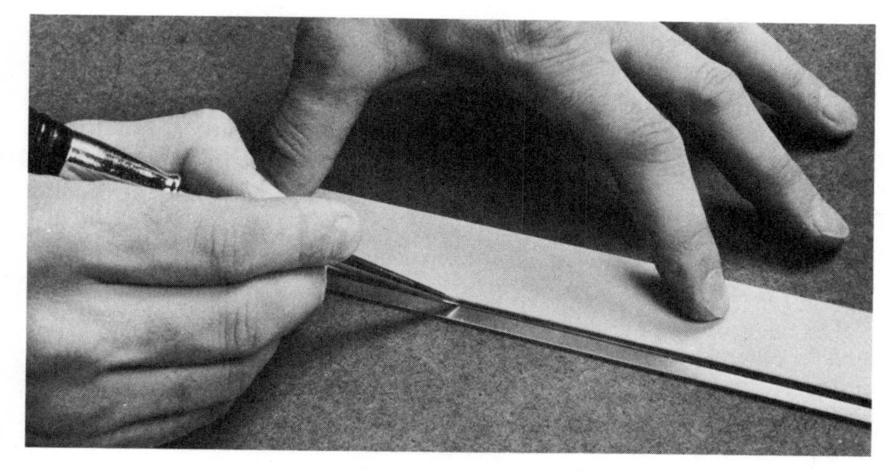

Photo 2–1—Scribe the lines in the brass, using a sharp point and a straightedge.

Two lines are then scribed the length of the piece ¼ inch in from each edge. As we show in photo 2–1, you can scribe against a straightedge. Clamp the straightedge and band firmly together. Keep the scribe against the edge of the straightedge. Your line does not need to be very deep, as it will darken in contrast to the polished brass. Repeat the operation for the other edge of the equator band.

Referring to illustration 2–2, lay out the hour and the half-hour lines on the band. We show 12 hours from 6:00 A.M. to 6:00 P.M. The first hour mark is 1 inch from one end of the band. Then each hour mark is 1½ inches from the previous one. Twelve noon is exactly in the center of the band. The half-hour marks lie centered precisely between, or ¾ inch from, the hour lines. The hour lines lie outside the lines you scribed to leave space for the numerals. Reinforce the marks with a cold chisel and hammer, working gently, or with a scriber and a straightedge.

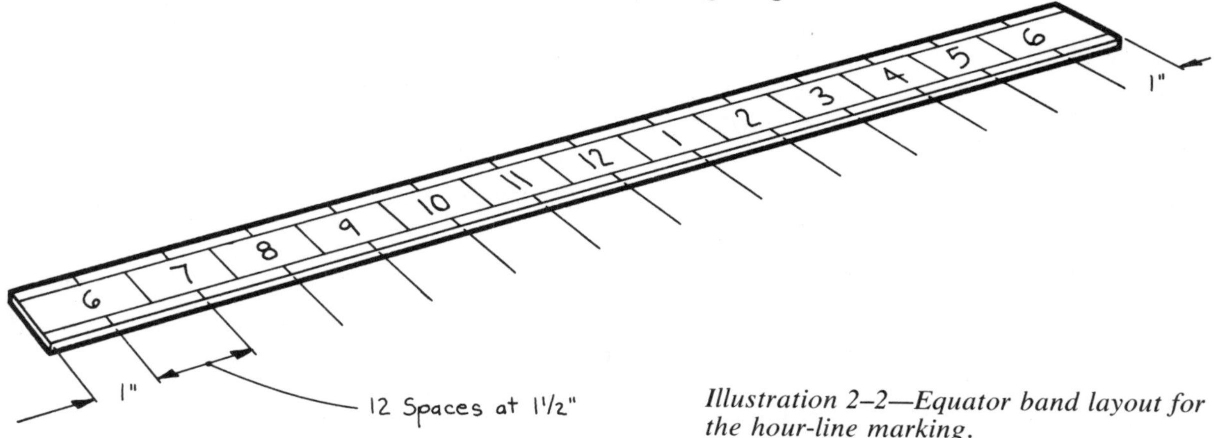

Illustration 2–2—Equator band layout for the hour-line marking.

Illustration 2–3—Crossbar holes.

Next, the numerals are put onto the equator band. This is done with a punch set of ¼-inch-high steel numerals. Or, you may take the band to a jeweler who can engrave it for a fee. As shown in illustration 2–2, the numerals are placed just at the hour lines. A third alternative in marking the equator band is to use the cold chisel or scriber with a straightedge to cut in Roman numerals. See pages 20–21 for Roman numeral design.

In the center of the band, at the marking for noon, center-punch and drill two 5/32-inch-diameter holes, 3/16 inch from each edge of the band. Countersink the holes to accept 6–32 brass screws, but be careful not to overenlarge the hole. At each end of the equator band, exactly at the six o'clock markings, drill two 3/32-inch-diameter holes, as shown in illustration 2–3, to hold the crossbars. *Note:* These holes are not countersunk.

Remove any burrs left after the drilling, either with a small file or by lightly touching the countersink to the reverse side of the hole. To obtain a mirrorlike finish on all the brass pieces, remove any scratches with #220 sandpaper and then 400- and 600-grit silicon carbide cloth. Polish the brass to a high-gloss finish before you do any bending or fastening.

Next, form the equator band into a half-circle with a diameter of 11½ inches, measured from the 6:00 A.M. to the 6:00 P.M. lines. If you have access to a form roller in a metal shop, use this machine to shape the brass band. If not, form the brass band around a jig that you cut from a scrap of ¾-inch plywood, as shown in illustration 2–4. Mark out a circle 8 inches in diameter on the plywood. Add a 2 × 2-inch square. Cut out the jig and cut a slot 1¼ inches long into the square, as shown. Place one end of the brass band in the slot, and gradually press the metal around this jig, as we show in photo 2–2. Alternating ends in the slot, work the band gradually into an 11½-inch-diameter half-circle. A 1-gallon paint can may also be used as a form to help shape the band.

Cut two crossbars, each 12 inches long, from the 3/32-inch-diameter brass rod. They will be silver-soldered across the equator band. Use the

Illustration 2–4—Plywood jig for bending the brass bands.

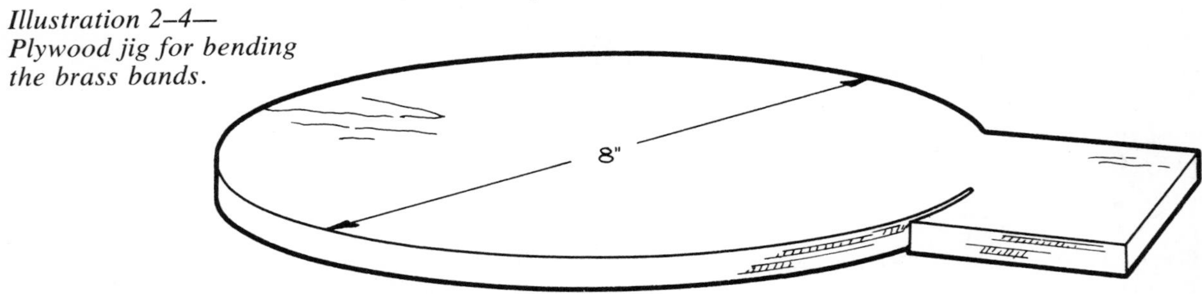

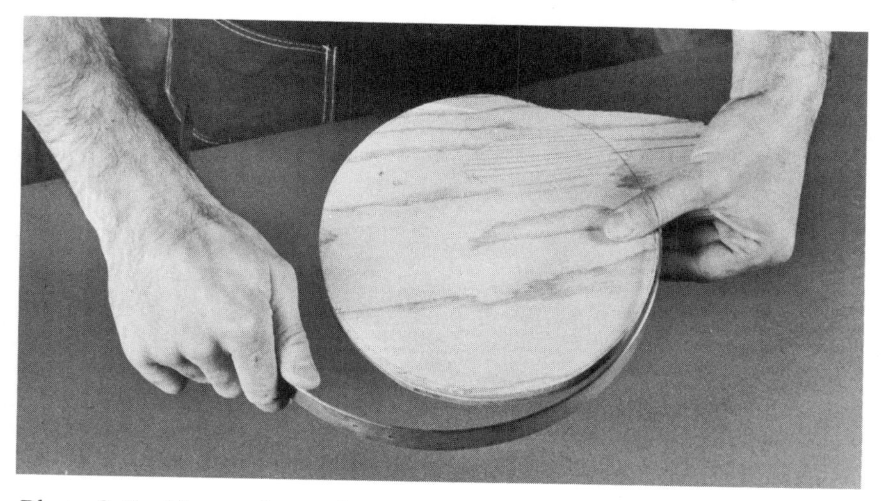

Photo 2–2—Use a plywood jig to bend the brass strip to 11½ inches in diameter.

silicon carbide cloth to clean the metal where it fits into the holes. Fit the rods into the drilled holes in the equator band, having them extend beyond the band. Be sure to maintain the 11½-inch diameter.

When you are silver-soldering or welding, follow recommended safety precautions for your equipment. Use goggles; remove flammable materials from your work area; place the work on firebrick or an asbestos pad; do not touch the heated metal; have a bucket of water and fire extinguisher close by.

If your silver solder contains a flux core, you will not need additional flux. Some form of flux is necessary to prevent oxides from forming on the surface of the metal, which would keep the metals from bonding. If your flux is separate, apply some to the joint area. Then heat the solder and dip it into the flux, also, to clean the solder. In silver-soldering, the flame is directed to the metal, not to the solder material. Concentrate the heat on the band, but keep the flame moving, being careful not to overheat or melt the rods. Touch the solder to the heated joint. When the temperature is right, the solder will flow into the joint between bar and band. Continue to solder all four rod ends in this way, being sure to maintain the 11½-inch spacing.

Cut off the ends of the crossbars and file them smooth. Use 220 sandpaper and 400- and 600-grit silicon carbide cloth to remove any scratches from the surface of the band, and repolish it.

You can now set aside the equator band and start to work on the second major part of your sundial, the meridian band. Cut a piece of 14-gauge brass 1¼ inches wide and 22 inches long for the meridian band, which is shown in illustration 2–5. Locate the center of the length of the band. Mark 7/16 inch on both sides of this center point and square lines across. On these lines, mark two holes in the center of the width, 5/8 inch from each edge of the band. Lay the equator band over the meridian and at right angles to it to check that the holes will align with the holes in the equator band. Center-punch and drill two 5/32-inch holes in the band.

Drill two additional holes in the ends of the meridian band to receive the brass gnomon bar. Measure 9 inches from the center of the meridian toward each end, and square lines across. Mark the position of each hole in the center of these lines, 5/8 inch from each edge of the band. Center-punch and drill two 1/8-inch-diameter holes. Smooth and polish the band after drilling.

Shape the meridian band in the same way that you formed the equator band. It should be a half-circle with a diameter of 11½ inches, measured at the holes.

The gnomon, which casts a shadow onto the equator band, is a 1/8-inch-diameter brass rod, 12 inches long. Cut this rod to length. Using a No. 6–32 die, cut threads into each end of the rod for a distance of 1/4 inch. See photo 2–3. Clamp the rod firmly in a vise. To help the die start more easily, file a slight bevel on the end of the rod. Fasten the die, which cuts the threads, into the diestock, or handle, so that the die-size markings are on top. Placing the die over the rod, apply pressure downward and turn the diestock clockwise. After two or three turns, back the die up half a turn to free the metal chips. Continue backing off every two turns until you have cut 1/4 inch of threads. Turn the die off the rod counterclockwise. You can have this step done for you for a minimal charge at a good hardware store or a machine shop.

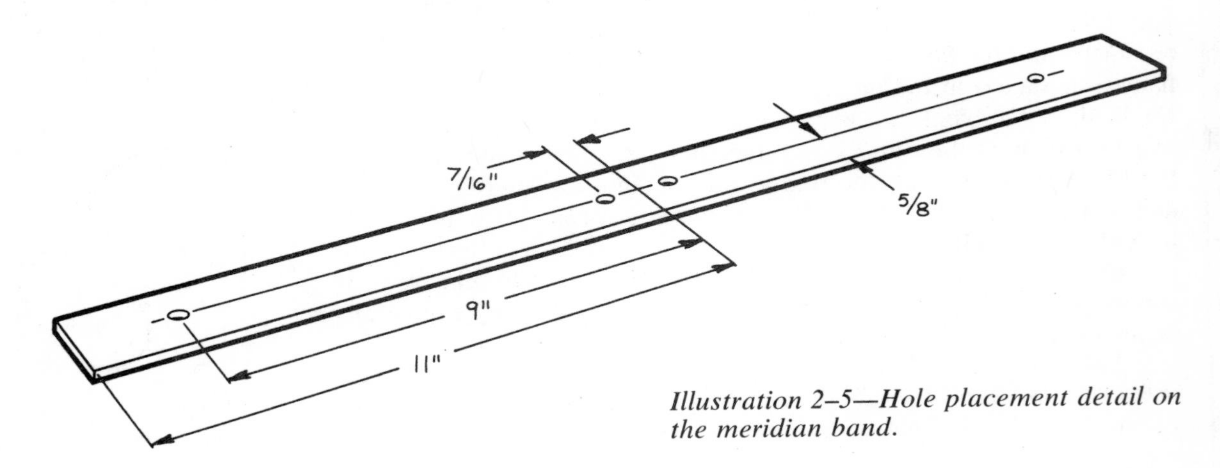

Illustration 2–5—Hole placement detail on the meridian band.

Photo 2–3—Cut threads into the ends of the gnomon rod with a tap set, or have a machine shop do this for you.

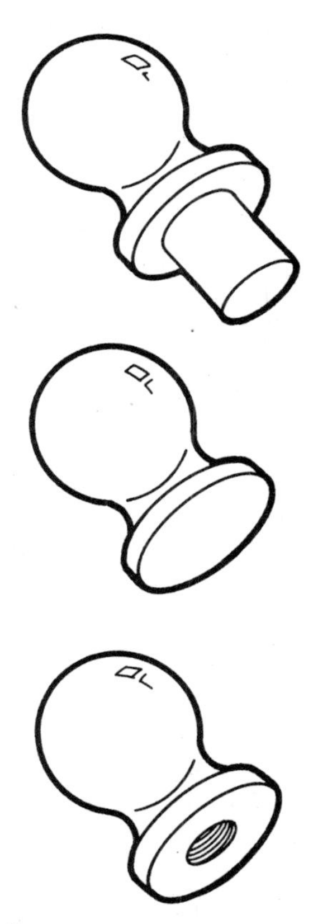

Illustration 2–6—Finial modification.

The two finials, which will be screwed onto the gnomon, should be filed flat, as they generally are manufactured with an extension. See illustration 2–6 for details. When the extension has been removed, it will probably be necessary to extend the threads. To do this, clamp the finial in a vise, then center-punch and drill a 7/64-inch-diameter hole in the center of the flat end. The hole should be ½ inch deep. With a No. 6–32 tap, cut threads into each finial. Interior threads are cut with a tap, which is fastened into a tap wrench. Place the tap into your drilled hole and gently press down on the wrench. Be careful to keep the tap vertical as you start to turn it clockwise in the hole. After two turns, check the position with a small square. Back the tap off to clear out the metal chips, then continue, backing off every two turns, until you have cut ¼ inch of threads. Again, you can have this done for you at a slight charge.

Insert the gnomon through the holes in the meridian band, and screw a finial onto each end of the gnomon. Adjust the finials until the diameter of the meridian band is 11½ inches.

The next step is to construct the sundial's mount. The sundial is supported by two brass flanges brazed to the meridian band. These flanges are bolted to a steel post that is welded to the steel base as shown in illustration 2–7. For the pivot flanges, cut two pieces of 14-gauge brass 1¼ inches wide and 2 inches long. On one end of each flange, mark out a ⅝-inch-radius curve. First measure ⅝ inch from the side to find the center of the width of the flange. Then measure on this line ⅝ inch from the end. With this as the center point, use a compass to draw the curve. To shape this curve, rough-cut it with the hacksaw and then refine it with a grinding wheel or file. Finish up with #220 sandpaper and 400- and 600-grit silicon carbide cloths.

Next, mark a concave radius on the opposite end of each flange to match the curve of the meridian band. Lay the curved meridian band over the flanges so that the band touches both corners of the flanges, and draw the curve onto each flange. File out these curves and smooth them.

Drill a ¼-inch-diameter hole in each flange at the center point you used in drawing the ⅝-inch-radius arc. This will hold the mounting bolt.

Cut a piece of ¼-inch steel 1¼ inches wide and 2¼ inches long for the mounting post, as shown in illustration 2–8. Mark a ⅝-inch radius at one end of the post the same way you marked the flanges. Shape this arc with a hacksaw and file. Drill a ¼-inch hole at the center point of the ⅝-inch radius.

Fasten the post and pivot flanges together with a ¾-inch ¼-20 bolt, washer, and acorn nut.

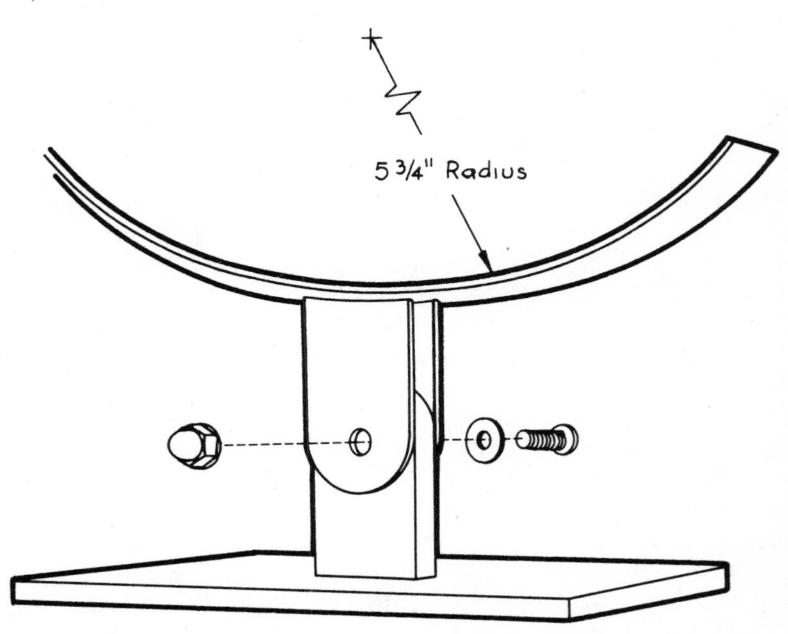

Illustration 2–7—Mounting flange attachment details.

The sundial is assembled by silver-soldering it into a position that corresponds to the latitude at which it will be used. Thus, the point at which you mount the flanges to the meridian band will vary. The circumference of the meridian band is 36 inches. Dividing the 360 degrees of a full circle by 36 inches shows that each inch of circumference equals 10 degrees of the circle. On the edge of the meridian band, mark the point where the center of the equator band will cross. Then measure out 1 inch for every 10 degrees of your latitude and mark this point. The flanges are centered and silver-soldered directly under this point.

5/8" Radius

1 5/8"

Illustration 2–8—Support post.

Clamp the flanges, with the post between them, and the meridian band together firmly, as shown in photo 2–4, so that they won't move during the silver-soldering. Using silver solder, join the flanges to the meridian band. Remove the post, and use sandpaper to clean the inside of the joint area.

For the base, obtain a ¼ × 6 × 8-inch piece of steel plate. File the edges smooth, and clean the whole piece with sandpaper. Measure and draw centerlines on the base, as shown in illustration 2–9. With the post detached from the flanges, position it so that it is in the center and at right angles to the base on all sides. Weld the post to the base.

If you do not have access to an arc welder, it is advisable to have this small job done at a welding shop. It is difficult to heat the ¼-inch-thick base plate with a home-size gas welding set.

If you will be fastening your sundial to a pedestal permanently, lay out mounting holes in the base. Measure in ⅝ inch from each side near the corners, and square lines across. At the intersection points at each

Photo 2–4—Solder the flanges to the meridian band, using the post for proper spacing. The clamps used to hold the pieces together were removed for clarity.

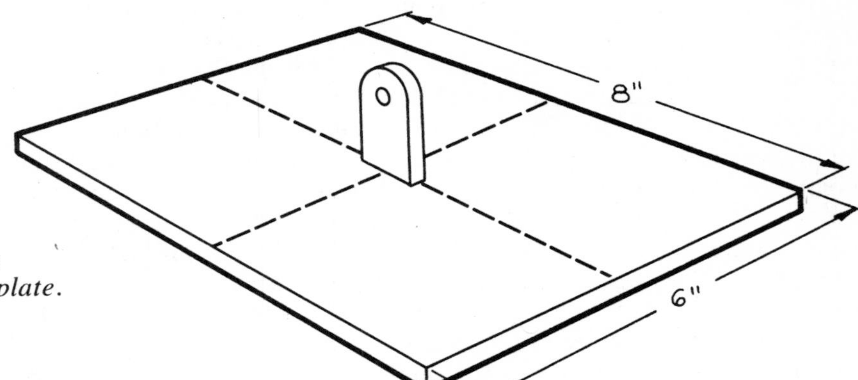

Illustration 2–9—Center the mounting post on the steel plate.

corner, drill a ¼-inch hole, and countersink these holes for ¼-inch flathead screws.

Next, form the direction indicators. Cut three pieces of 3/32-inch-diameter brass rod, each 1½ inches long. These are the east, west, and south indicators. Cut one piece of rod 3 inches long and one piece 2¼ inches long. These two will form the north arrow. Following illustration 2–10, bend the brass rod. The three short pieces are ¾ inch long, the arrow 2¼ inches long, and the arrowhead, as shown, is bent in a 60-degree angle with each leg of the angle ¾ inch long. There is a 90-degree angle bent ⅜ inch from each end of each indicator, so the indicators can be inserted into holes in the base.

On the base, lay out the direction indicators on the centerlines, keeping each indicator ¼ inch from the edge. The north indicator is positioned ½ inch from the end of the base, and the arrowhead holes are 1 inch from the end. Use the indicators themselves to mark the locations, then center-punch and drill ten 3/32-inch-diameter holes, each ⅛ inch deep. Spray-paint the base and post flat black. Put a small amount of epoxy in each hole, and insert the direction markers in the holes.

Now you are ready for the final assembly of the sundial. Fasten the equator band to the meridian band, using two ¼-inch 6–32 brass flathead screws with washers and acorn nuts. Mount the sundial on the post so that the open area faces the south indicator, and the upper end of the gnomon faces the north indicator. Be sure the meridian band is vertical. Bolt the flanges to the post with the ¾-inch ¼-20 bolt, washer, and acorn nut. If the flanges align with the sides of the post, the sundial should be calibrated to your location.

In the event you, or your sundial, should ever move to a different latitude, you will have to recalibrate the sundial. To do this, simply loosen the bolt on the post and flanges, and align the angle of the gnomon

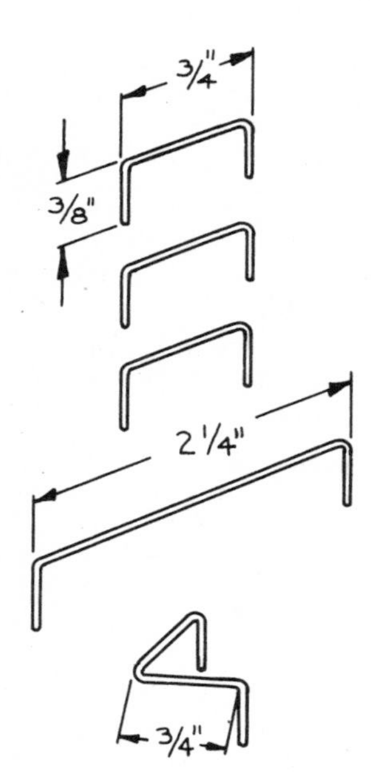

Illustration 2–10—Direction indicators.

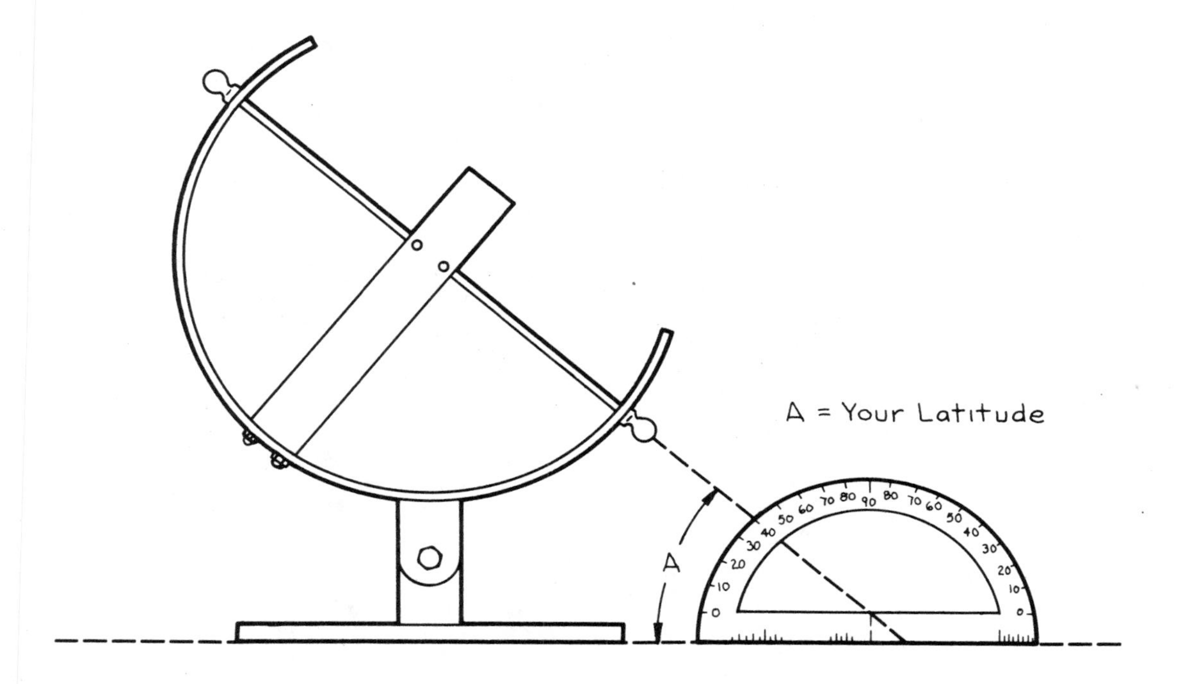

to correspond with the latitude where the sundial will be positioned. It is somewhat tricky to get an exact alignment, and you most likely will have to eyeball it. Illustration 2–11 shows where to position a straightedge and protractor to recalibrate the sundial.

Illustration 2–11—Adjust the sundial to your latitude by holding a protractor in line with the gnomon bar.

Before the sundial is mounted in its permanent position on a post, be sure the post is plumb and level. Check for plumb by testing in two directions.

The sundial must be oriented to true north (not magnetic north). A simple procedure for finding true north at your location is to determine the hour of sunrise and sunset on any particular day. Ask your local weather station or check your local newspaper for this information. The halfway point between the two is known as solar noon. At that time the gnomon's shadow must be on the midpoint, twelve o'clock on the equator band. Fasten the sundial permanently in this position. Use 1¼-inch #10 brass flathead wood screws to mount the sundial.

Throughout the year there will be varying differences between solar time on the sundial and the time shown on your watch, since the earth orbits at a slightly uneven rate. Your location within your time zone also may result in a discrepancy in times. Thus, we appreciate the need for watches but enjoy the sundial in our growing appreciation of the sun's influence on our lives.

3 Site Evaluator

This is a do-it-yourself version of the Solar Site Selector, developed and patented by Sheri Lewis. Complete units are available from Lewis and Associates, 105 Rockwood Drive, Grass Valley, CA 95945, 916: 272-2077. Their manufactured version has both winter and summer grids and covers the entire world.

A site evaluator is fun, instructive, and downright essential for anyone involved with solar projects. The evaluator is used to determine the hours of sunlight and shade that a site will receive during any time of the year. In any solar project, the orientation of your building, or solar device, in relation to sun and shade is often your most critical decision. A site evaluator will help you decide whether a proposed project should be oriented to solar south or somewhat to the east or west, if shade will be a problem, during what time of year shade will occur, and if an entirely different location is needed.

A site evaluator is very helpful in understanding how the sun's path

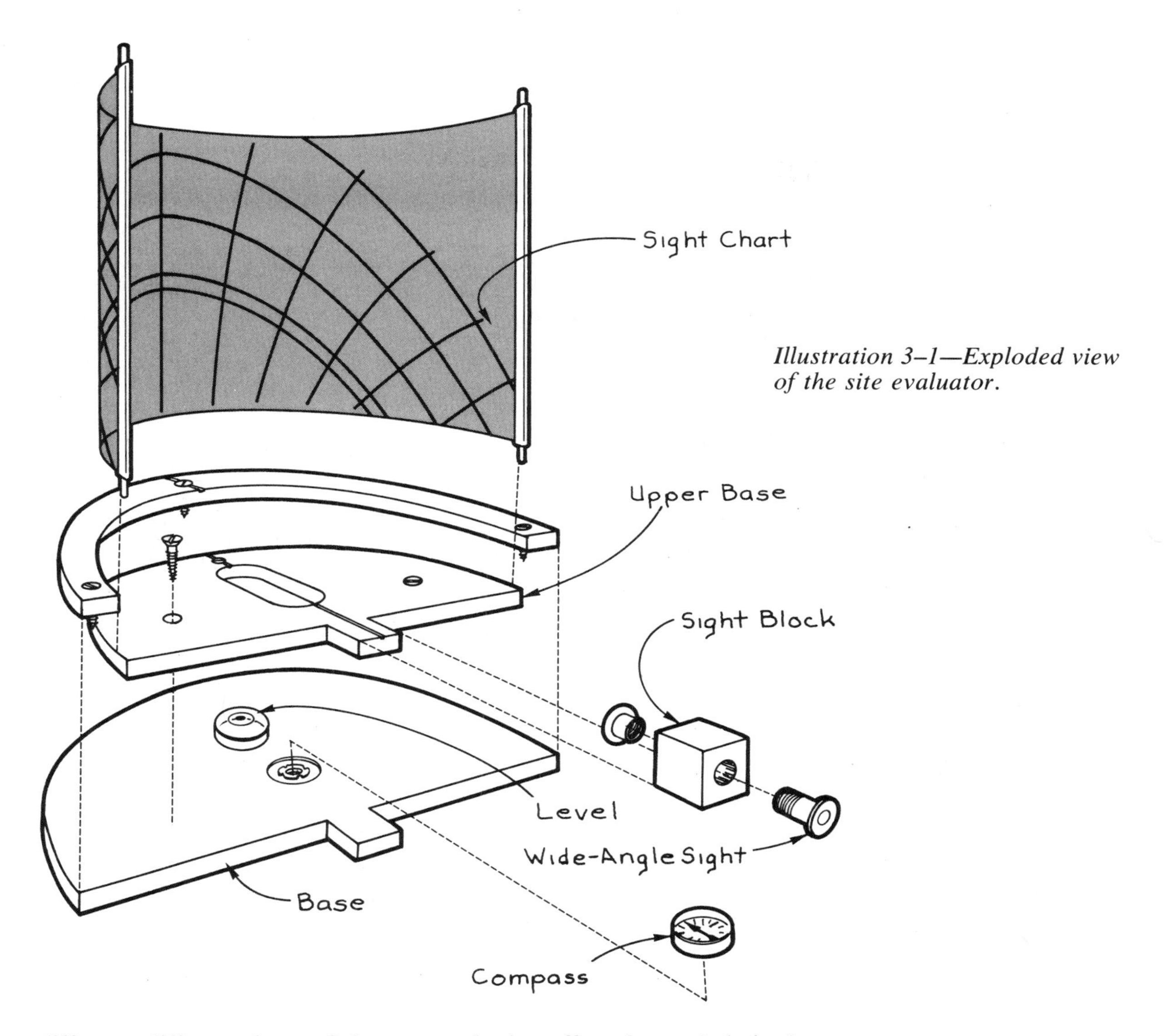

Illustration 3–1—Exploded view of the site evaluator.

differs at different times of the year and what effect the sun's latitude has on your proposed site. Whether you're planning a solar collector, a new garden site, an add-on greenhouse, or a new house, with the site evaluator you can determine how much sunlight the proposed site will receive during any month of the year.

Our site evaluator is made from commonly available materials, and by making it yourself you will enjoy considerable savings over purchasing a ready-made evaluator. It is made with two pieces of plywood, which hold instruments identical to those found in the most expensive professional site evaluators: a wide-angle door sight, a level, and a small compass. Because these are all mounted on the site evaluator, all you need to do is level it, aim through the sight, and trace the top edges of the trees and buildings as you see them. A sheet of acetate provides the surface on which to sketch the skyline. It may then be wiped clean for a new site and used many times over.

We show an exploded view of the site evaluator in illustration 3–1. Chart 3–1 lists the necessary materials for this project, and chart 3–2 is the tools list. Charts 3–3 to 3–8 are the sun charts to be used with the site evaluator.

CHART 3–1—
Materials

DESCRIPTION	SIZE	AMOUNT
	Lumber	
#2 Pine	1⅛″ × 1⅜″ × 1⅜″	1
A-C Exterior Plywood	½″ × 10″ × 15″	2
Hardwood Dowels	3⁄16″ × 10¾″	2
	Hardware	
#8 Brass Flathead Wood Screws	¾″	6
#4 Brass Roundhead Wood Screws	½″	3
¼–20 Tee Nut	½″	1
	Miscellaneous	
Fish-Eye Wide-Angle Door Sight	...	1
Bull's Eye Level	1¾″ dia.	1
Round Liquid-Filled Compass	1¾″ dia.	1
Wood Glue	...	2 oz.

CHART 3–2—
Tools

Tools
Saber Saw
Drill and Bits
Square
Clamps
Utility Knife
Compass

Construction of the site evaluator begins with the two-part base. It is designed in two layers to eliminate the need for a router to cut holes for the compass, level, and acetate sheet for the sun chart. Begin by cutting two pieces of ½-inch plywood, each 9⅜ × 14¼ inches.

Lay out one of the pieces for the lower, solid base, according to illustration 3–2. Square a line across the piece 1⅜ inches from one of the longer edges. Designate this edge *north*. Measure 2¼ inches from the same edge, and draw a second line parallel to the first. Mark a centerline 7⅛ inches from one side. With a compass set to 7⅛ inches, scribe the curved front edge of the site evaluator.

Square lines 9⁄16 inch on both sides of the centerline from the edge of the piece to the first line. This defines the area for the sight support. Carefully cut out the pieces with a saber saw.

A tee nut in the base bottom allows the site evaluator to be attached to a tripod. The tee nut is positioned 2¼ inches from the center point and on the centerline, as shown in illustration 3–2. With a spade bit, drill a ¾-inch-diameter hole 1⁄16 inch deep in the top of the lower base piece. This is called a counterbore, and it will hold the tee nut. Next, drill a 5⁄16-inch-diameter hole through the center of the counterbore. Insert a ¼–20 tee nut in this hole from the top, and tap it in place.

The shape of the upper base is identical to that of the lower base. It has additional holes cut in to hold the evaluator equipment. Start by

CHART 3–1—*Continued*

DESCRIPTION	SIZE	AMOUNT
Miscellaneous—*continued*		
Epoxy	...	1 oz.
Flat Black Spray Paint	...	1 can
Wood Putty	...	4 oz.
Clear Acetate	0.015 × 10″ × 20″	1
Report Cover Clips	...	2
Flat Black Artist's Tape	4 pt. width	1 roll
Permanent Marking Pen	0.4 mm. width	1
Drafting Paper	10″ × 20″	1
Washable Marking Pen	...	1

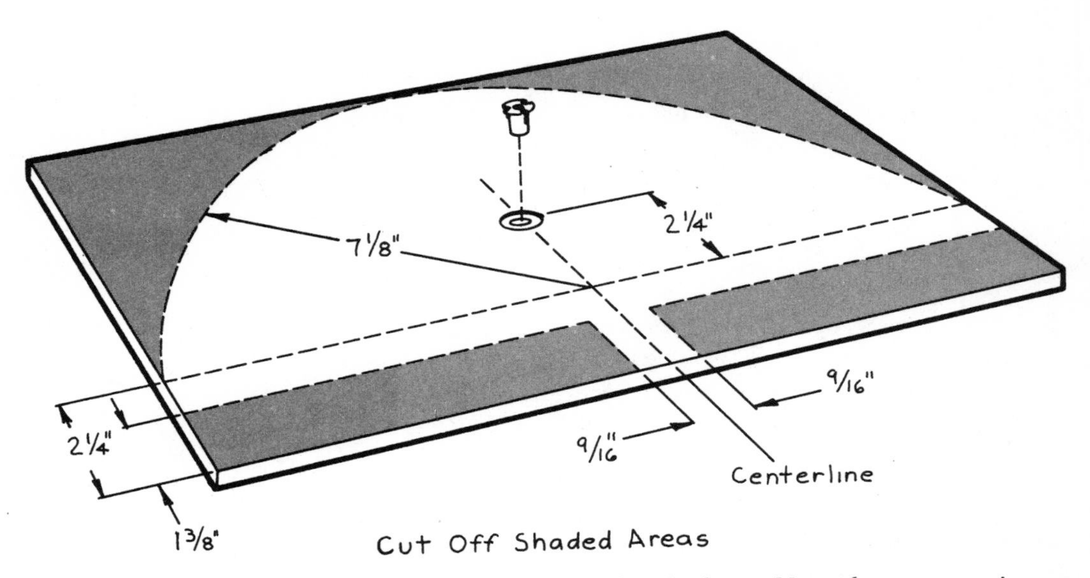

Illustration 3–2—Layout details for the base. Note the center point location for the 7⅛-inch radius.

duplicating the measurements, lines, and shape of the lower base on the second piece of plywood. Then, referring to illustration 3–3, mark the half-circle. From the center point on the second line, draw a curve with a radius of 6³⁄₁₆ inches. Stop the curve at the second line. With a square, continue the line to the edge of the base. Cut the curve with the blade of your saber saw centered on the line. The top base piece will separate into two pieces with approximately ¹⁄₁₆ inch between them for a sheet of acetate. Sand the edges smooth so that the sheet will fit easily later.

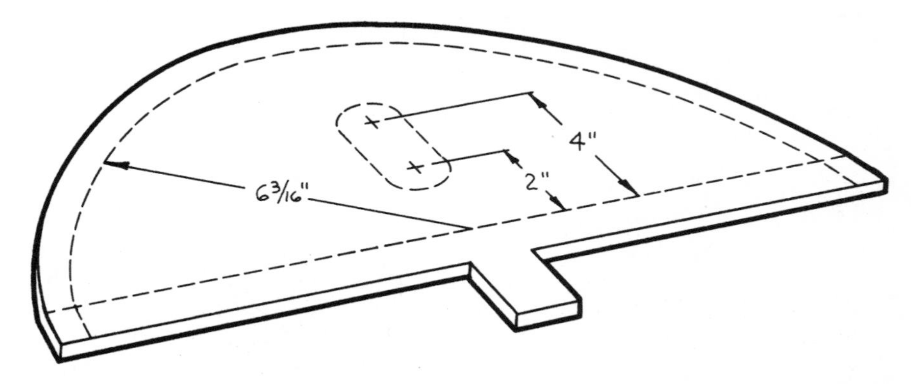

Illustration 3–3—Cutting details for the acetate sheet groove and the recess for the compass and level.

Form the oval shape for the center cut-out by drawing two circles equal in diameter to your level and compass along the centerline of the top base piece. Measure 2 inches and 4 inches from the center point toward the south edge of the piece. Scribe the circles and connect them with a straightedge, and the oval shape will be formed. Cut out the slot with a saber saw after drilling an entry hole for the blade.

Place the top pieces and the bottom base piece together, aligning the edges and maintaining the slot for the acetate. Lay out six screw holes for the brass screws that will join the unit; three in the outer arc and three in the center section. Drill 1/16-inch-diameter pilot holes. Enlarge the holes to 3/16-inch diameter in the top pieces only. Countersink for the screw heads. Use wood glue and six 3/4-inch #8 brass flathead wood screws to fasten the base pieces together.

The sight block is a small scrap of pine that is fastened to the base and holds the sighting device. Refer to illustration 3–4, noting the direction of the wood grain. Cut a block 1⅛ × 1⅜ × 1⅜ inches, being careful to keep the sides straight. On one face measuring 1⅛ × 1⅜ inches, draw diagonal lines from corner to corner. At the center point, drill a 9/16-inch-diameter hole to hold the sight device, as shown in photo 3–1. It is very important to have this hole truly square through the block

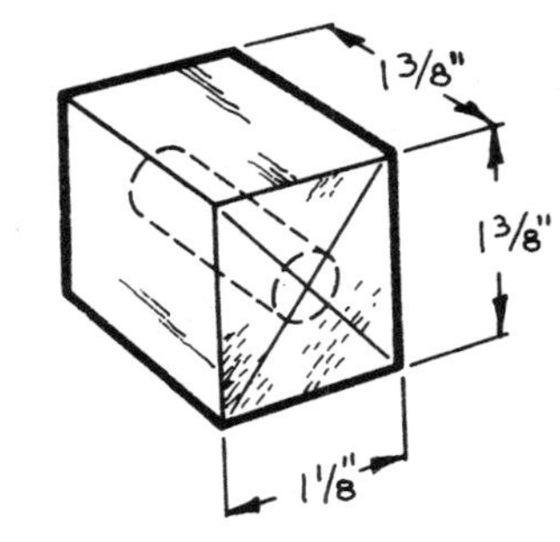

Illustration 3–4—Sight block.

Photo 3–1—Drill a hole in the sight block to hold a wide-angle door viewer.

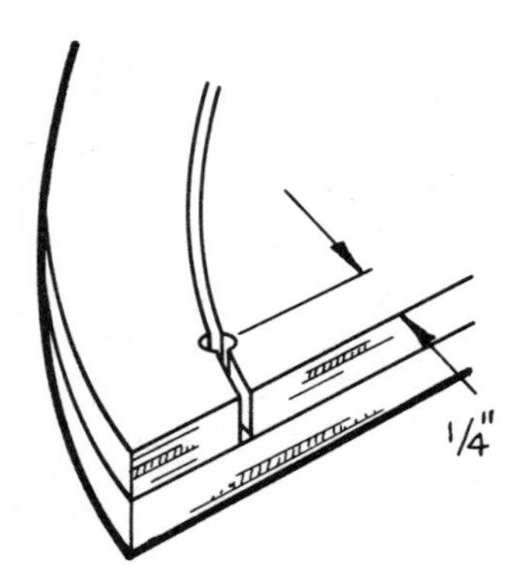

Illustration 3–5—Dowel holes.

for accurate sighting. It is a good practice to mark both sides of the block, then drill from both sides to produce a straight hole. Sand the sight block and glue it to the base with wood glue. Make sure the sight hole is properly aimed and level.

The acetate sheet is held in position by two wooden dowels inserted in the base. Drill two 3/16-inch-diameter holes in the base, 3/4 inch deep. The holes are located within the groove for the chart, 1/4 inch from the sight edge of the base. See illustration 3–5. To finish the base, fill any cracks in the plywood with wood putty, and then sand the unit.

Cut two 3/16-inch-diameter dowels, each 10 3/4 inches long, to serve as uprights for holding the acetate sheet. Paint the base and the uprights with flat black spray paint.

When the paint is dry, use a utility knife and a straightedge to cut a slight groove in the base for a reference mark on the centerline. Cut into the wood enough to make a contrast against the black paint, as we show in photo 3–2. This groove will be useful in aligning the evaluator to a north-south axis when it is in use.

Now you are ready to attach the site evaluator equipment. Use epoxy to attach the compass directly above the tee nut. North points directly at the sight, and south, the direct opposite. Place the bull's-eye level in the slot's south end, and mark the screw holes. Drill 1/16-inch-diameter pilot holes, then fasten the level into the slot with three 1/2-inch #4 brass roundhead wood screws. The sight, which is a fish-eye wide-angle door viewer, is assembled in the sight block by screwing its two components together until they are flush with the block.

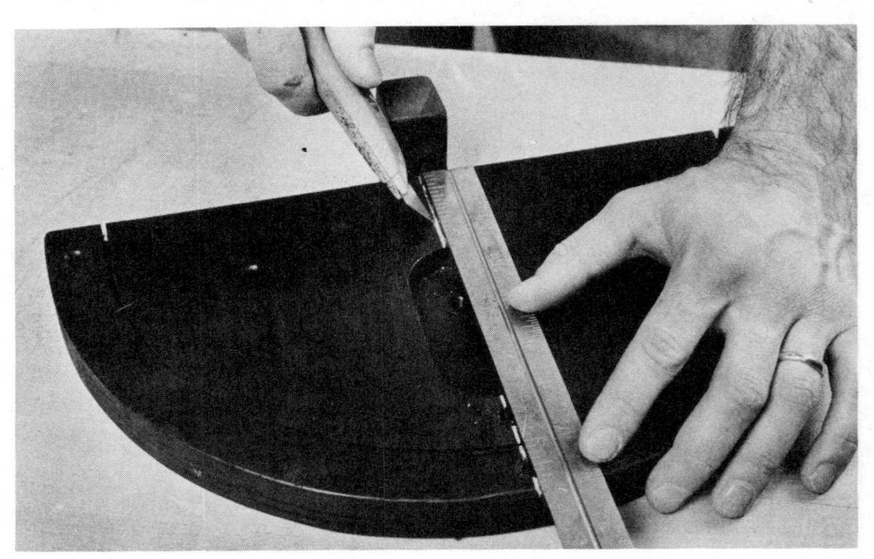

Photo 3–2—Cut the centerline reference mark into the base with a utility knife.

Your latitude determines the amount of sun you will receive throughout the year and how to set up your evaluator. On a road map, find the latitude of your location. You can also call the local airport for this information. Then, select the sun chart for your latitude from charts 3–3 to 3–8, pages 47–49. We have included six sun charts to cover the United States, with the exception of Alaska and Hawaii.

You have to transfer the pattern we show for your location to a 10 × 20-inch piece of drafting paper. Lay out a grid pattern of 1-inch squares on the drafting paper to match the chart. Using the chart that corresponds most closely with your latitude, transfer the curved lines that show the sun's path at different times of the year. Also, transfer the straight lines showing the times of the day.

Place a 10 × 20-inch sheet of 0.015 clear acetate over the drafting paper grid. Duplicate the sun path lines onto the acetate, using black, 4-point-wide artists' tape. Add the straight lines, also with tape, to represent the hour lines. Label the sun chart dates and hours, using a permanent marking pen.

The position on the sun's horizon is duplicated for several months of the year, and thus these pairs of months share the same line. For

Photo 3–3—Attach the sun chart to the dowels with the clips, then put the unit into the slot in the base.

example, the sun will be at the same height in October as it is in February.

Note: For some geographical latitudes, the sun paths for some months have been omitted from the sun charts. At this time of the year, specifically May, June, and July, the sun will be very high, and, in most cases, plotting the shadows for these months is not necessary.

The sun chart is attached to the evaluator base with two plastic report cover clips and the dowels. Cut the clips to 9½ inches. Place the dowel uprights into the drilled holes in the base, push the sun chart into the clips, and then slide the clips over the dowels, as shown in photo 3–3.

Take the site evaluator to the prospective site. The evaluator may be fastened to a tripod or any steady mount that is convenient for you. It must be level to operate properly. Also, try to position the site evaluator at a spot that corresponds to the center of the glazing of the proposed project.

Using your compass, orient the reference groove so that it runs north and south, with the sight edge to the north. Your compass shows magnetic north and does not indicate true solar south. To adjust for the

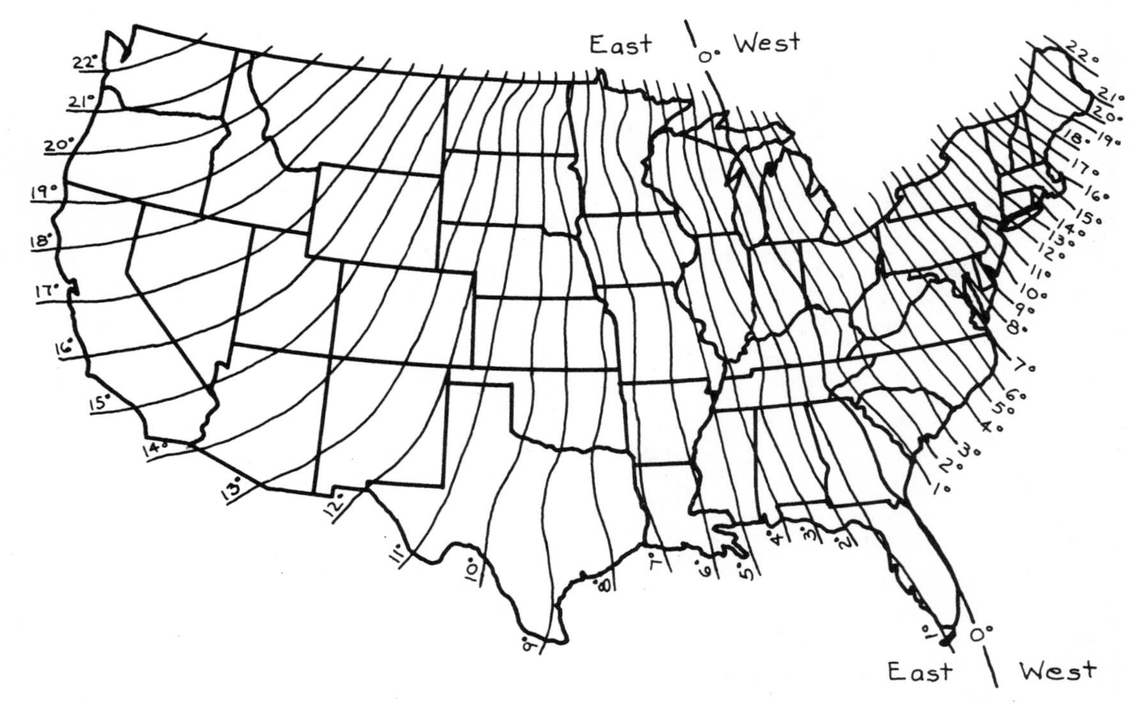

Illustration 3–6—Use this map to determine your location and magnetic variation from solar south.

variation, which depends on your location, use illustration 3–6 or consult your local geological office.

This map enables you to determine the necessary adjustment for the earth's magnetic influence. Along a line of 0 degrees variation, from Michigan to the eastern Florida coast, you need to make no adjustment. To the east of the 0-degree line, true south is west of the compass-indicated south by the degrees shown. To the west of the line, true south is east of magnetic south. See illustration 3–7 as an example. The illustration indicates the variation between magnetic south and true south in San Francisco, where there is a 17-degree difference. Set the site evaluator to face true solar south, taking your local variation into account. *Note:* You need to be aware of local conditions, such as power lines, metal buildings, and so forth, to get a true reading.

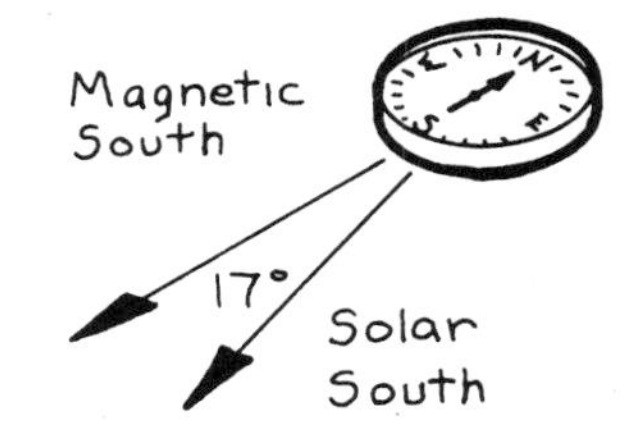

Illustration 3–7—Compass correction.

To use the site evaluator, look through the sight and draw the objects you see on the skyline onto the sun chart, as shown in photo 3–4. Be sure to use washable ink and to draw on the unlettered side of the sun chart to make it easy to wash the chart later. Draw the roofline of buildings and the outline of trees exactly where you see them on the acetate screen. You may want to draw in deciduous trees with a dotted line to indicate that they will provide less shade in the winter. When you have finished drawing your skyline, remove the sun chart and analyze the results. In illustration 3–8 we give a sample skyline drawn onto a sun chart.

Photo 3–4—Draw the outline of all objects in view on the outside of the sun chart with removable ink.

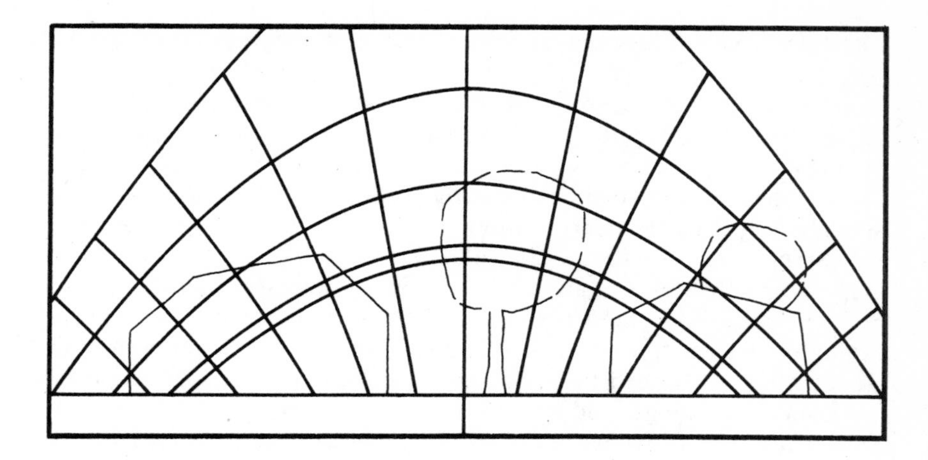

Illustration 3–8— A typical site evaluation shows trees and buildings along the skyline.

The arc lines indicating the sun's path on different days of the year are guides to which objects will shade your site during each season. If a tree or building is below any given line, it will not provide shade; the sun will shine above it at that time of year. Since the sun rises and sets in a much wider arc in the summer than in the winter, a tall object 45 degrees from south may give shade in summer but will not interfere with winter sun.

If, for instance, you discover that an object in your chosen location for a collector presents shading problems, consider orienting the collector up to 30 degrees east or west of solar south, or move the shading object. A variation from true south of over 30 degrees in either direction will result in a substantial loss of solar efficiency.

For the most complete picture of your proposed building site, use the site evaluator at several locations: southeast corner, southwest corner, center of south-facing wall, and two spots at the height of your planned second-story windows. For a small solar installation, one reading is sufficient. There is usually a significant increase in available light with increased height. If you are planning a roof collector, it is especially important to take a reading at the proposed height. Record the results from the sun chart at each location, using drafting or tracing paper, so that you have a record to compare locations.

Remember that the evaluator can be used to place shading objects in front of windows, doorways, and even entire homes in sunnier climates where solar cooling is desired. Take the same siting procedures, but place the objects so that their shadows will fall where intended. The analysis is the opposite of removing the shade from an area, and the shadows of the objects will fall above the sun path lines.

CHART 3-3—28 Degrees

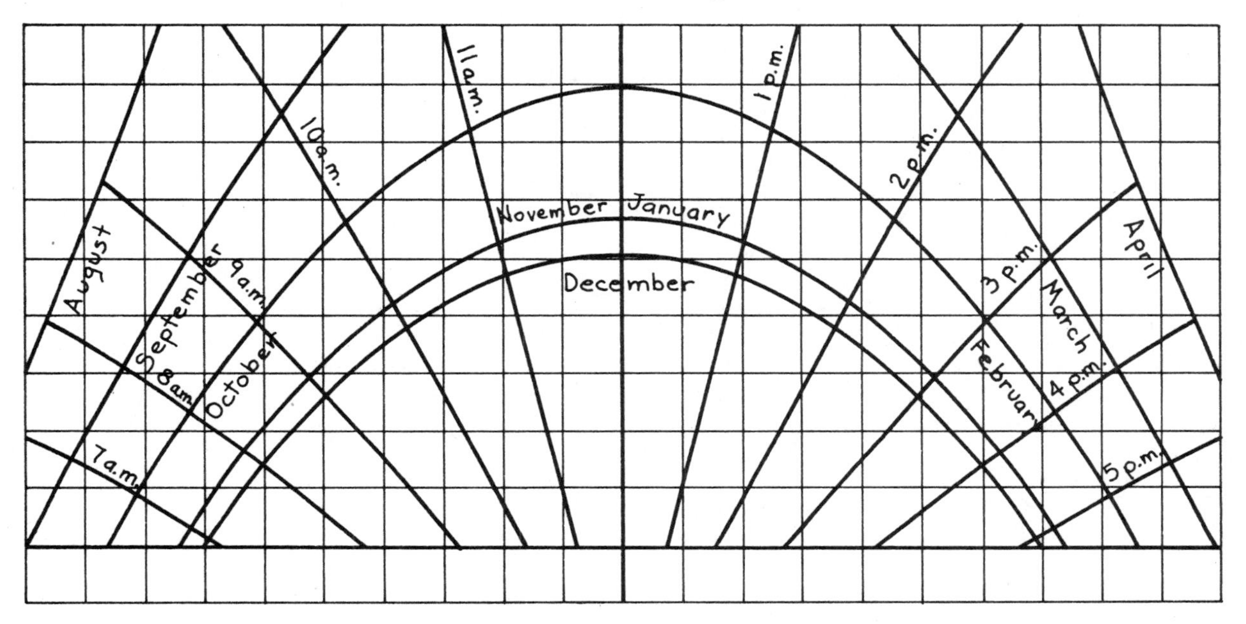

CHART 3-4—32 Degrees

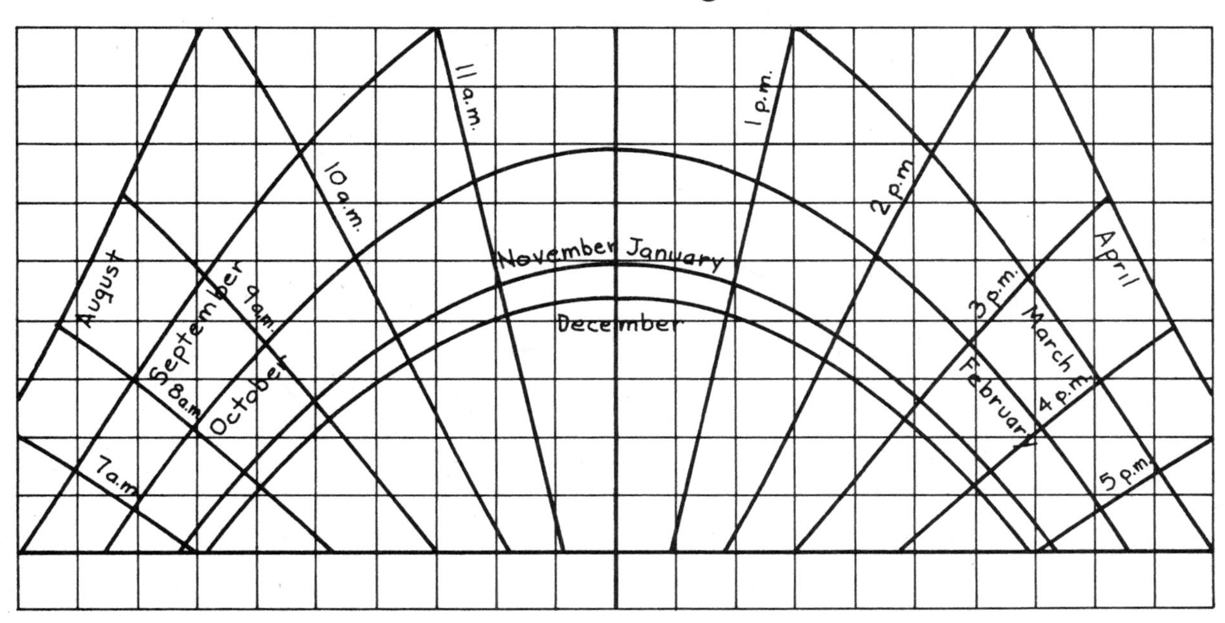

CHART 3-5—36 Degrees

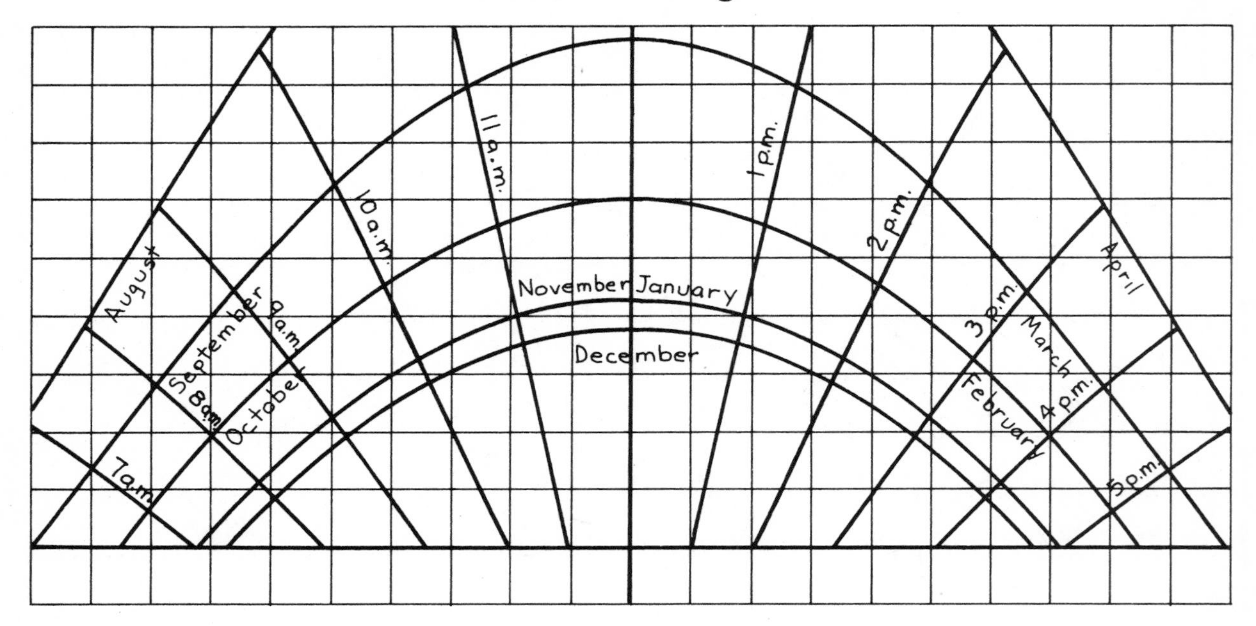

CHART 3-6—40 Degrees

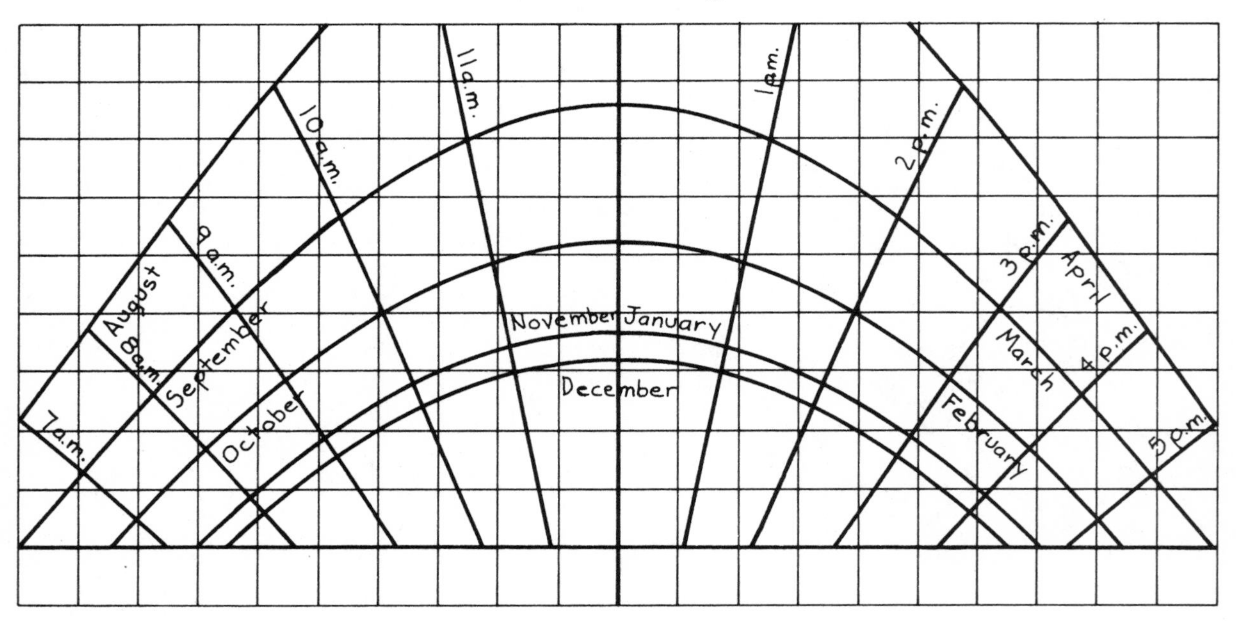

CHART 3-7—44 Degrees

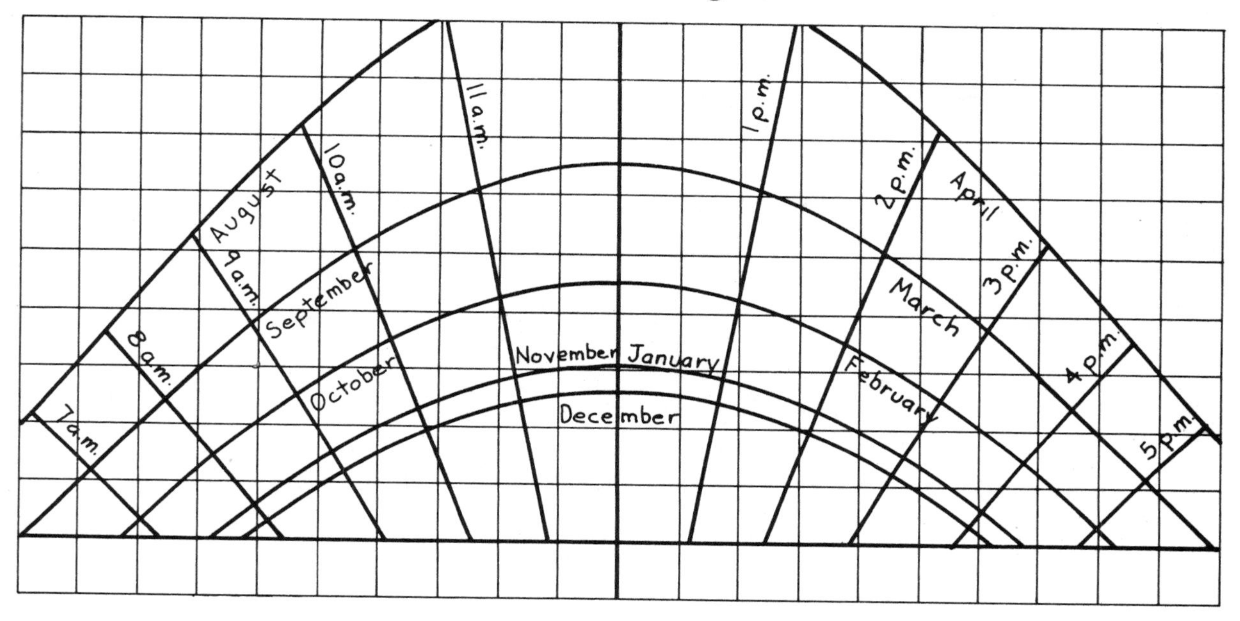

CHART 3-8—48 Degrees

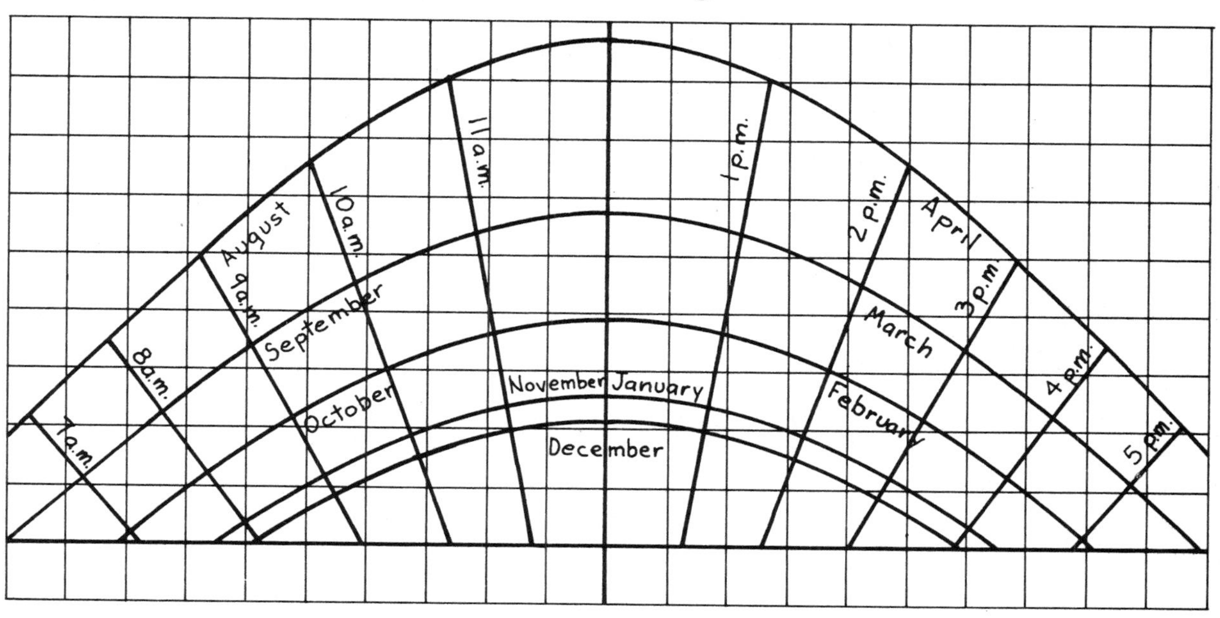

4 Window Greenhouse

This window greenhouse is a handsome yet inexpensive alternative to kits and ready-made units. It can be constructed in a weekend with hand tools and will convert an ordinary window into a rewarding growing area. Houseplants, as well as herbs and salad greens, do well in this unit.

A curved front gives this small greenhouse its distinctive shape. This design works particularly well in a second-floor situation, because it doesn't limit downward visibility as much as most greenhouses. The added visibility is achieved by using wraparound acrylic glazing.

The frame is very simply constructed of plywood and pine. It fits

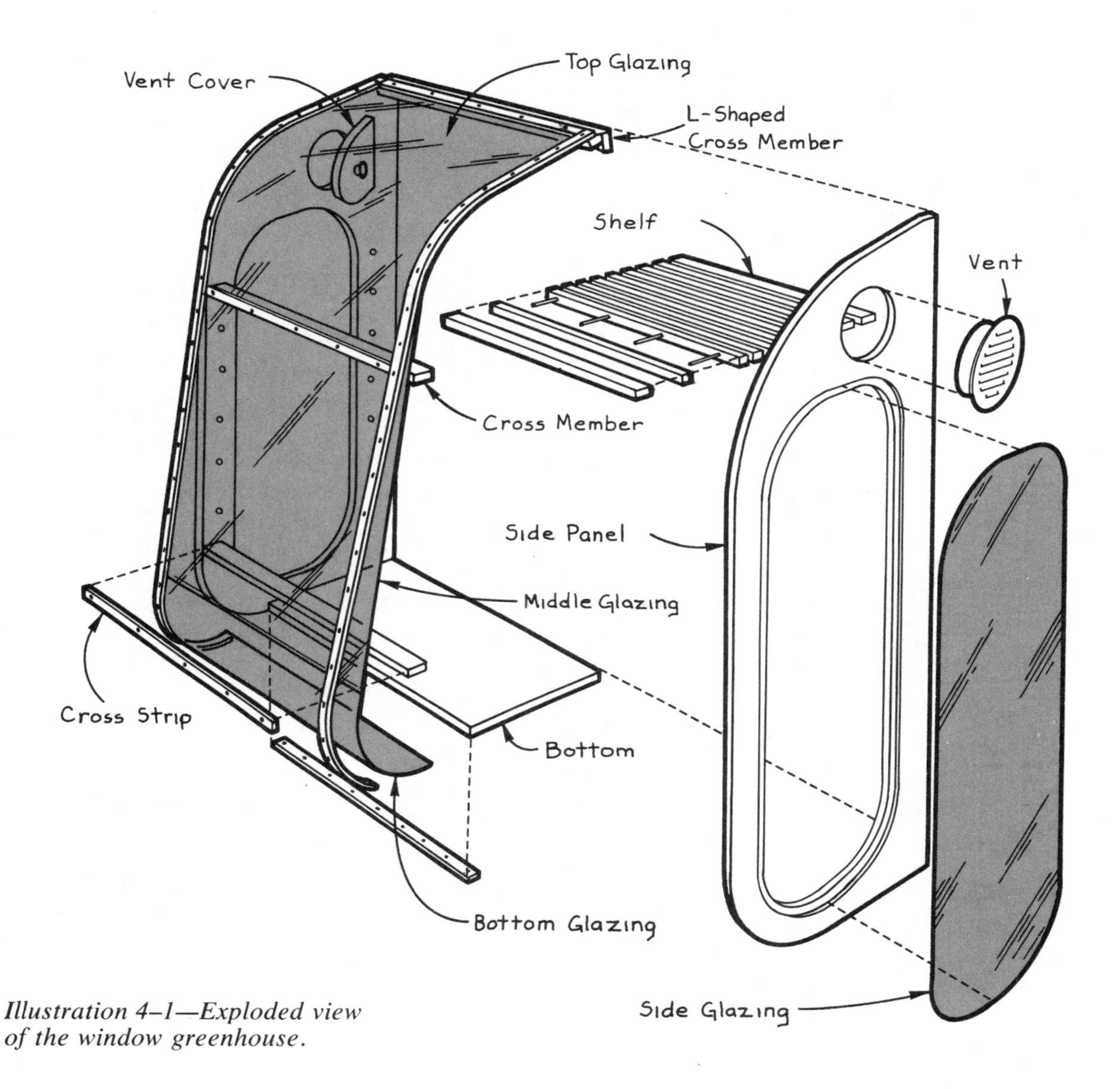

Illustration 4–1—Exploded view of the window greenhouse.

into the window jamb and is held in place by eight screws. A single bead of caulk around the edge of the greenhouse after installation makes it completely weathertight. Vents are provided in the sides, and slatted shelves allow good air circulation between the plants. Illustration 4–1 shows an exploded view of the greenhouse.

The dimensions we give are for a greenhouse that will fit a specific window size, 31½ inches wide by 54 inches tall. Illustrations 4–2 and

4–3 show where to take measurements for your unit. Measurement A is the height; B is the width. Take your own window measurements and adjust the dimensions according to chart 4–3. A materials list, chart 4–1, details your necessary supplies (you may have to adjust the quantities, depending on your window size). A tools list, chart 4–2, follows.

CHART 4–1—
Materials

DESCRIPTION	SIZE	AMOUNT
	Lumber	
#2 Pine	1 × 2 × 8′	4
Baluster Stock	¾″ × ¾″ × 8′	7
A-C Exterior Plywood	¾″ × 4′ × 8′	1
Hardwood Dowel	¼″ × 3′	4
	Hardware	
18-Gauge Brads	½″	1 box
Cement-Coated Box Nails	6d	20
#8 Flathead Wood Screws	1½″	8
#6 Aluminum Flathead Wood Screws	¾″	52
#4 Aluminum Flathead Wood Screws	½″	40
Flush Cabinet Hinges, Self-Closing, with Screws	2½″ × 1½″	4
Wooden Knobs	1¼″ dia.	2
Plastic Adjustable Shelf Brackets	¼″	8
Aluminum Strips	⅛″ × ¾″ × 6′	4
Screened Aluminum Louvered Vents	6″ dia.	2

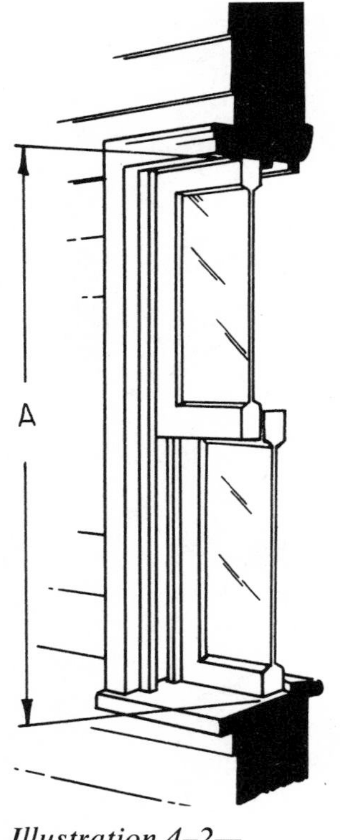

Illustration 4–2—
Measurement A.

Before shopping for materials, take the measurements of your window according to illustrations 4–2 and 4–3. The amounts of materials we give in our chart are for a greenhouse that will fit in an average-size window with a height (A) of 54 inches and a width (B) of 31½ inches. We have figured in ⅛-inch clearance on all sides.

CHART 4–1—*Continued*

DESCRIPTION	SIZE	AMOUNT
Miscellaneous		
Closed-Cell Foam Weather Stripping	⅛″ × ¼″	4′
Waterproof Wood Glue	...	4 oz.
Primer	...	1 pt.
Exterior Enamel	...	1 qt.
Silicone Caulk	...	1 tube
Acrylic Sheet	1⁄16″ × B − ½″ × A + 16″	1
Acrylic Sheet	1⁄16″ × 14¼″ × A − 15″	2

CHART 4–2—
Tools

- Hacksaw
- Saber Saw
- Circular Saw (optional)
- Drill and Bits
- Router
- Plane
- Caulking Gun
- Compass
- Countersink

CHART 4–3—
Sizing Chart

DESCRIPTION	SIZE	AMOUNT
Side Panels	19¾″ × A − ¼″	2
Bottom	11¾″ × B − 1¾″	1
Cross Members	B − 1¾″	3
Shelf Edges	B − 2	2
Shelf Slats	B − 1¾″	10
Cross Strips	B − 1¾″	4
Side Strips	A + 16¾″	2
Front Glazing	See text	See text
Side Glazing	14¼″ × A − 15″	2

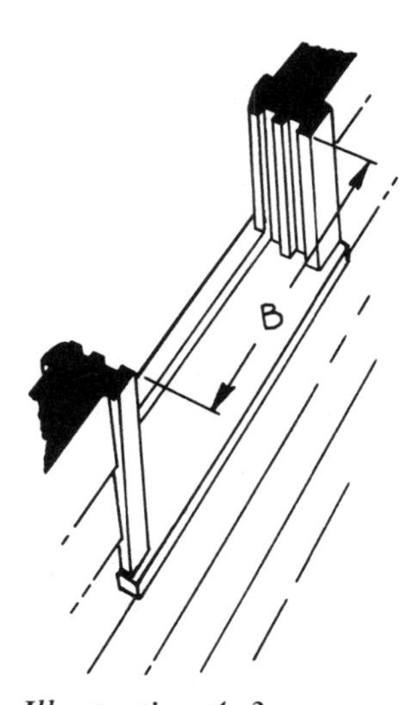

Illustration 4–3— Measurement B.

Keep a few items in mind when choosing a window for the greenhouse. First, the window must have double-hung sash that open both top and bottom. Second, if there is a storm window, it must be removed to make space for mounting the greenhouse. Take the measurements after you remove the storm window.

Start the construction of the greenhouse by cutting the sides and bottom from ¾-inch plywood. Lay out two side panels, each 19¾ inches wide and measurement A minus ¼ inch high. If you lay out several pieces before cutting, be sure to allow space between them for the width of the saw blade. Lay out the bottom piece 11¾ inches wide and B minus 1¾ inches long.

The side panels are the heart of the greenhouse project, since they determine the front glazing curve and hold the side glazing, shelves, and vents. As you mark these panels, work closely with illustration 4–4 as well as with our written directions.

Refer to illustration 4–4 and lay out the curved top and bottom edges, vent opening, and side panel opening on one side panel. First, measure 2¼ inches up from the bottom to mark the bottom edge of the side opening. Then mark the front edge, also at 2¼ inches. Square these two lines across the side panel for the lower and outer edges of the

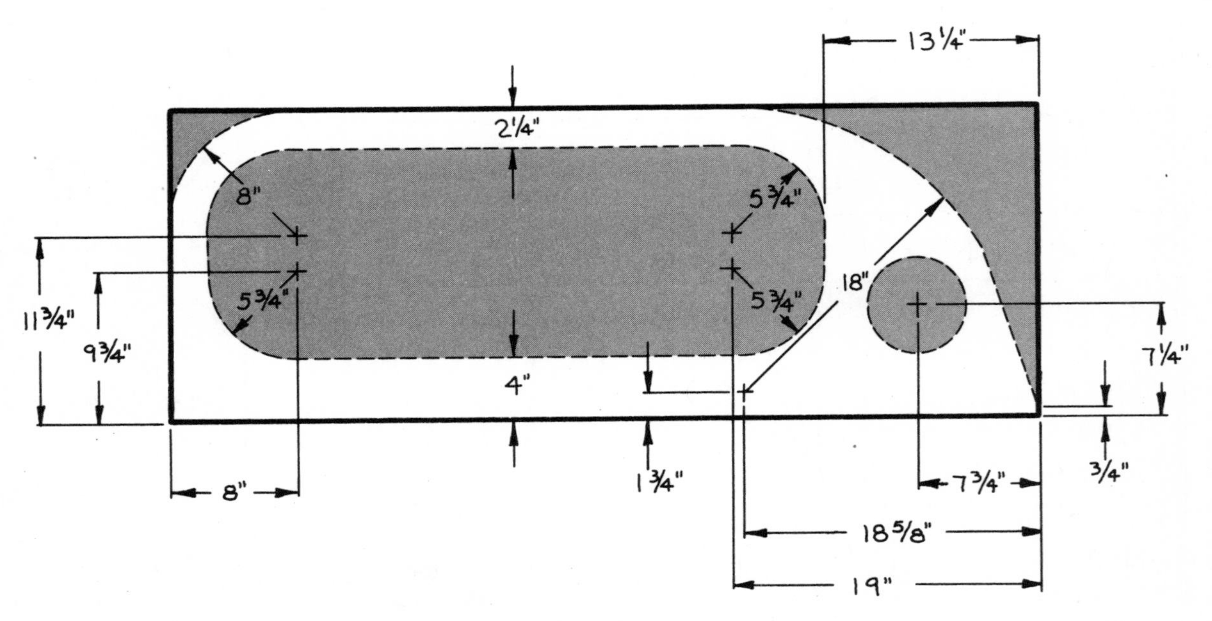

Illustration 4–4—Layout details for the greenhouse side. Note the center points for the curves.

opening. From the top of the side panel, measure down 13¼ inches and square a line across. In the same way, draw the line for the back edge of the side opening 4 inches from the back edge of the panel.

Referring to illustration 4–4, next mark the two center points for the bottom curves. The outer curve has an 8-inch radius; the inner curve, a 5¾-inch radius. Using a compass, scribe these curves as shown.

At the top of the panel, measure down 19 inches and square a line across the panel. Measure in 9¾ inches and 11¾ inches from the straight edge. Using these two center points and a radius of 5¾ inches, draw the two inner corner curves.

To shape the outer top of the greenhouse sides, start by making a mark on the top edge, ¾ inch in from the back corner of the piece. Next, mark a point 18¾ inches down from the top of the side panel and 1¾ inches in from the edge. Using this point, with an 18-inch radius, draw a large curve from the outer edge up the panel. See photo 4–1 for a homemade compass large enough for this step. Use a straightedge to draw a line connecting the ¾-inch mark with the arc. Form a smooth transition from the curved edge to the top.

Next is the vent opening. From a point 7¾ inches from the top edge and 7¼ inches from the side panel's back edge, draw a circle with a 3-inch radius, then cut the piece to shape.

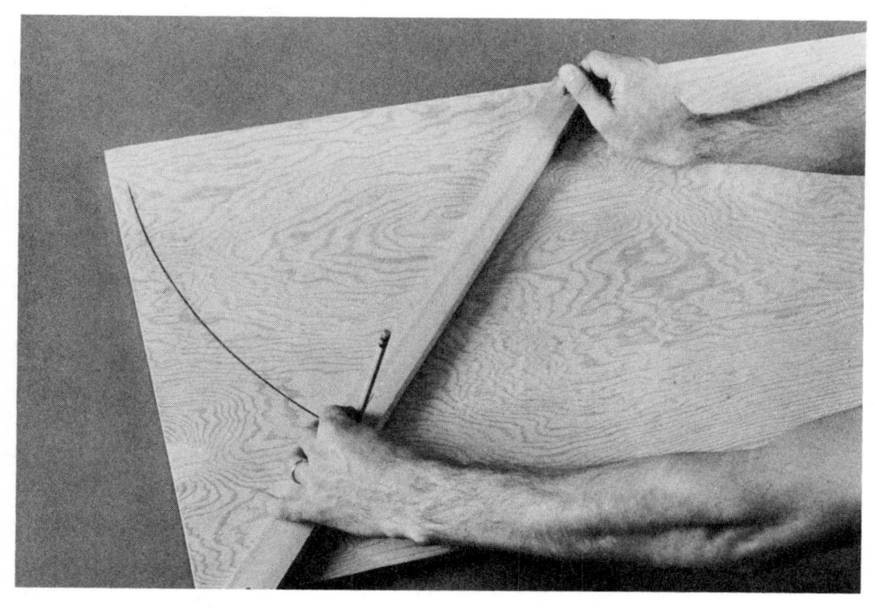

Photo 4–1—Use a shop-made compass to draw the large-diameter curves for the sides of the window greenhouse.

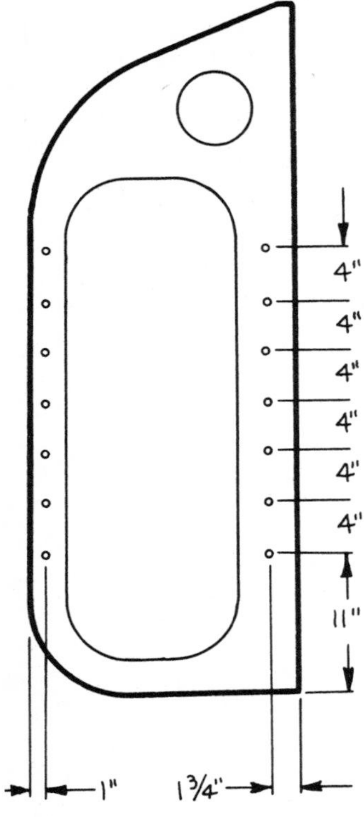

Illustration 4–5—Shelf bracket hole placement.

Use the first side panel after it is cut out to shape the second side panel. After cutting out the second side panel, fasten the two panels together temporarily with finishing nails or clamps, and shape them together so they match exactly. Remove the nails and proceed with the remaining work.

Holes are drilled into the inside of the side panels to hold adjustable shelf brackets, as shown in illustration 4–5. On a line 1¾ inches in from the back edge of the side panel, mark the holes every 4 inches, starting 11 inches from the bottom. The second set of holes are placed 1 inch from the outside edge, on the same level as the first set. Use a strip of tape wrapped around a ¼-inch drill bit to indicate ½-inch depth. Drill the bracket holes only ½ inch into the inner side of each panel. Be sure to put the holes on the inner side of the panels.

To set the glazing into the sides, a rabbet must be cut around the side openings. Set a router with a ⅜-inch rabbeting cutter to cut 1/16 inch deep. Cut the rabbet ⅜ inch wide and 1/16 inch deep, on the outside face of the panels, as shown in photo 4–2.

Once you have completed the two side panels, you are ready to prepare the greenhouse framing assembly. Crosspieces will hold one side to the other and support the front glazing. From the pine 1 × 2s, cut four cross members, each measurement B minus 1¾ inches. Take one of the cross members and plane off a long edge 15 degrees to fit the slope of the front edge of the side panels, as in photo 4–3. Refer to

Photo 4–2—A router is the best tool to use for cutting the rabbets in the sidepieces for the glazing.

Photo 4–3—Plane a 15-degree angle on one cross member to fit the sloped front glazing.

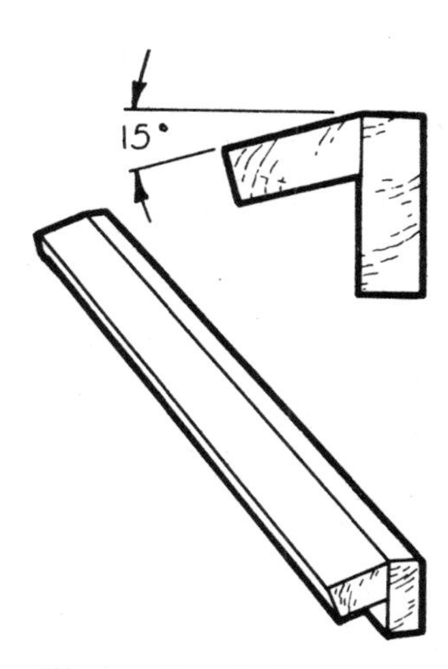

Illustration 4–6—Top L brace.

illustration 4–6 to assemble the trimmed crosspiece with another crosspiece, using waterproof wood glue and 6d cement-coated box nails, to make an L-shaped top cross member.

Now start to assemble the greenhouse frame. Using the waterproof glue and 6d box nails, fasten the bottom and the top cross member to one side panel. Space the two remaining cross members evenly along the front edge and fasten them in place. Then glue and nail the second side panel in place. Be sure each of the two middle cross members is square with the sides before fastening the second side.

Inside the greenhouse there will be covers to close off the vent openings. Lay out and cut the vent covers from ¾-inch plywood, each 7¾ × 7¾ inches. See illustration 4–7 to shape the front curve on the covers. First, find the center point of the plywood square. Draw a half-circle with a radius of 3½ inches. Cut the cover with a saber saw. In the center of each cover, drill a ¼-inch hole for the vent handle. Mount a vent handle on the inside of each cover, as shown in illustration 4–8.

You are now ready to paint the assembled greenhouse unit and the vent covers. Fill and sand all voids in the plywood edges. Use a coat of primer and two coats of exterior enamel to protect the greenhouse both inside and out. You may also want to use a wood sealer before painting, as the greenhouse humidity makes it important to seal the wood.

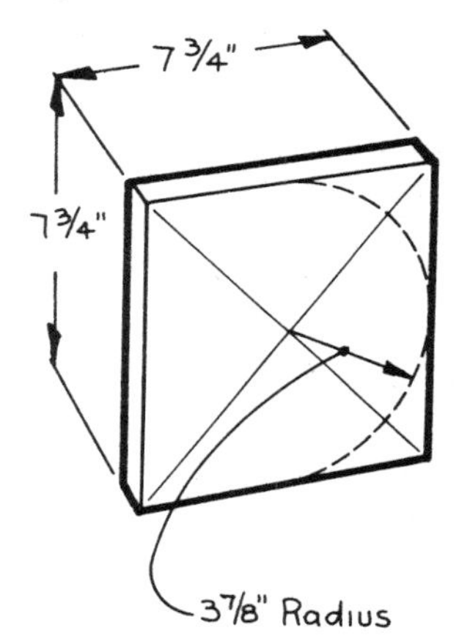

Illustration 4–7—Vent cover.

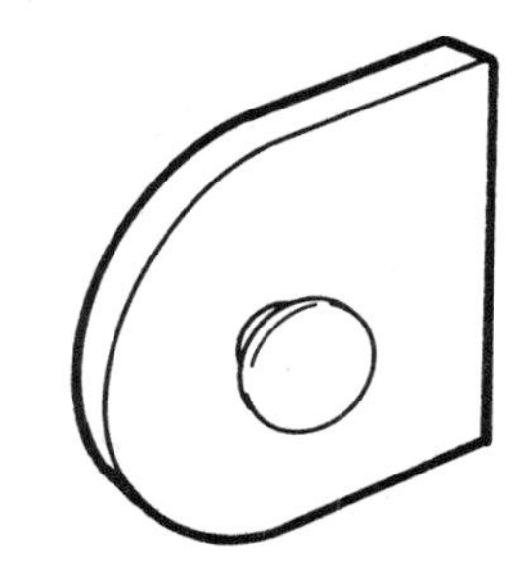

Illustration 4–8—Completed vent cover.

Using the center of the vent handle attachment hole as the center point, lay out a circle with a 3½-inch radius on what will be the outside face of the vent cover. Attach closed-cell foam weather stripping along the circumference of this circle on both vent covers. Next, mount self-closing hinges on the vent covers and attach them to the inside of the side panels so that the weather stripping seals the opening.

Snap a 6-inch aluminum louvered vent into the vent holes on each side from the outside to check the fit. Push the vent out again, and apply caulk to the flanges of the vent, then push the vent back into the hole. That completes work on the vents for now.

The front glazing, shown in illustration 4–9, gives the greenhouse its distinctive look. It is made of three pieces of 1⁄16-inch acrylic sheet. Before you cut the glazing to length, place it on the curved front and check the measurements carefully. Be sure the glazing and the frame are square. Lay out and cut the bottom piece of glazing first. The glazing should cover the lower cross member completely and overlap the bottom by ¾ inch. The glazing should be about ⅛ inch narrower than the greenhouse to provide a space for caulk under the aluminum side strips.

The middle section of glazing extends from the top edge of the upper cross member to the bottom of the lower cross member and overlaps the bottom glazing piece by ¾ inch. The top piece covers the L-shaped

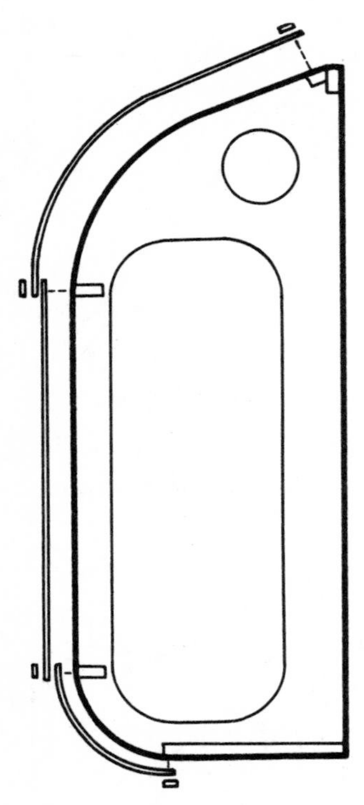

Illustration 4–9—Glazing pieces and battens.

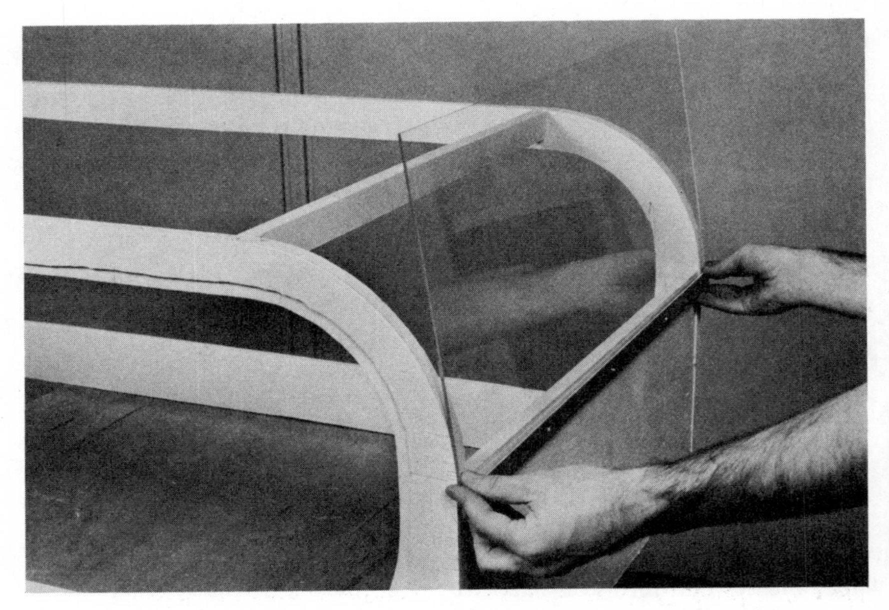

Photo 4–4—Align and mark the holes in the glazing, using the cross strips for guides.

cross member and overlaps the middle glazing piece by ¾ inch. The overlaps allow for expansion and contraction of the glazing and will shed water if installed as they were cut, from the bottom to the top.

After cutting the glazing to size, the next step is to cut the aluminum cross strips. The strips are used to hold the glazing in place. Use a hacksaw to cut four cross strips, each measurement B minus 1¾ inches. Drill five evenly spaced 3⁄16-inch-diameter holes in each of the four strips. Designate one strip as the bottom strip, and drill four additional holes, evenly spaced, in that strip. Countersink all the holes with a ¼-inch drill. Test your first countersink with a #6 screw to be sure the screw doesn't pull through when tightened. Remove all burrs and label the strips *bottom, lower, upper,* and *top*.

To install the glazing, simply start at the bottom of the unit and hold the glazing in place, with a ¾-inch overlap on the bottom, and check the fit, as shown in photo 4–4. Then, using the bottom cross strip to position the holes, drill 3⁄16-inch-diameter holes to align with the holes in the cross strip. The holes are slightly oversized to allow for slight contraction and expansion of the acrylic sheet. With this done, remove the glazing and use the cross strip to mark 1⁄16-inch-diameter pilot holes in the wood. This way, all holes should align.

Tighten the bottom cross strip and the glazing in place with ¾-inch #6 aluminum flathead wood screws. Then, holding the top of the glazing in place, check the fit of the middle section of glazing. Again, use the cross strip to position holes and drill the holes in both pieces of glazing and the pilot holes in the brace. Then attach the middle piece of glazing and the top of the bottom piece, being sure the middle piece overlaps the top of the bottom piece. Follow the same procedure for the top piece of glazing.

With the glazing firmly in place, use a hacksaw to cut the two aluminum side strips. Measure the outside edge of a side panel, using a measuring tape to bend around the curved edge. Then cut the two side strips. Drill and countersink fourteen 3⁄16-inch-diameter holes in each of the side strips. If your greenhouse is taller than ours, drill holes every 4 to 5 inches. Screw the side strips in place, as you did the cross strips. The only difference is that you will have to hold the pieces in place, mark the hole positioning on the glazing, and drill the oversize glazing holes and the pilot holes with the glazing in place. Use care not to extend the glazing hole into the wood.

On the acrylic sheet, lay out the two side glazing pieces, each 14¼ inches by measurement A minus 15 inches. Follow illustration 4–10 to mark the corner curves with a 6⅛-inch radius. Cut out the glazing with a saber saw with a fine-toothed blade, and fit it into the rabbet around the side opening. If you need to shave the glazing to make it fit, a file

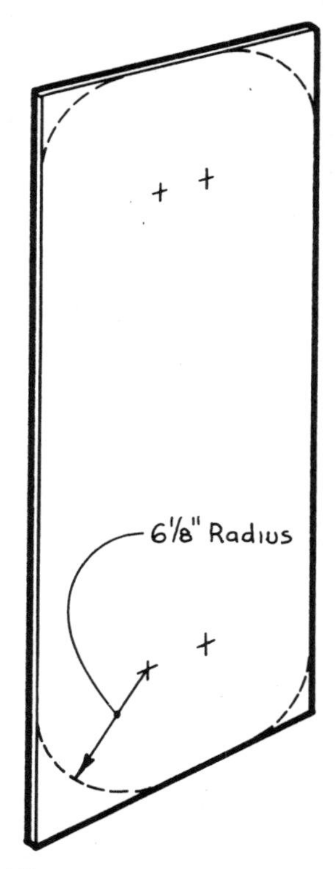

Illustration 4–10—Side glazing layout details.

will work well. With the glazing in the opening, mark the location of approximately 20 holes around the edge of each sheet. With the glazing on your workbench, drill ⅛-inch-diameter holes where marked. Hold the acrylic firmly on both sides of the hole as you drill, as it tends to catch the drill bit and splinter. Drill very shallow countersink holes in the glazing. Replace the glazing in the side to mark the pilot holes.

Lift the glazing out of the rabbet, drill 1/16-inch-diameter pilot holes, and lay a bead of caulk in the rabbet. Replace the glazing and screw it in place with ½-inch #4 aluminum flathead wood screws.

The vents and aluminum strips should be primed with metal primer and then painted with exterior paint. Caulk the top edges of the cross strips and the outside edges of the side strips.

Make one or more shelves to fit the greenhouse with ¾ × ¾-inch baluster-stock slats spaced evenly on ¼-inch-diameter hardwood dowels, as shown in illustration 4–11. The front and back shelf edges are cut from lengths of 1 × 2. Cut the slats and shelf edges to a length of measurement B minus 1¾ inches. This allows ⅛-inch clearance on each side of the shelves. Drill four 5/16-inch-diameter holes in each piece. Each piece should be matched with the others so that the slats slide

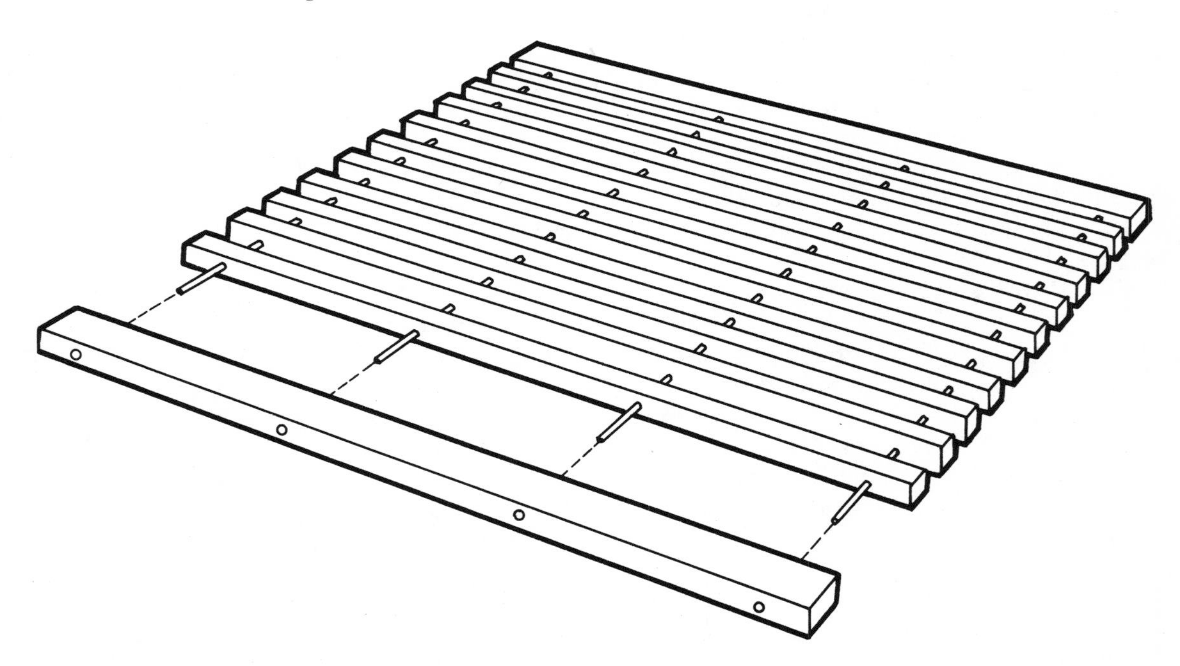

Illustration 4–11—Shelf construction, using dowels and baluster stock.

easily onto the dowels. Fit the slats onto the dowels and space them evenly, with the shelf edges at the ends. Glue any slats that are loose-fitting. Check the fit of the shelves in the greenhouse, and trim the slats where needed so they can be easily adjusted in the greenhouse. Use ¼-inch shelf brackets pushed into the holes in the sides of the greenhouse to hold the shelves.

To mount the greenhouse to your window, drill eight 3/16-inch-diameter holes through each side panel, close to the house edge. Set the greenhouse in place, and mark for pilot holes in the window frame. Drill 1/16-inch-diameter holes in the wooden frame, then fasten the greenhouse in place with 1½-inch #8 flathead wood screws. Caulk around the greenhouse frame.

If the greenhouse is oriented properly, it is possible it may help to heat your house. Open both the top and bottom window sashes about 6 inches, and an air current will develop. Cool room air will be pulled into the greenhouse, and, as it is warmed, it will rise and circulate out over the top window into the room. At night the flow will reverse so that the greenhouse will receive heat from the house.

5 Solar Barbecue

This solar barbecue is somewhat of a breakthrough in using solar energy at home. It is easy to use, lots of fun, efficient, durable, and handsome. It can be the center of attraction in your outdoor living space. The barbecue is a unique, high-temperature cooker that can easily broil meat or bake potatoes in the same length of time as a conventional barbecue—but without the use of any energy except the sun.

Five reflectors focus the sun's rays onto a griddle. Each reflector is set at a slightly different angle and radius to concentrate the sunlight on the griddle properly. By concentrating all the energy that strikes the

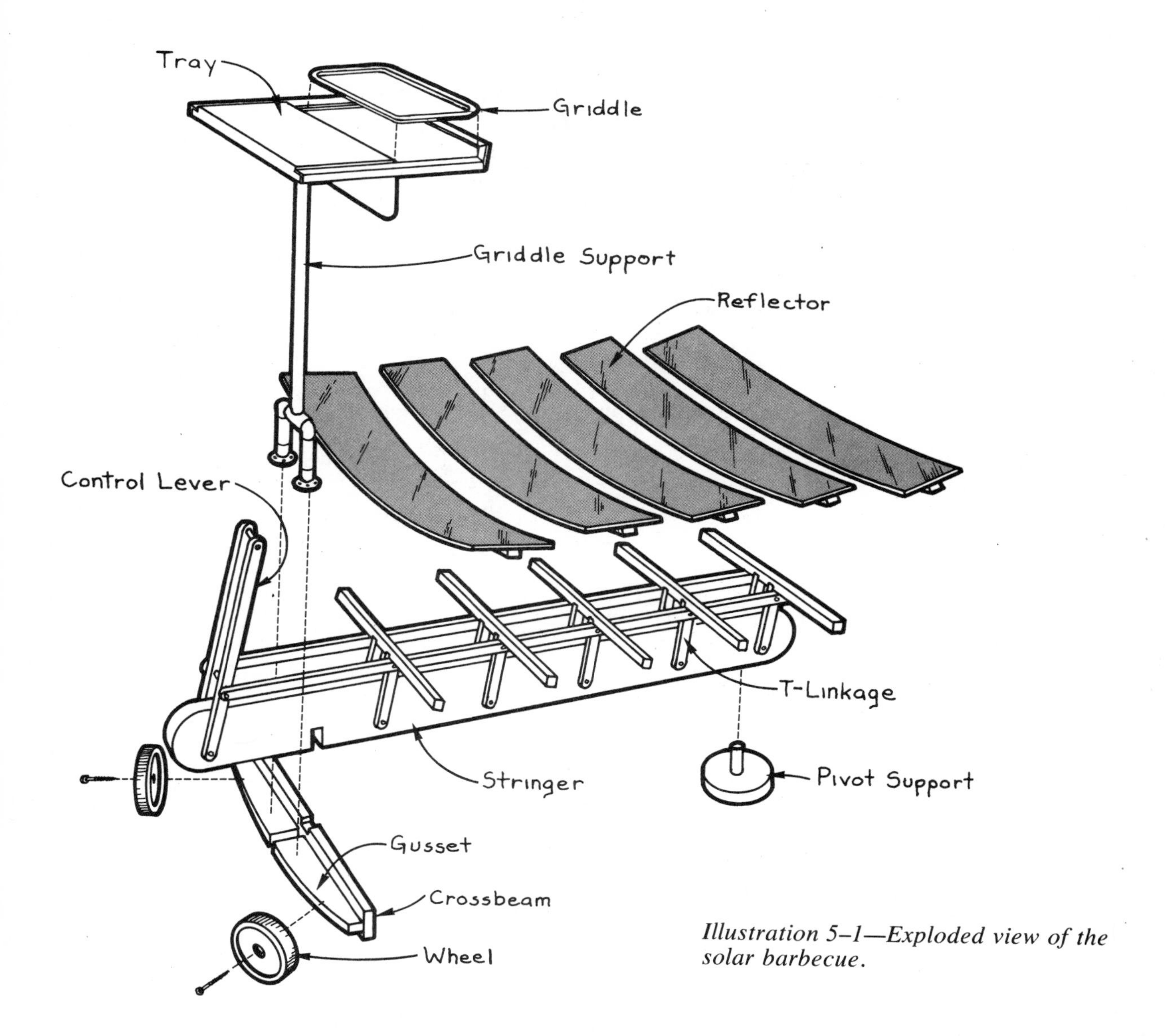

Illustration 5–1—Exploded view of the solar barbecue.

large reflector area onto the smaller griddle, high temperatures are reached. The barbecue is sized to handle a big steak or several pieces of chicken at one time. An added plus is that the grilling surface is at a convenient working height.

The solar barbecue is constructed of redwood, 1-inch pipe, and steel reflectors. It can easily be partially dismantled for winter storage.

CHART 5–1—
Materials

DESCRIPTION	SIZE	AMOUNT
	Lumber	
Redwood	2 × 6 × 8′	4
Redwood	2 × 4 × 8′	1
Redwood	2 × 2 × 8′	1
Redwood	1 × 2 × 8′	4
A-C Exterior Plywood	¼″ × 4′ × 4′	1
Hardwood Dowel	⅞″ × 8″	1
	Hardware	
Finishing Nails	8d	6
Aluminum Nails	1¼″	5
#14 Flathead Wood Screws	1¼″	12
#10 Flathead Wood Screws	3½″	4
#10 Flathead Wood Screws	1½″	1
#8 Aluminum Flathead Wood Screws	1½″	16
#8 Aluminum Flathead Wood Screws	1″	60
Lag Bolts with Nuts	½″ × 4″	2
Hex Head Bolts with Nuts	⅜″ × 4″	6
Hex Head Bolts with Nuts	¼″ × 5½″	6
¼–20 Toggle Bolts	3″	10
¼–20 Hex Nuts	...	30
Flat Washers	½″	4

CHART 5–1—*Continued*

DESCRIPTION	SIZE	AMOUNT
	Hardware—*continued*	
Flat Washers	1/4″	20
Fender Washers	3/8″	12
Fender Washers	1/4″	12
Loose-Pin Hinges with Screws	1 1/2″ × 2″	10
#8 Cup Hook	...	1
Screw Eyes	3/8″ × 3/4″	10
Aluminum Angle	3/4″ × 3/4″ × 1/8″	58″
Aluminum Flashing	8″ × 22″	1
	Miscellaneous	
Pipe Flanges	1″	3
Pipe Nipples	1″ × 5 1/2″	2
Pipe Nipples	1″ × 2″	2
Pipe Elbows	1″	2
Pipe Tee	1″	1
Pipe	1″ × 30″	1
Wheels	7″	2
Waterproof Wood Glue	...	1 pt.
Trim Adhesive	...	1 can
Metal Primer	...	1 can
Aluminum Paint	...	1 can
Exterior Urethane	...	1 qt.
Lead Solder	...	trace
Flux	...	trace
Reflective Mylar	8″ × 20′	5
Cast-Iron Griddle	10″ × 17″	1

CHART 5–2—
Tools

Hacksaw
Circular Saw
Backsaw
Saber Saw with Rip Guide
Drill and Bits
Bar Clamps: 12-inch
Utility Knife
Wood Chisel
Pipe Wrenches
Protractor
Trammel Points
Grinding Wheel (optional)
Propane Torch
Angle Finder

At first the solar barbecue may appear complicated to build, but the construction methods are not difficult. Our directions are detailed for the novice builder. We include a complete list of materials, chart 5–1, and of the tools you will need, chart 5–2. An exploded view of the barbecue is shown in illustration 5–1.

The design that follows is for a wheeled version of the solar barbecue. The unit could be mounted permanently, and the modifications needed for a permanent mount are discussed at the end of this chapter.

Begin construction of the solar barbecue with the support stringer, shown in illustration 5–2. Cut the 2 × 6 stringer to a length of 74 inches. Measure 4 inches, 16 inches, 28 inches, 40 inches, 52 inches, and 70 inches from one end of the stringer. Drill ⅜-inch-diameter holes 1 inch from an edge, designated as the lower edge, at the six points. Five holes are for the reflector pivots and one, at the 70-inch mark, is for the control handle pivot. Draw a 2¾-inch-radius half-circle at each end of the stringer, and round off the ends with a saber saw.

The stringer is also notched to fit a 2 × 4 crossbeam, set on edge. Measure 15 inches and 16½ inches from the control handle end to mark the sides of the notch. Referring to illustration 5–2, measure 2½ inches from the lower edge for the depth of the notch. Cut the notch and clear away any excess wood with a wood chisel.

The solar barbecue is designed to pivot around a fixed point at the front of the stringer so that the reflectors can easily be turned to face the sun. The pivot pin is glued into a ⅞-inch-diameter hole drilled 2 inches deep into the bottom edge of the stringer. Drill the hole 11 inches

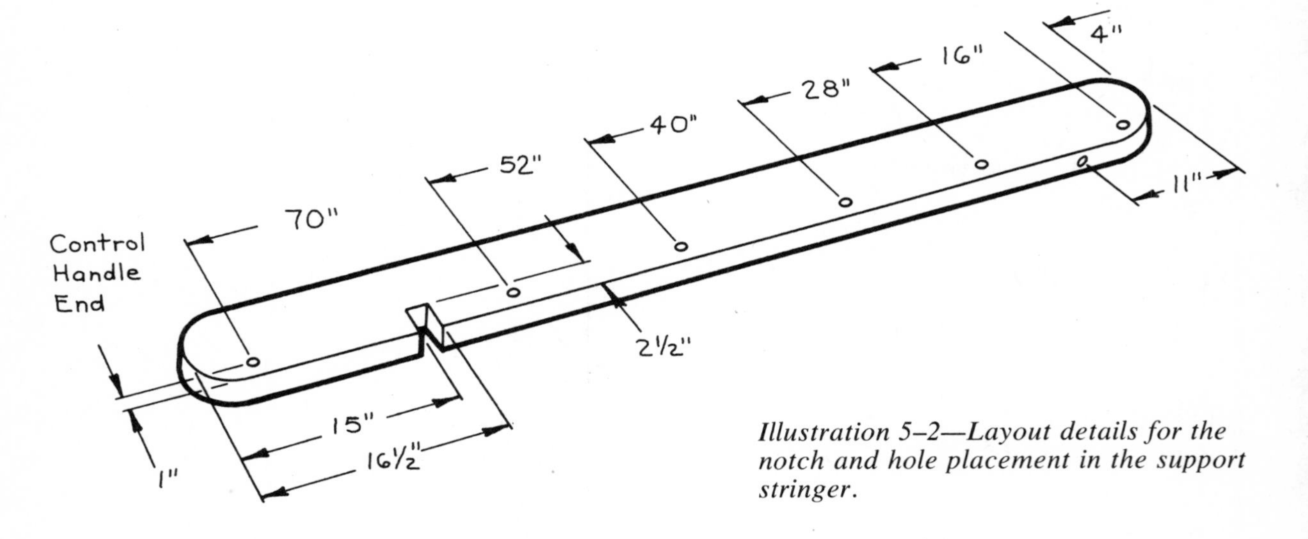

Illustration 5–2—Layout details for the notch and hole placement in the support stringer.

from the reflector end of the stringer. Cut a 3¼-inch-long piece of ⅞-inch-diameter hardwood dowel to make a pivot pin. Chamfer the bottom end of the pivot pin by rasping or sanding a 45-degree bevel around the edge. The chamfered end will rotate inside the pivot support. Glue the pivot pin into the hole in the stringer with waterproof glue.

Make a circular pivot support, shown in illustration 5–3, from scrap 2 × 6 redwood, to hold the pivot pin. Be sure to save three 8-foot 2 × 6s for the reflector bases and the gusset. Cut a 5½-inch-diameter circle, using the saber saw. Drill a 1-inch hole ¾ inch deep in the center to fit the pivot pin. Then drill a ⅛-inch-diameter pilot hole through the center of the larger hole for the fastening screw.

The crossbeam, illustration 5–4, supports the stringer and, with the gusset, provides a stable support for the wheels that allow the grill to be turned to face the sun. Cut the crossbeam from a 2 × 4, to a length of 31¼ inches. The crossbeam receives a notch 1 inch deep and 1½ inches wide, centered in the crossbeam. Cut this notch as you did the one in the stringer.

The wheels of the barbecue are attached to two gussets, made from a single piece of 2 × 6. The gussets are curved to set the wheels at an angle so they easily move around the pivot point. The gusset shape is drawn onto a single board, then the board is cut to make two gussets, as shown in illustration 5–5. First, cut the 2 × 6 to 29¾ inches and clamp it to a flat surface where you have several feet of work space. At the midpoint, measure 4 inches across the width and make a mark. Draw a line on your work space perpendicular to the midpoint of the 2 × 6,

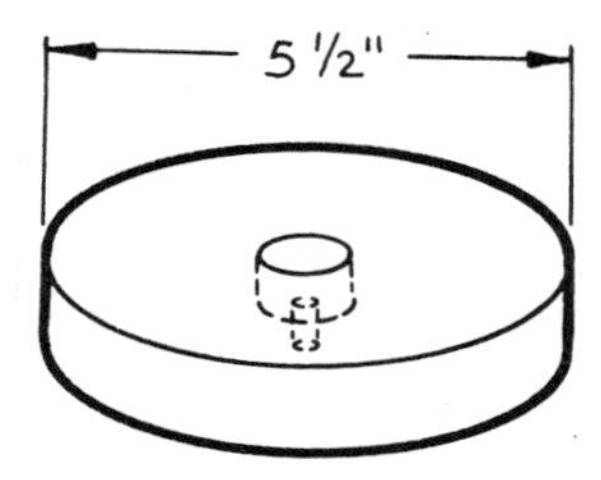

Illustration 5–3—Pivot support.

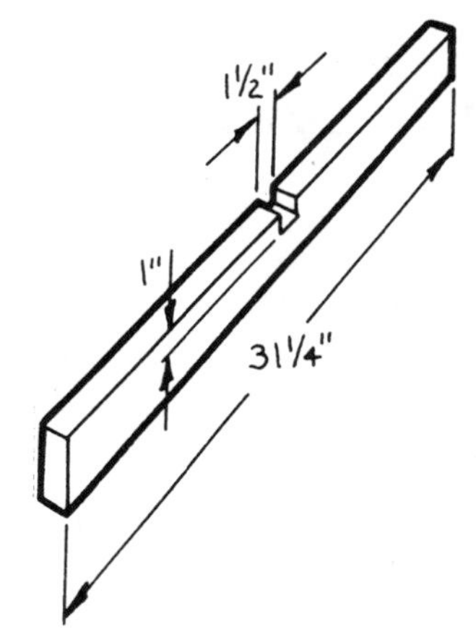

Illustration 5–4—Crossbeam layout.

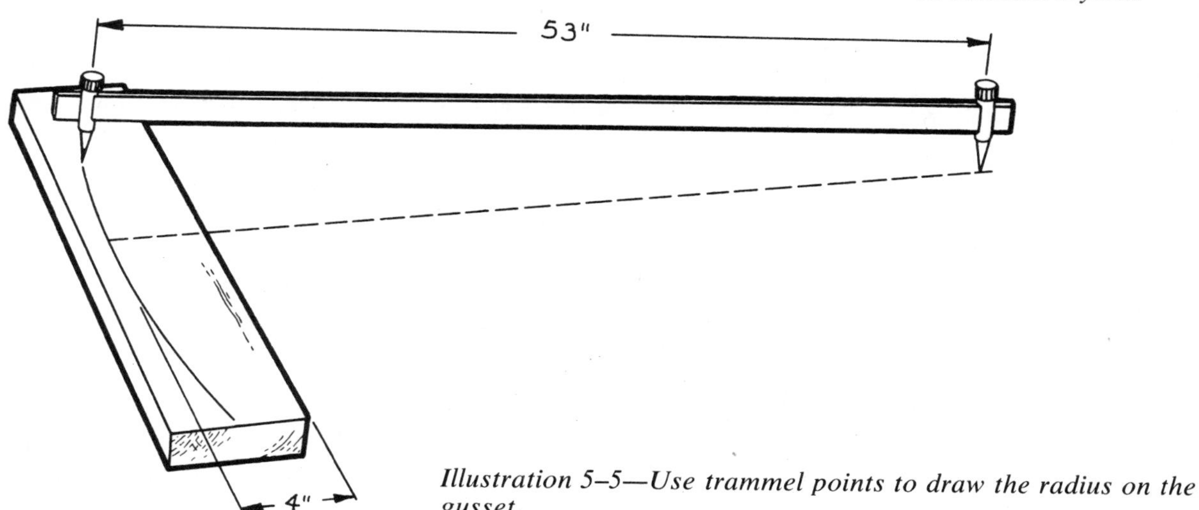

Illustration 5–5—Use trammel points to draw the radius on the gusset.

Photo 5–1—Draw the 53-inch gusset radius, using a shop-made compass pivoting from the floor.

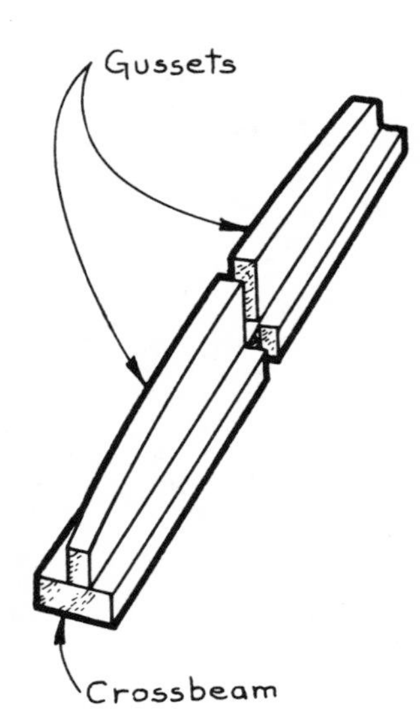

Illustration 5–6—Gussets and crossbeam.

extending 53 inches from the mark. Set trammel points, which are shown in photo 5–1, for a 53-inch radius. With one end on the perpendicular line and the other end on the 4-inch mark at the midpoint of the gusset board, draw a 53-inch radius lengthwise onto the gusset, as shown in illustration 5–5. Cut the curved edge first, then cut the board in half at the midline to make the two gussets.

Fasten the gusset pieces to the crossbeam with four 3½-inch #10 flathead wood screws and waterproof glue. Set the crossbeam on its face, and place the gussets on edge in the center of the crossbeam, ends flush. Be sure to leave a space to fit the stringer between the gusset halves, as shown in illustration 5–6. Glue the pieces, clamping them together, then drill ⅛-inch-diameter pilot holes and 3/16-inch-diameter shank holes 4 inches and 10 inches from the center of the crossbeam. Screw the crossbeam to the gussets.

When the glue has set, attach the crossbeam-and-gusset assembly to the stringer, using waterproof glue and 8d finishing nails. Then sand the assembled unit while the surfaces are still accessible.

Drill ⅜-inch-diameter pilot holes 3 inches from the ends of the gussets for fastening the two 7-inch-diameter wheels, as shown in illustration 5–7. The next step is to attach the wheels. For each wheel, use two ½-inch flat washers as wheel spacers between the wheel and gusset. Slip a ½ × 4-inch lag bolt through the wheel and two washers, then screw it into the gusset.

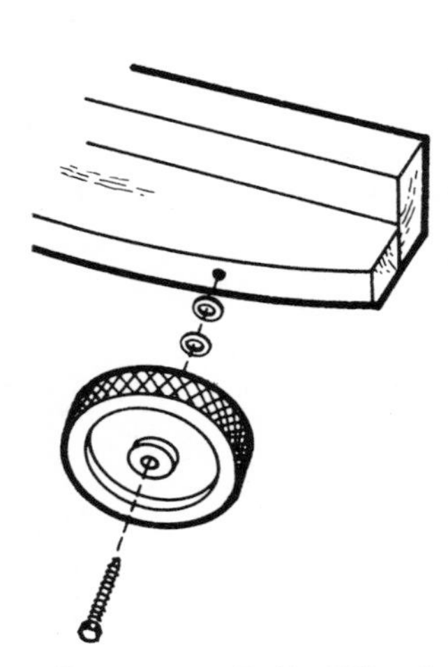

Illustration 5–7—Wheel assembly.

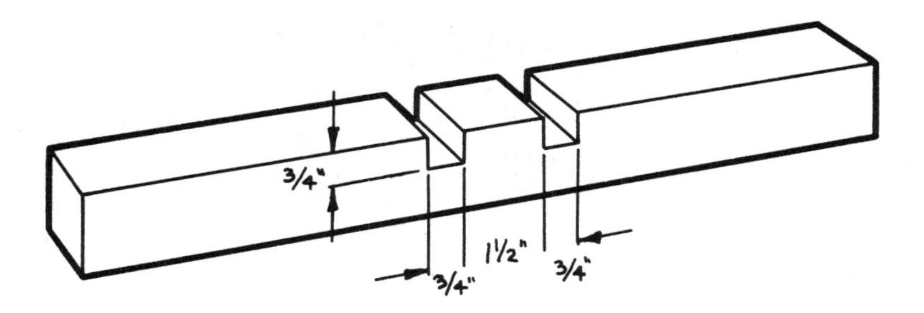

Illustration 5–8—
T-linkage crosspiece.

The adjustment system for your solar barbecue is the T-linkage, which enables you to move all five reflectors with just one lever. In this way, maximum sunshine can be directed to the griddle at any season of the year or any time of the day. The T-linkage enables you to capture the sun as its angle of altitude changes due to seasonal variations, and the pivot point allows you to swing the entire unit to track the sun's path during the day!

The T-linkage is constructed from redwood. First cut five pieces of 2 × 2s, each 16 inches long, for the cross members. As shown in illustration 5–8, each crosspiece receives two dadoes. Each dado is ¾ inch wide and ¾ inch deep, and starts ¾ inch on either side of the centerline of the crosspiece. This positioning allows a 1½-inch space for the stringer. Mark the lines and cut the dadoes, then clean the cuts with a chisel.

Next, for the vertical pieces, cut ten pieces of redwood 1 × 2s, each 8 inches long. Measure ¾ inch from one end of each piece, and drill a ⅜-inch-diameter hole as the pivot. Mark a point 6¼ inches from the same end, and drill a ¼-inch-diameter hole for the control arm. Cut and shape a ¾-inch radius at the bottom of each vertical piece, as shown in illustration 5–9.

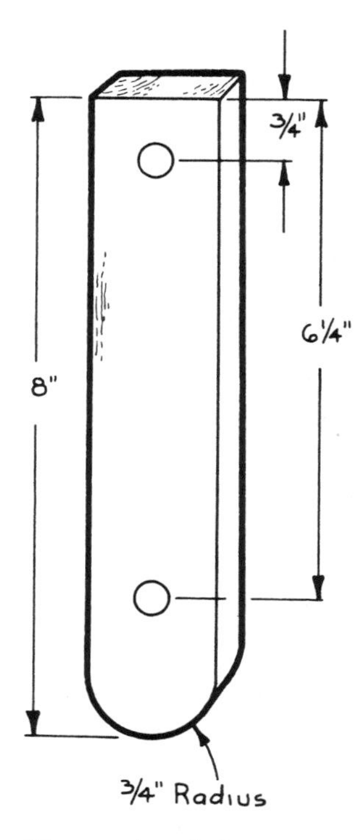

Illustration 5–9—
T-linkage vertical piece.

Assemble the pieces of the linkage with waterproof glue and 1½-inch #8 aluminum flathead wood screws, as shown in illustration 5–10. Drill a 3/16-inch-diameter shank hole and a 1/16-inch-diameter pilot hole, and screw down through the crosspieces into the vertical pieces. Put two ¾-inch-long, ⅜-inch screw eyes into each crosspiece to serve as adjustment guides for the reflectors. Position the screw eyes 1 inch from each end and ½ inch above the lower edge of the crosspiece. Drill ¼-inch-diameter pilot holes, and insert the screw eyes. Tighten the screw eyes and position them so they are parallel to the long axis of the crosspiece. Sand the completed T assembly, and move on to the control linkage.

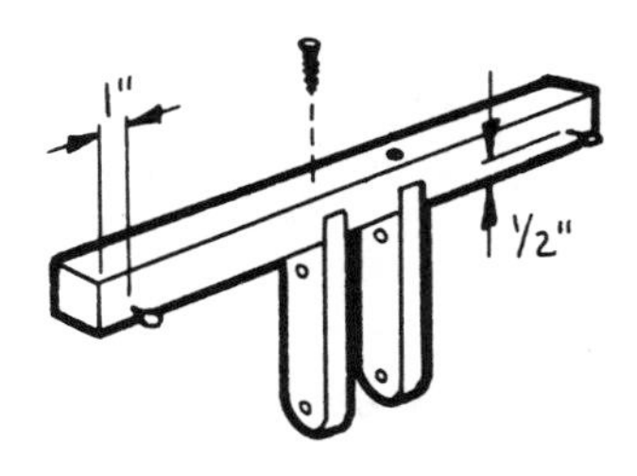

Illustration 5–10—
T-linkage assembly.

The T-linkage is controlled through two arms extending the length of the barbecue. The arms are fastened to the Ts and to the control lever. Refer to the exploded illustration 5–1 to see the entire control assembly. First, cut two pieces of clear redwood 1 × 2s, each 68 inches

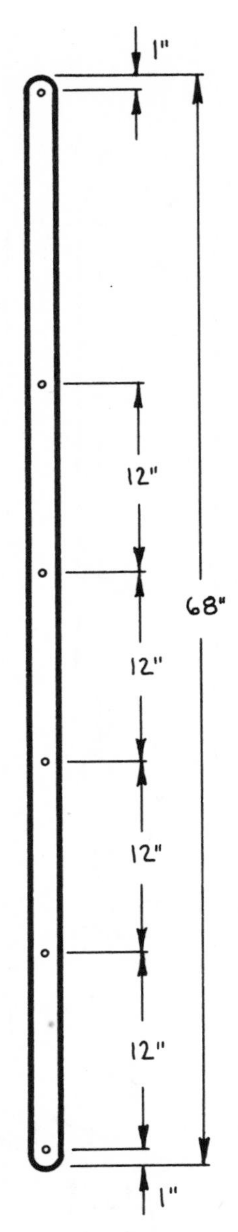

Illustration 5–11—Adjustment rod layout.

Photo 5–2—Rip 1-inch-wide T-linkage strips, using a saber saw and the guide arrangement shown here.

long. Rip each piece to 1 inch wide. If you have only hand-held power tools, use the method shown in photo 5–2.

Drill six ¼-inch-diameter holes, as shown in illustration 5–11, starting at a point 1 inch from an end, and repeating at 12-inch-on-center intervals from the first hole for four more holes. The last hole goes 1 inch from the other end. At each end of the arms, cut and shape a 1-inch-diameter radius.

The lever controls extend to a convenient height so that you can easily adjust the reflectors. Cut two pieces of redwood 1 × 2, each 24 inches long, for the levers. Round both ends of each piece by cutting a 1½-inch-diameter radius. Then, ⅜ inch from the end designated as the top, drill a ⅞-inch-diameter hole in each control lever to fit the dowel handle. The ⅜-inch-diameter pivot holes are ¾ inch from the bottom end of the control lever. The ¼-inch control arm holes are 6¼ inches from the bottom end.

Finally, for the handle, cut a ⅞-inch-diameter hardwood dowel 3¾ inches long. Glue the handle between the lever controls, keeping them parallel and spacing them so that they easily slip over the stringer. Sand the parts of the control system, rounding or "breaking" the edges.

The power of the solar barbecue lies in the reflectors. Each panel has been designed with a different curve to best concentrate the sun's rays on the griddle. Since we have already calculated these arcs, all you need to do is use the figures we give here to lay out the reflector bases. Label each one, as it is important to assemble them in a particular order. Each reflector consists of a redwood base, plywood backing, and a laminated Mylar and aluminum reflective surface.

Illustration 5–12 shows the layout for the reflector bases. Cut five base pieces, each 48 inches long, from redwood 2 × 6s, and label the pieces *A–E*. On each piece, measure 24 inches to the center point, and square a line across. Measure ⅞ inch from the edge, and make a mark on the centerline. Use this point to strike off each radius. As you did for the gusset, clamp the base pieces to your table, and draw a line from the center point perpendicular to the base. Draw the radius lines with trammel points, giving each reflector base a dished shape, ⅞ inch wide at the narrow center point. The radii of the reflectors are as follows: reflector A, 64 inches; reflector B, 68 inches; reflector C, 79 inches; reflector D, 97 inches; and reflector E, 114 inches.

Lay out the bottom edges of the reflector pieces before cutting the arc radii. Mark 7 inches in from each end, then measure 1½ inches down from the point at which the arc crosses the end of the base. Draw a diagonal line from the 7-inch point to the end point, and cut along this line. Then cut along the curved radius lines of the upper edges.

Making the reflector backing is next. Cut five pieces, each 8 × 48 inches, from ¼-inch plywood. Round each corner by cutting a ½-inch

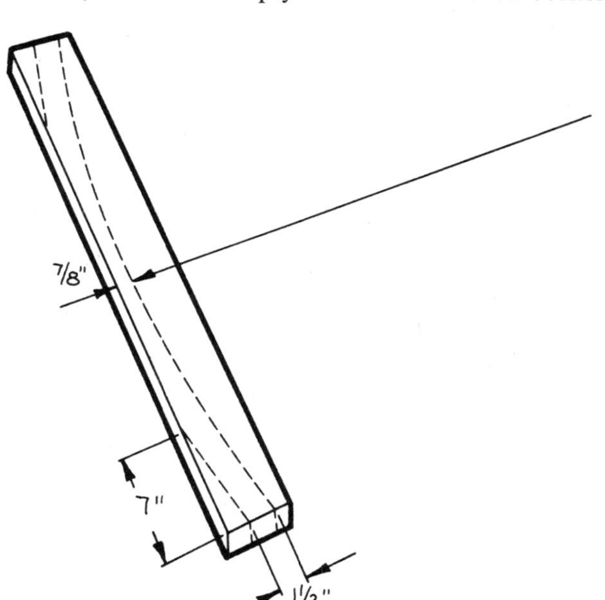

Illustration 5–12—Lay out and label each reflector support individually.

radius. Center the backing pieces on the bases, and fasten them with waterproof glue and 1-inch #8 aluminum flathead wood screws. Drill 1/16-inch-diameter pilot holes and 3/16-inch-diameter shank holes, and countersink the screws. Use ten screws in each reflector. Sand the completed reflector bases to remove any sharp edges.

At this stage, finish all the wood parts of the solar barbecue with three coats of exterior urethane, or the finish of your choice.

The next step in constructing the reflectors is to add a surface of reflective material. Use Mylar and aluminum flashing, as described below, or substitute heavy-duty aluminum foil for the Mylar. Aluminum foil will need to be replaced periodically. A very durable reflector material is polished stainless steel, although it is somewhat expensive initially, harder to work with, and not always available in some locales. The description that follows pertains only to the aluminum flashing and Mylar construction.

Cut five pieces of aluminum flashing material to 8 × 48 inches each. Fasten the flashing pieces to the plywood reflector backing pieces with automotive trim adhesive. Round the corners of the flashing to match the plywood backing pieces with a utility knife. The flashing reflects a good amount of light to the griddle, but this amount can be increased with the addition of reflective Mylar. Cut the Mylar pieces slightly oversize, apply them to the aluminum flashing with automotive trim adhesive, and trim them to size with a utility knife.

Now that you have built all the parts of the solar barbecue, you are ready to assemble them. Start the assembly by fastening each reflector

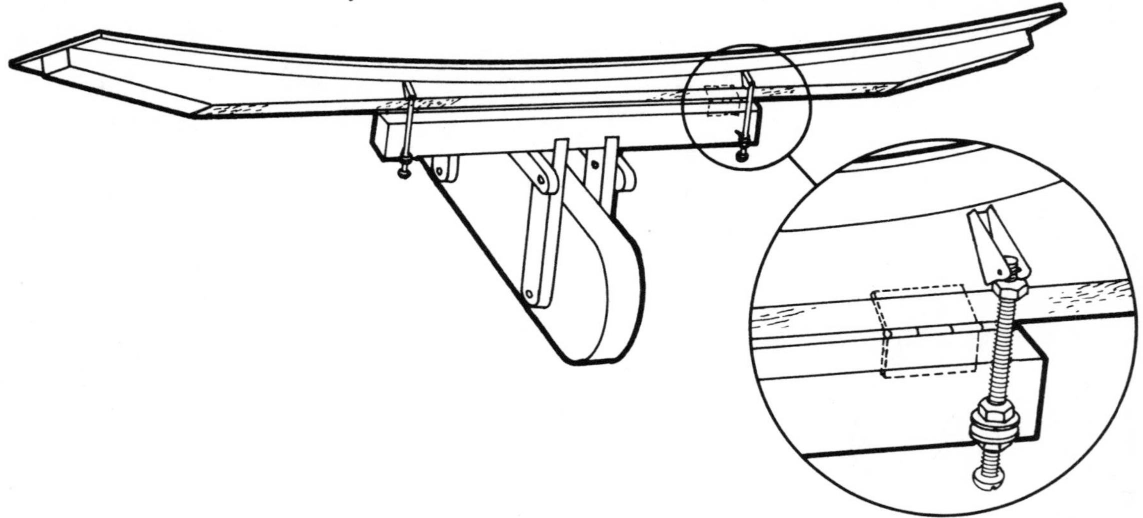

Illustration 5–13—Reflectors are adjusted with modified toggle bolts.

to the T-linkage with two loose-pin hinges. Using loose-pin hinges will allow you to remove the reflectors when you want to store the barbecue. Center the reflectors over the T-linkage. Place the hinges ½ inch in from each end of the horizontal member, and screw them to it and to the reflector base. Check illustration 5–13 to be sure you are mounting the hinges correctly.

Adjusting rods are necessary on each reflector to set the reflector to the correct angle. The rods are made from 3-inch-long ¼–20 toggle bolts, as shown in illustration 5–13. Remove one ear from each toggle bolt, and drill a 3⁄16-inch-diameter hole in the remaining ear. Trim the ear if necessary to fit the base. Using 1-inch #8 aluminum flathead wood screws, screw the toggle bolt ear to the reflector base directly above the adjustor screw eye on the T-linkage crosspiece, as shown in photo 5–3. Use three ¼–20 hex head nuts to secure the adjustor in position, one at the ear and two around the screw eye, as shown in illustration 5–13. Also use ¼-inch flat washers between the nuts and the screw eye. Repeat this procedure for each reflector, two adjustors for each one.

Photo 5–3—Modified toggle bolt.

Proceed to assemble the remaining parts of the barbecue and then make the final adjustments. *Note:* Each nut-and-bolt joint is a pivot, so do not overtighten them. Begin by fastening the control lever to the stringer with a 4-inch-long, ⅜-inch-diameter hex head bolt. Use two large washers and a hex nut with the bolt.

Next, attach the control arms to the control lever with a 5½-inch-long, ¼-inch-diameter hex head bolt. Use a hex nut to match and two ¼-inch fender washers at the head and nut. Finally, fasten the reflector units to the stringer and control arms. Be careful to keep the reflectors in order, with reflector A closest to the griddle and reflector E at the far end of the T-linkage. For the connection at the stringer, use 4-inch-long, ⅜-inch-diameter hex head bolts with two ¼-inch washers and a hex nut. *Note:* Mount reflector A, the reflector closest to the grill, with its adjustor rods on the grill side, opposite the other four reflectors, as shown in illustration 5–19, on page 77. At the control arm connection, use 5½-inch-long, ¼-inch-diameter hex head bolts with two ¼-inch fender washers and a hex nut.

The griddle, which is the cooking surface, is supported by a 1-inch galvanized pipe assembly. Following illustration 5–14, assemble the parts of the griddle support. Join the two 5½-inch nipples to the flanges and elbows. Use two pipe wrenches to tighten these joints, as shown in photo 5–4. Trim two of the flanges with a hacksaw to a width of 3¼ inches, as shown in illustration 5–15, so they will fit onto the gusset. The next four joints will need to be permanently fixed by soldering, brazing, or a mechanical connection. In preparation for soldering or brazing, clean the ends of the 2-inch nipples, the elbows, and the tee

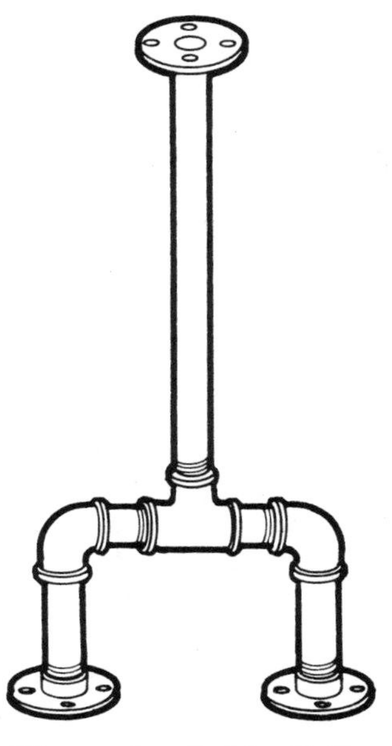

Illustration 5–14—Grill support.

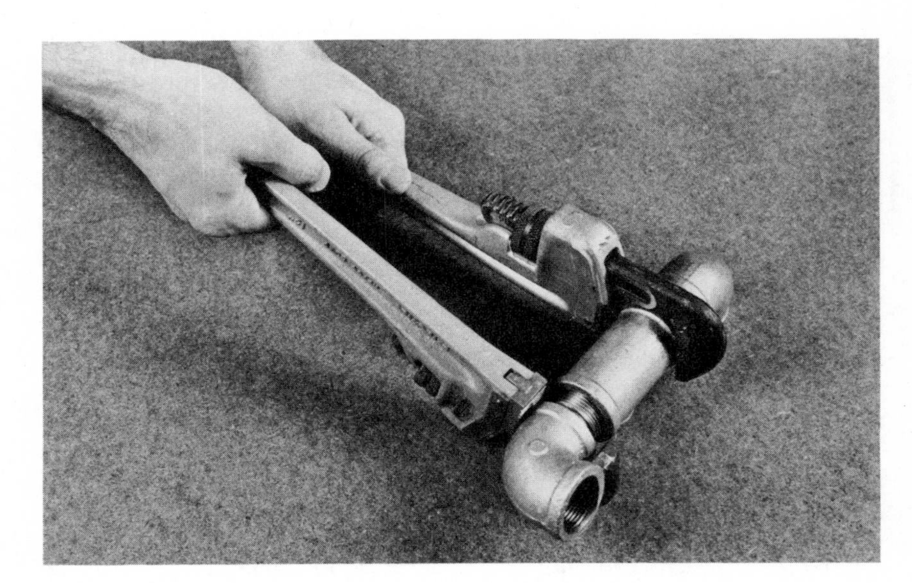

Photo 5–4—Use two pipe wrenches to tighten the fittings for the griddle support.

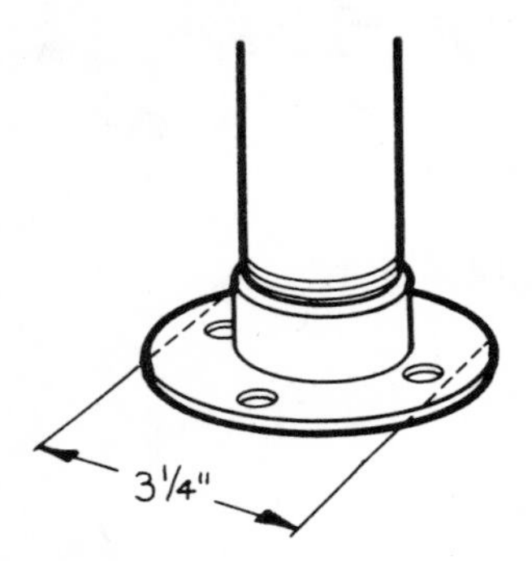

Illustration 5–15—Cutting details for the flange.

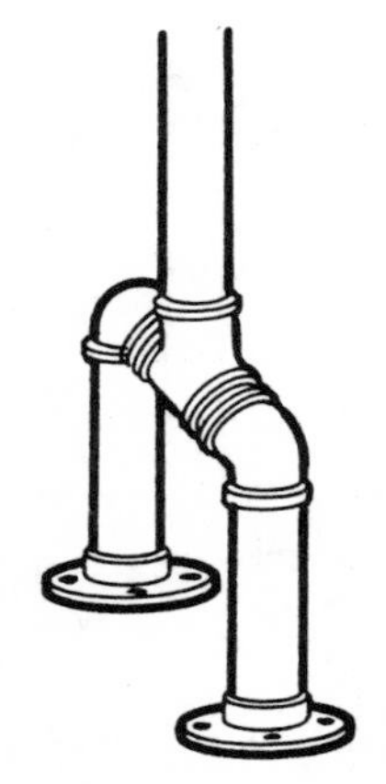

Illustration 5–16—Supports must be plumb.

with solvent to remove any grease. Cover the joint areas with flux or soldering paste. Tighten these pieces together and be sure the tee is in the correct vertical position in relation to the flanges, as shown in illustration 5–16. Using a propane torch (or Mapp gas in the case of brazing), flow lead solder into the four joints so that the tee cannot change position.

If you prefer, these four joints can be secured with ¼-inch rolled pins. Drill a ¼-inch hole through each joint to match the pin diameter, tap in the pin, and cut it flush with the pipe.

Finish the support assembly by adding a 30-inch section of pipe to the tee. Paint the griddle support, using a metal primer and finish coat of aluminum spray paint.

Fasten the griddle support to the gusset with 1¼-inch #14 flathead wood screws. First set the support in place to mark the holes. Drill ⅛-inch pilot holes in the gusset, then screw down the flanges.

At the top of the support we have designed a tray of redwood to help hold the griddle and to provide an additional work surface. Our dimensions are calculated to fit a 10 × 17-inch cast-iron griddle (including the handles, which rest on the aluminum frame). Adjust your dimensions for the tray and frame if your griddle is a different size. The tray is made of three pieces of redwood 2 × 4, cut to 18¾ inches long. Using waterproof glue, fasten the three pieces together on edge lengthwise, as shown in illustration 5–17. Hold the pieces with pipe clamps until the glue has set.

The ends of the tray are rabbeted to fit the aluminum angle that holds the grill. To mark the rabbet, draw a line 9/16 inch from each end of the tray. Mark ¾ inch deep at each end of both lines. With a backsaw or circular saw, cut just to the line, as shown in photo 5–5. Remove excess wood with a wood chisel, and sand the tray, rounding the top edges and the corners, then finish it with three coats of exterior urethane.

Fasten a pipe flange underneath the tray to fit the pipe support. The outer edge of the flange is ¼ inch from the forward edge of the tray and centered on the width.

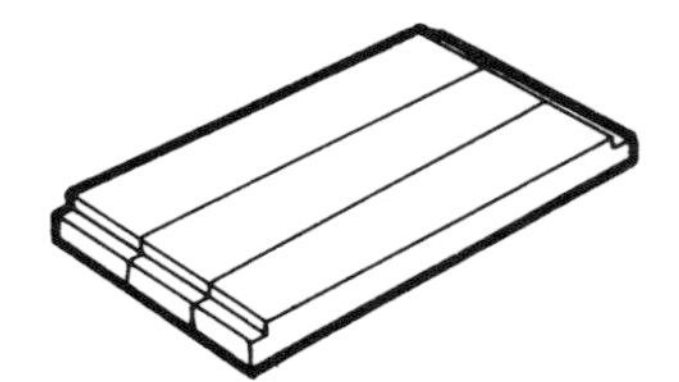

Illustration 5–17—Grill tray.

The griddle is held by an aluminum angle frame, which is cut and shaped to fit. As illustration 5–18 shows, the aluminum is cut with a hacksaw through one side only so that the bends can be made. Cut a piece 58 inches long, and measure 18 inches, 21½ inches, 36½ inches, and 40 inches from one end of the aluminum. Cut a 45-degree angled V at these points. Bend the angle so that it will rest on the rabbets in the tray. Lay the shaped angle in place against the tray. Drill ³⁄₁₆-inch-diameter countersunk holes in the aluminum and ¹⁄₁₆-inch-diameter pilot holes in the tray. Be sure the countersink is shallow so the screw can't slip through the aluminum. Fasten each side with three 1½-inch #8 aluminum flathead wood screws.

A tray shield serves to protect the tray and the person cooking if the reflectors are accidentally focused incorrectly. Cut the 8 × 18¾-inch shield from aluminum flashing. Fasten it to the front edge of the redwood tray between the griddle and the tray, using 1½-inch aluminum nails. Paint the flange as you did the support, then fasten the tray unit in place onto the support pipe.

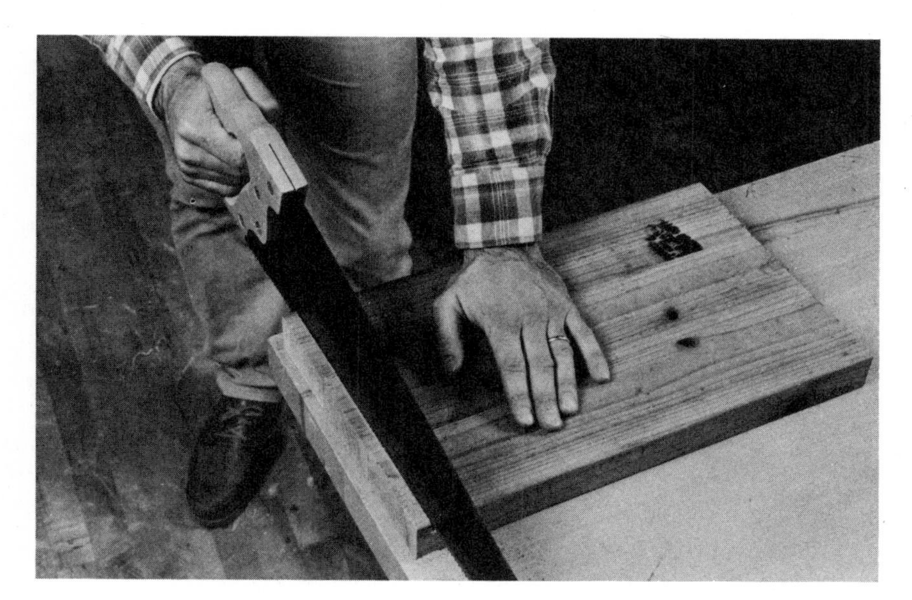

Photo 5–5—Cut the rabbets in the ends of the tray with a handsaw or a circular saw.

Adjustments need to be made so the barbecue operates efficiently. Start by moving the control lever so that the T-linkage is perpendicular to the stringer. Use an angle finder, a device for accurately determining the angle in relation to a vertical or horizontal surface. *Note:* The stringer must be level to adjust the reflectors. Start with reflector A, the one closest to the grill, and adjust the adjustor bolts under the reflectors. These angles are approximate and will be slightly changed to fit your particular location when the barbecue is facing the sun. Set reflector A at 10 degrees from the stringer. Set the second reflector, B, approximately parallel to the stringer. Reflector C is set at 9.5 degrees, reflector D at 15 degrees, and reflector E at 19 degrees, as shown in illustration 5–19.

Now for the final adjustments: In the middle of the day, take the barbecue into the sun and be sure the unit is level. Align the stringer directly toward the sun. Cover four of the reflectors with tarps. Adjust the one uncovered reflector so that sunlight is beamed onto the bottom of the grill. Then secure the adjustor bolt by tightening all three of the lock nuts. Repeat this procedure for the other four reflectors.

When you are using the barbecue, the stringer should be kept in line with the sun and the control lever adjusted every 15 minutes or so for the maximum heat you require on the grill. It is easy to "turn it down" by a less-efficient orientation or angle of reflection.

Food should be covered while it is cooking to reduce heat loss. The best method is to cover the food with aluminum foil when cooking—this gives the highest heat—or you can make a cover by attaching a cabinet handle to an 8 × 10-inch baking pan. Put a #8 cup hook into

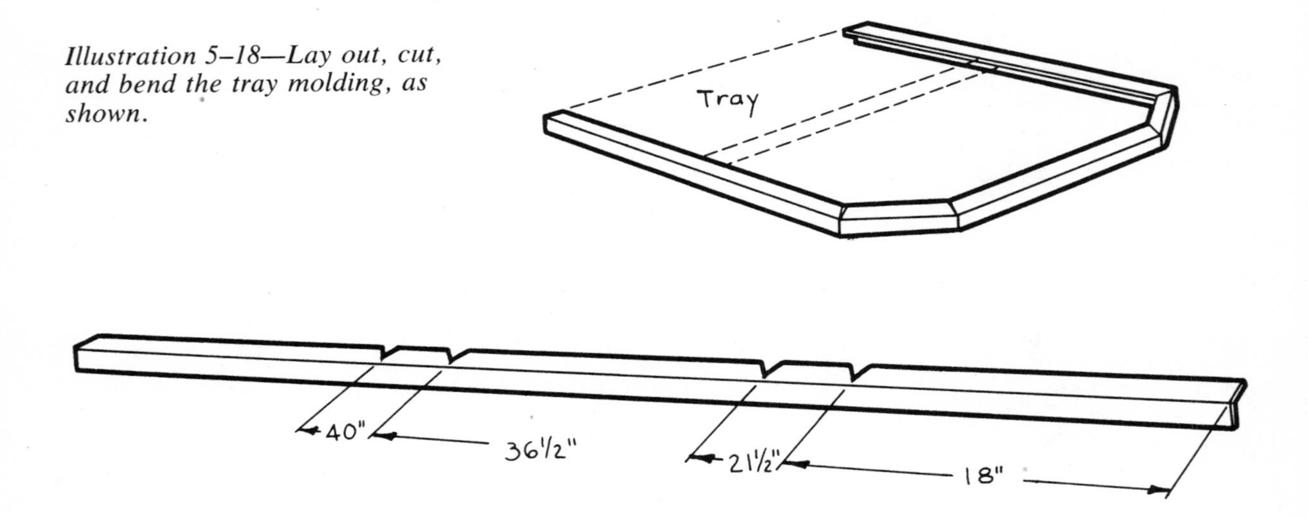

Illustration 5–18—Lay out, cut, and bend the tray molding, as shown.

the bottom of the redwood tray so you can hang up the baking pan cover or other cooking utensils.

Be careful when the solar barbecue is facing the sun. Use pot holders to handle the grill and cover. Remember, that grill gets hot enough to broil a steak within a very short time after the sunlight is focused.

To mount the barbecue permanently, find the balance point along the stringer, which should be close to 32 inches from the control end of the stringer. Cut a 3½ × 1½-inch notch at this point, and fasten a 5-inch piece of redwood 2 × 4 as a crosspiece. Use waterproof glue and 4-inch #10 flathead wood screws. Connect a 1-inch pipe flange at the stringer balance point, and fasten a 1-foot length of 1-inch pipe to the flange.

Embed and plumb-level a 3-foot length of 1½-inch pipe 2 feet in the ground with concrete at the permanent mounting site. Lift the completed barbecue onto the 1½-inch pipe, inserting the smaller pipe into the larger. Your barbecue is now swivel mounted and yet can be lifted off the pipe and stored away for the winter.

The permanently mounted barbecue requires the short crosspiece, but the wheels and pivot peg can be eliminated. Make a short version of the crossbeam and gusset to hold the griddle support. Make the crossbeam 12 inches long, and the gusset pieces 5⅜ inches long. Make the gussets square, not arced as the wheeled versions need to be.

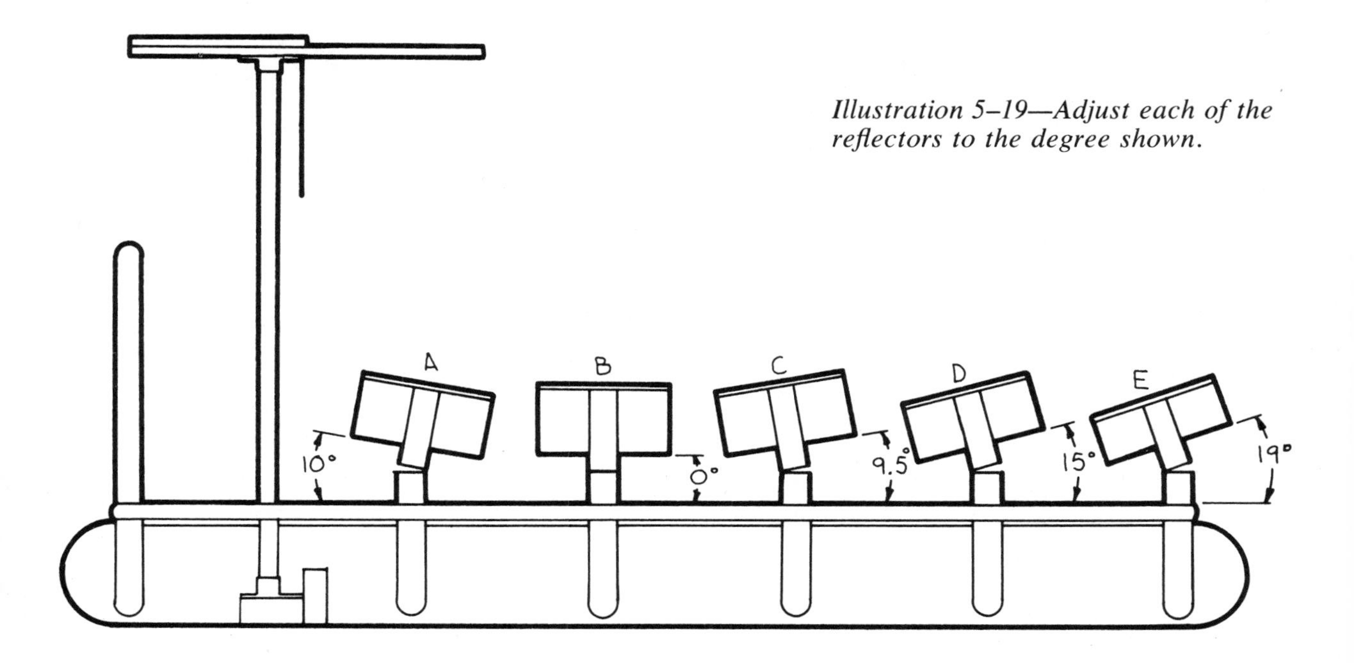

Illustration 5–19—Adjust each of the reflectors to the degree shown.

6 Solar Oven

It really works! Organic Gardening magazine tested many solar oven cookers and chose this design as the most effective. The temperatures normally achieved are higher than those in a slow cooker, yet the foods retain their natural juices and are very flavorful. Solar oven users are finding these cookers successful for nearly every food that is usually baked—from turkey to breads. The oven is sturdy, good-looking, and inexpensive.

The solar oven has large, flaplike reflectors made of thin plywood covered with either shiny Mylar or heavy-duty aluminum foil. The reflectors direct sunlight into the plywood oven section, which is insulated

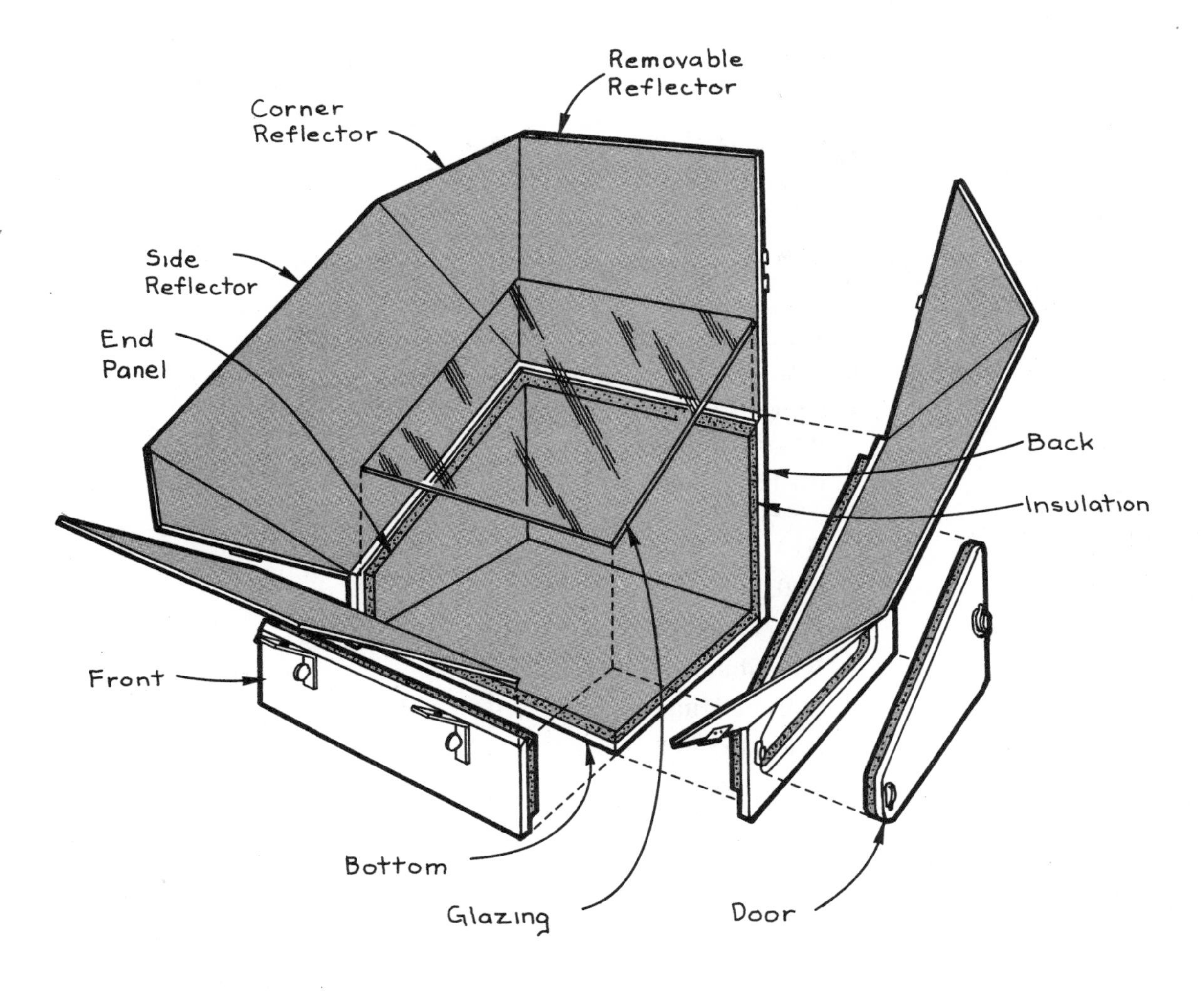

Illustration 6–1—Exploded view of the solar oven.

with ¾-inch foil-clad rigid foam insulation. The oven may be positioned for either low or high sun, so it will be useful in the winter as well as summer. The opening is covered with a layer of double-strength glass, and a removable door on one sidewall of the oven allows easy access. The reflectors may be folded in and two of them removed, so that the oven is easy to move and store. The exploded view, illustration 6–1, shows the principal parts of the oven. A materials list, chart 6–1, and a tools list, chart 6–2, follow.

CHART 6–1—
Materials

DESCRIPTION	SIZE	AMOUNT
Lumber		
A-C Exterior Plywood	½″ × 4′ × 4′	1
A-C Exterior Plywood	⅛″ × 4′ × 8′	1
Hardware		
Finishing Nails	4d	30
#4 Flathead Wood Screws	½″	8
#10 Tee Nuts	½″	4
#10 Tee Nuts	⅜″	4
#10 Thumbscrews	½″	4
#10 Thumbscrews	⅜″	4
Shutter Hinges	1½″ × 1½″	4
Butt Hinges	1½″ × 1½″	4
Sash Locks with Screws	…	2
Straight Braces	3″	2
120-Degree Braces	3″	2
Miscellaneous		
Foil-Clad Rigid Foam Insulation	¾″ × 2′ × 8′	1
Nontoxic Wood Glue	…	4 oz.
Epoxy	…	1 tube
Trim Adhesive	…	1 can
Primer	…	1 pt.
Exterior Enamel	…	1 pt.

CHART 6–1—*Continued*

DESCRIPTION	SIZE	AMOUNT
Miscellaneous—*continued*		
Nontoxic Flat Black Enamel	...	1 pt.
Double-Strength Glass	⅛″ × 18″ × 18″	1
Aluminized Mylar or Heavy-Duty Aluminum Foil	...	18 sq. ft.
Heat-Resistant Aluminum Tape	...	1 roll
Duct Tape	...	1 roll

CHART 6–2—
Tools

Saber Saw
Circular Saw
Drill and Bits
Utility Knife

Begin construction of your solar oven with the oven box itself. Working with ½-inch plywood, first cut out the bottom, 15⅞ × 18 inches. Next, lay out the back, 14⅛ × 18 inches. The top edge must be beveled to match the slope of the side, and it is best to make this angle cut first. Set your saw to cut a 30-degree angle. Working on the good side of the plywood, with the board clamped to your bench, cut the top edge on an angle, as shown in photo 6–1. Then, resetting the saw, check your original measurements and cut out the rest of the back.

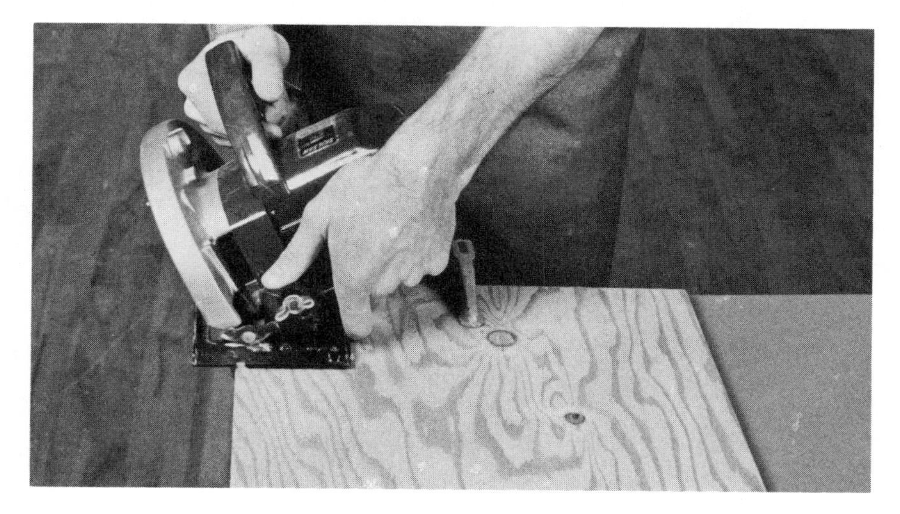

Photo 6–1—Clamp the back piece firmly to a workbench, and cut the 30-degree angle with a circular saw.

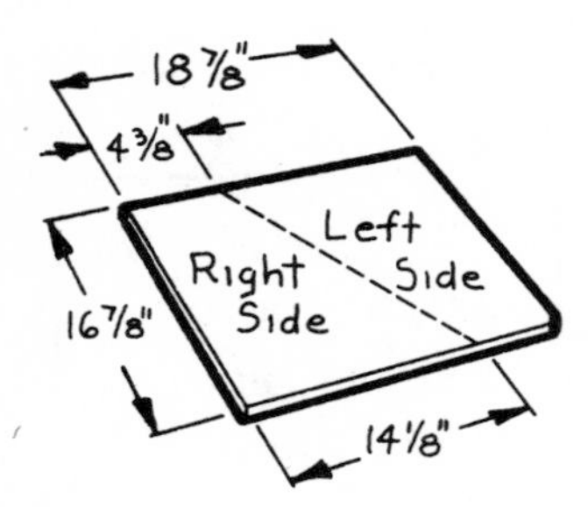

Illustration 6–2—Side layout.

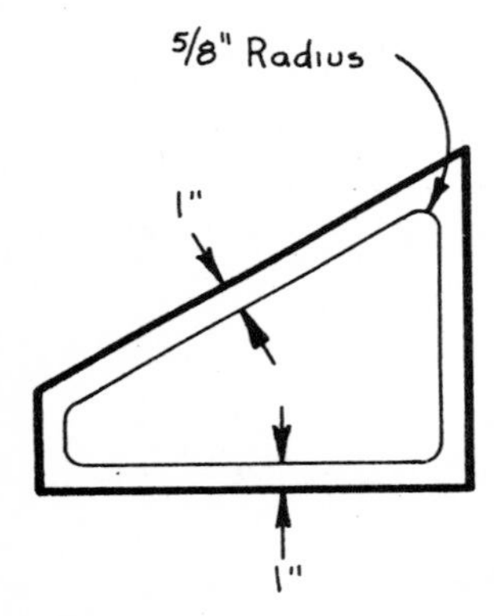

Illustration 6–3—Oven door layout.

In the same way, measure out the front, 4¾ × 18 inches. However, this time make the bevel cut along the top edge from the back of the plywood. Then check the measurements and cut out the three remaining edges of the oven front.

Cut one 16⅞ × 18⅞-inch piece of ½-inch plywood for the ends of the oven box. Illustration 6–2 shows the measurements to use when laying out the diagonal cut which separates the two end pieces. The angle from 14⅛ inches to 4⅜ inches will match the beveled front and back. *Note:* One end piece is a little higher than the other to accommodate the folding reflectors.

Use the shorter end piece, the right side, to make the oven door. Measure in 1 inch on all sides of the panel. Lay out a ⅝-inch radius at each corner, as shown in illustration 6–3. Start the cut by pressing the front edge of the saber saw base firmly against the wood. Tip the saw forward so that the blade does not touch the wood. Turn on the saw and gradually lower the blade into the plywood on the cut line. This is called plunge cutting. Photo 6–2 shows this useful technique. If you drill a hole to start the blade, you will have to cut a replacement door panel. Cut the door free from the panel.

Now you are ready to put the oven together. Assemble the oven box with wood glue and 4d finishing nails. First attach the front and back to the bottom, then add the ends, fastening them to the bottom and to the front and back.

Photo 6–2—Plunge-cut the door panel by starting the saber saw on an angle. Lower the blade slowly and accurately into the cut line.

Two sash locks, the type used on windows, hold the door in place. Lay out the locks on the door and end panel, drill pilot holes, and screw them in place.

With the box constructed, but before you put in insulation, fasten two reflector braces on the back and two on the front, with the hardware shown in illustration 6–4. Use straight braces on the back and 120-degree braces on the front. If you are unable to buy 120-degree braces, clamp straight braces in a vise and hammer them to the necessary angle. Mark and drill 3/16-inch-diameter holes in the plywood and the braces to accept the 1/2-inch #10 tee nuts, one for each brace. Hammer the tee nuts into the front and the back from the inside of the oven. Brace the sides of the oven on your workbench to absorb the pounding.

Insulation maintains the oven's heat. Use foil-clad rigid foam insulation. Cut the insulation to fit 1/8 inch below the top edges to leave room for the glass top. Use a utility knife and a straightedge, as shown in photo 6–3, to cut the insulation.

Cut the bottom piece of insulation 15 7/8 × 18 inches, and cover all

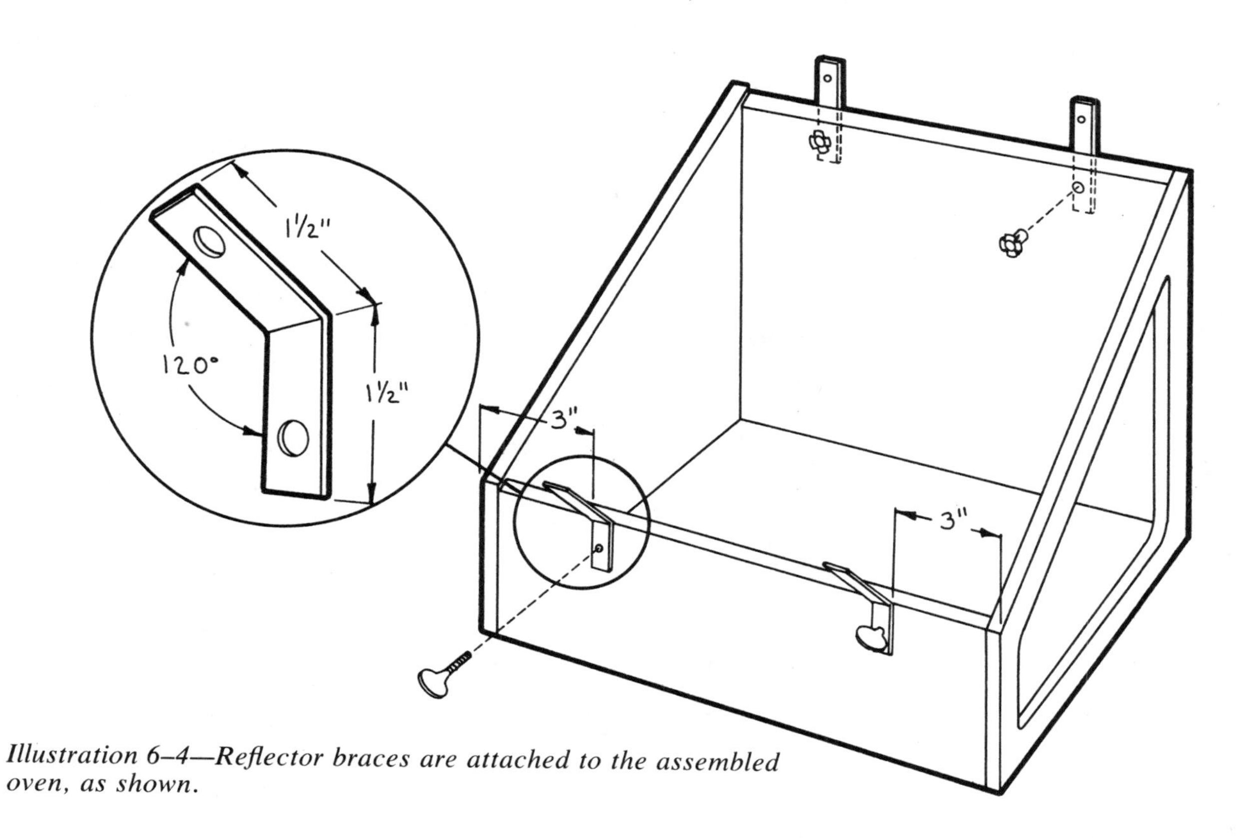

Illustration 6–4—Reflector braces are attached to the assembled oven, as shown.

Photo 6–3—A straightedge and a utility knife will help you cut the insulation accurately.

the edges with heat-resistant aluminum tape. It is important to tape the edges so that the insulation will not deteriorate from exposure to light. Install the bottom insulation panel with nontoxic wood glue. Next, cut the back insulation to fit, approximately 12½ × 18 inches. Bevel the top edge of the panel to 60 degrees, cover all the edges with aluminum tape, and glue it in place.

Cut the front insulation panel to fit, about 3¾ × 18 inches. Bevel the top edge of the insulation to 60 degrees. Tape all the edges and install this front panel as you did the back.

Measure, cut, and tape insulation to fit the left side of the oven. Use wood glue to fasten it in place. Cut a second piece of insulation to fit the door side. Cut the door opening and install the pieces on the door and space around the door. Tape all edges of the pieces and then fasten them to the door end with wood glue. Be sure that the insulation is ⅛ inch below the edge of the oven all the way around so the glass will fit.

Paint the interior of the oven flat black so it collects the maximum heat available. We suggest that you use a flat black paint that is nontoxic. This paint is sold for use on barbecues and engines, and as solar collector paint.

Having finished the oven box assembly, you are ready to make the reflector panels from ⅛-inch plywood. Paneling is a good, inexpensive alternative. Illustration 6–5 shows the dimensions and pattern of the reflectors. First, cut the two side panels. The left-hand panel is 17¾ × 19⅜ inches and the right-hand panel is 17½ × 19⅜ inches. For the corner sections, cut four triangles, each 12¾ × 17½ × 17½ inches. Cut the removable front and back panels, each 18 × 19 inches.

Set the removable front and back reflector panels in place on the braces, flush with the ends of the oven, and mark the hole placement of the braces on the panels. Drill two 3⁄16-inch-diameter holes, and insert two ⅜-inch #10 tee nuts in each removable panel.

Next, position two corner pieces on each side panel, as shown in illustration 6–6. With the two triangular pieces in place, apply aluminum tape the length of the joint on the outside of the panels to make a flexible union. Carefully open the panels and tape the inside of the joint. The triangular sections should be capable of being folded onto the side panels, which in turn can be folded onto the oven, for convenient storage.

Cover each of the two side reflectors and the front and back reflectors with reflective material. Cut aluminized Mylar or aluminum foil to fit each reflector area. Spray a panel with trim adhesive, and carefully smooth the reflective material from one side across the panel, as shown in photo 6–4. It is important to keep it as wrinkle-free as possible. Repeat this operation on all of the panels.

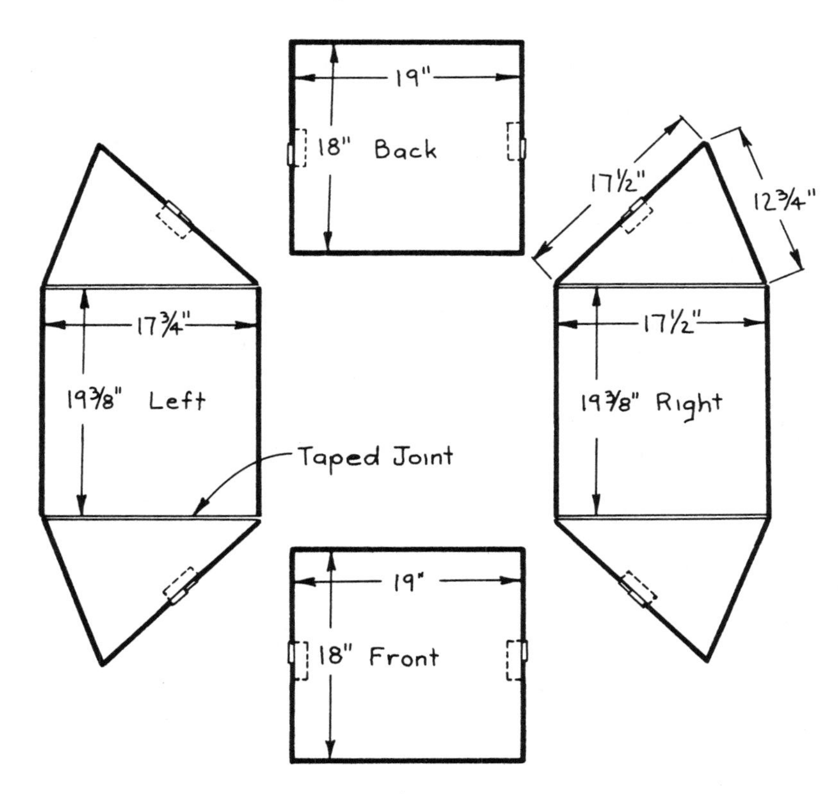

Illustration 6–5—Layout details and placement of the reflectors.

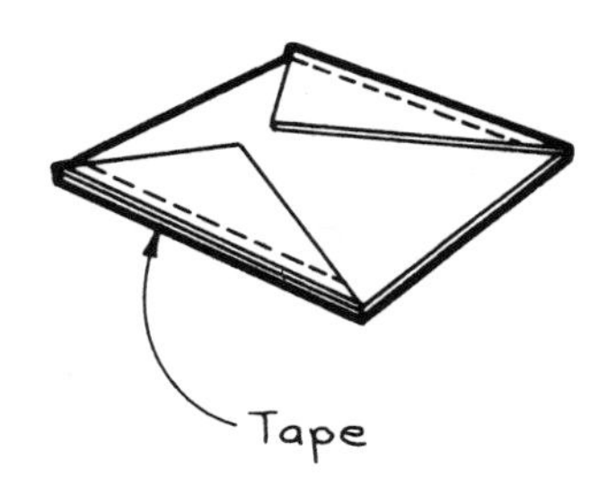

Illustration 6–6—Tape hinge.

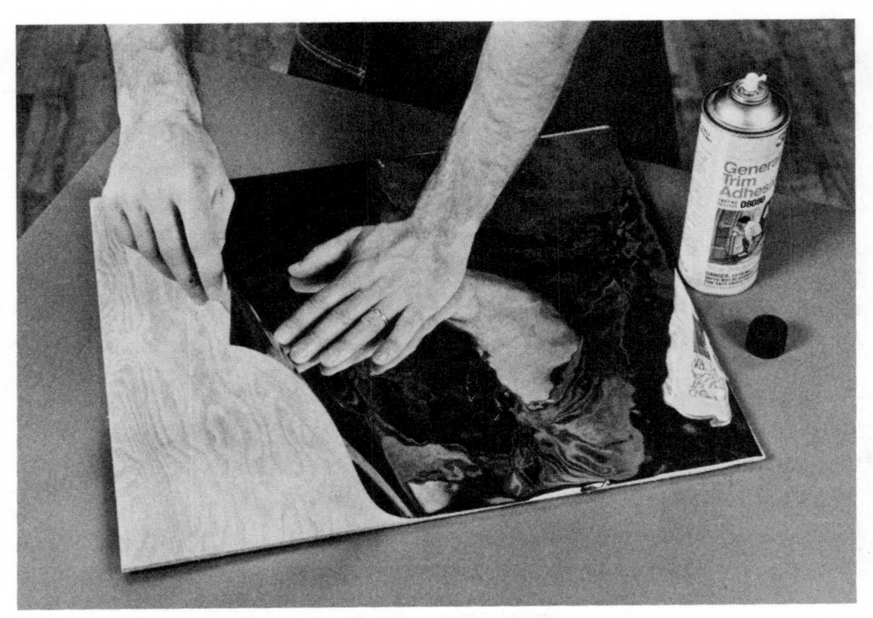

Photo 6–4—Carefully smooth the reflective Mylar onto the plywood reflector backs, using trim adhesive.

Prime and paint the outside of the reflectors and the oven any color you choose.

Attach the reflectors to the oven by attaching two butt hinges at the lower edge of each reflector side panel, using epoxy. Then screw the hinges into the outer face of the oven end pieces using ½-inch #4 flathead wood screws.

Set the removable reflectors into the front and back braces. Fasten them in place with ⅜-inch #10 thumbscrews turned into the tee nuts.

The corner reflectors are joined to the removable panels with loose-joint shutter hinges, shown in illustration 6–7, which are easily disassembled. If shutter hinges are not available, butt hinges with a removable pin are satisfactory. Using epoxy, cement the male half of a shutter hinge about 4 inches below the top edge of each corner panel. Then, cement the corresponding female half with epoxy to the adjacent removable panel. This will allow the removable panels to be slipped in and out of place for assembly and easy storage.

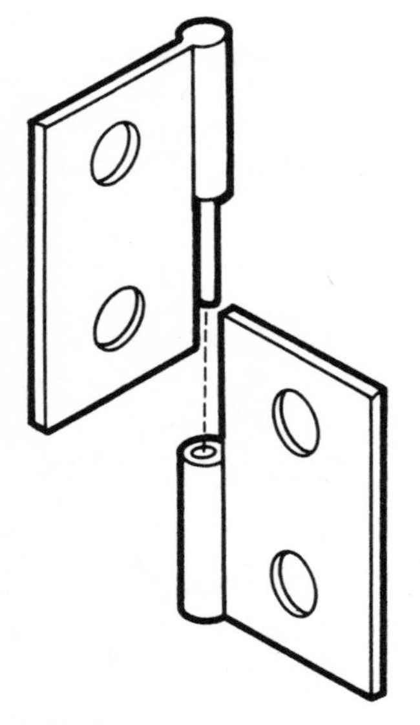

Illustration 6–7—Loose-joint hinge.

Glazing is necessary to complete the oven enclosure. Cut or buy a ⅛ × 18 × 18-inch piece of double-strength glass. Use aluminum tape to cover the edges for safety. Lay the glass on the edges of the insulation to close off the top of the oven. Tape the glass in place with aluminum tape to seal the oven fully.

To use your oven, place a dark cookie tin on the bottom and rest a small cooking rack on it to allow air circulation around your pot. Use dark-colored cookware, such as cast iron or dark blue enamel. Be sure the pot has a tight-fitting lid. Light-colored cookware will reflect light and diminish the heat.

For maximum heat, orient the oven into the sun every 20 minutes or so. However, if you wish to leave it unattended, place the cooker in mid-position for the time it will be cooking. An oven thermometer will let you monitor cooking temperatures.

Don't underestimate the heat achieved by this small oven. Use potholders to handle the equipment, and expect the cooker to operate effectively. Illustration 6–8 shows the two positions of the oven. When the sun is high, set the oven in position A. When the sun is lower, flip the oven so that it is in position B to capture heat more effectively.

Before cooking with your oven for the first time, allow it to heat up for several days with the door open so the paint residue can burn off. To keep food warm in the oven after sundown, heat a clean, dark rock in the oven with your cookware, and cover the oven with a blanket when the sun is low.

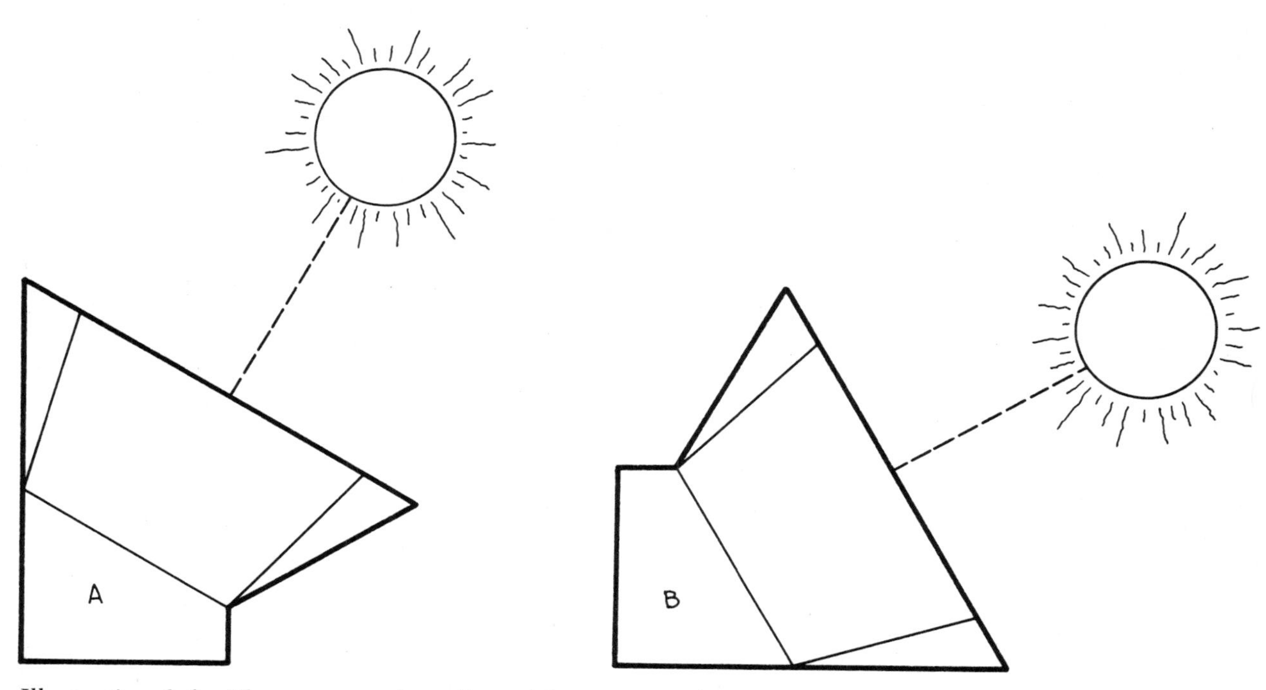

Illustration 6–8—The oven can be adjusted for summer (A) and spring/fall (B) use.

7 Stovepipe Food Dryer

Instead of turning on your oven or buying an electric dryer, thus adding to your utility bills, you can use this stovepipe dryer. The drying chamber is fashioned from 8-inch-diameter stovepipe. A curved reflective surface concentrates sunlight onto the drying chamber. More than a pound of food can be dried at a time on trays inside.

It is difficult to give exact times for solar drying, as conditions vary from one climate to another. Rodale researchers determined that here in Pennsylvania most solar food dryers require about two days of strong sun to dry fruits and vegetables, whereas in Colorado, one full day of bright sun and low humidity might actually toast your foods. A small

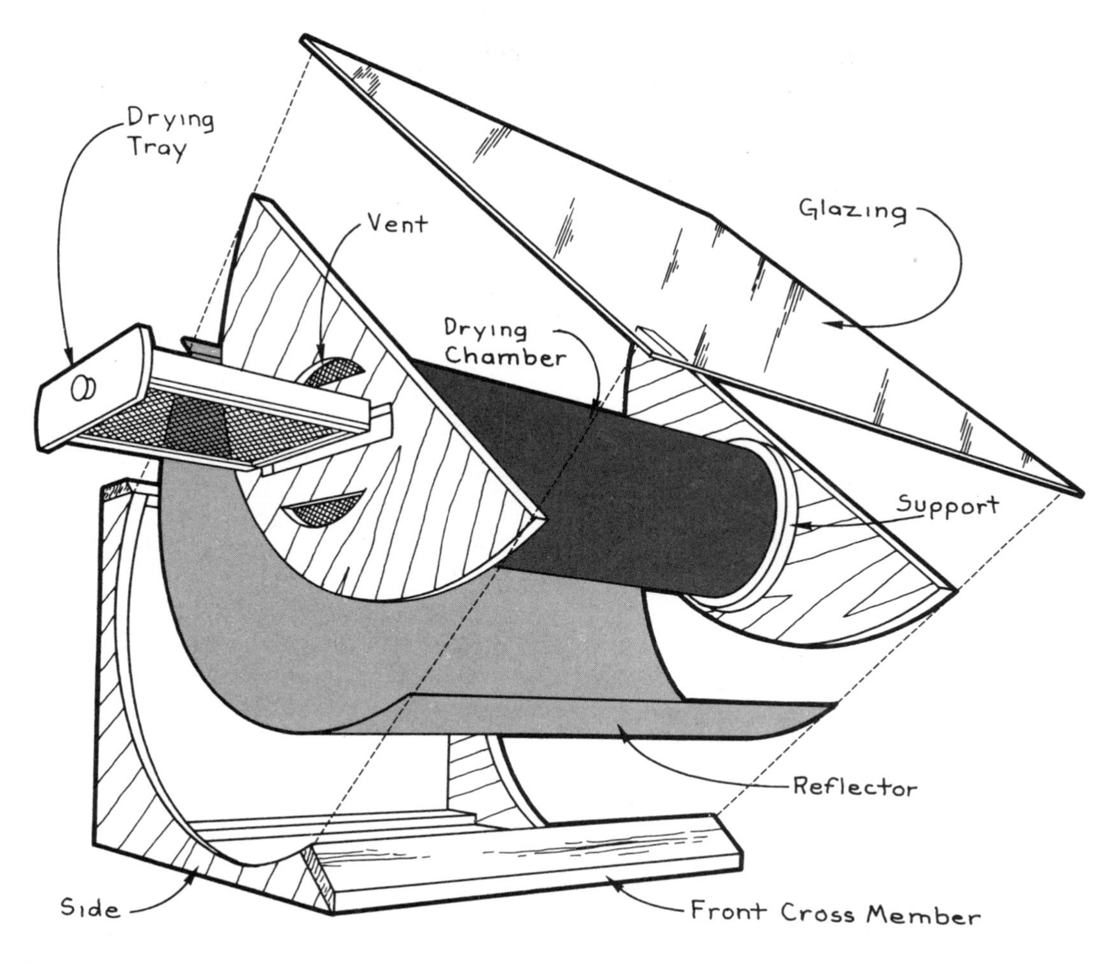

Illustration 7–1—Exploded view of the stovepipe food dryer.

oven thermometer will help to monitor the dryer and avoid excess heat. Usual air temperatures required for drying are 95°F to 130°F. *Note*: We do not recommend the use of a solar dryer to dry meat. Blanching is required as a pretreatment for drying most foods, as it is for freezing them. A more complete discussion of methods of drying foods may be found in Rodale's *Solar Food Dryer* Plans Book, *Stocking Up*, and other books.

An exploded view of the dryer is shown in illustration 7–1. The materials and tools you will need are detailed in the materials list, chart 7–1, and a tools list, chart 7–2, which follow.

CHART 7–1—
Materials

DESCRIPTION	SIZE	AMOUNT
	Lumber	
#2 Pine	3/8″ × 3/4″ × 8′	1
#2 Pine Trellis Stock	1/8″ × 1¼″ × 8′	1
A-C Exterior Plywood	3/4″ × 2′ × 4′	1
A-C Exterior Plywood	3/8″ × 6″ × 8½″	1
	Hardware	
16-Gauge Brads	1¼″	16
16-Gauge Brads	5/8″	1 box
#8 Flathead Wood Screws	2″	4
#6 Brass Roundhead Wood Screws	3/4″	12
Cement-Coated Box Nails	4d	1 box
Aluminum Pop Rivets	1/8″ × 1/8″	6
Staples	3/8″	...
Wooden Knobs	1″ dia.	2
Aluminum Flashing	2′ × 4′	1
Fiberglass Screen	1′ × 2′	1
	Miscellaneous	
Stovepipe	8″ dia.	22″
Closed-Cell Foam Weather Stripping	1/8″ × ½″	8′
Waterproof Wood Glue	...	4 oz.
Trim Adhesive	...	1 can
Primer	...	1 qt.

Begin construction with a 24 × 48-inch piece of ¾-inch exterior plywood. First, cut a 22 × 22-inch square for the sides of the dryer. As shown in illustration 7–2, draw diagonal lines from corner to corner. Using trammel points with the intersection point of the lines as the center point, draw a circle 20 inches in diameter. Cut along one diagonal line, to give you two triangular sidepieces, each with a half-circle marked on it. Do not cut out the marked half-circles at this time.

Mark the vent openings and tray opening on both sidepieces, using illustration 7–3 as a guide. Be sure to mark the pieces so you end up with a pair, with the good side of each piece facing out. First, designate one 22-inch edge as the bottom of the dryer. The other 22-inch side is the back edge, and the diagonal edge is the front.

Measure 7⁷⁄₁₆ inches from the 90-degree corner along the back edge and along the bottom edge, and square lines across from the back and bottom edges toward the front edge. Using the intersection of these lines as the center point, draw an 8-inch-diameter circle. Measure ¾ inch into the circle from both the top and bottom of the circle along the vertical line from the bottom. Square lines across the marks, perpendicular to the back edge, to mark the vent openings, as shown in photo 7–1. For the tray opening, measure ¾ inch both above and below the center point of the circle, and square lines across the circle.

Cut out the large half-circles from the sides, using a saber saw. Work carefully, and save the whole board, since both pieces of each half are needed. Clamp the plywood to your workbench, readjusting it as necessary for the saw cut. Mark the pieces so that you can later

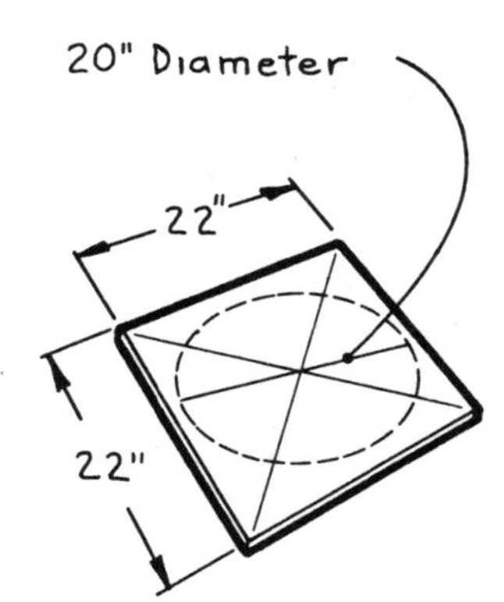

Illustration 7–2—Side layout.

CHART 7–1—*Continued*

DESCRIPTION	SIZE	AMOUNT
	Miscellaneous—*continued*	
Exterior Enamel	...	1 qt.
White Exterior Enamel	...	1 qt.
Flat Black Absorber Paint	...	1 can
Wood Putty	...	4 oz.
Acrylic Sheet	⅛″ × 23½″ × 22″	1
Heavy-Duty Aluminum Foil	23½″ × 32⁷⁄₁₆″	1

CHART 7–2—
Tools

Saber Saw
Circular Saw (optional)
Drill and Bits
Pop Rivet Tool
Heavy-Duty Staple Gun
Tin Shears
Wood Chisel: ½-inch
Trammel Points or Compass

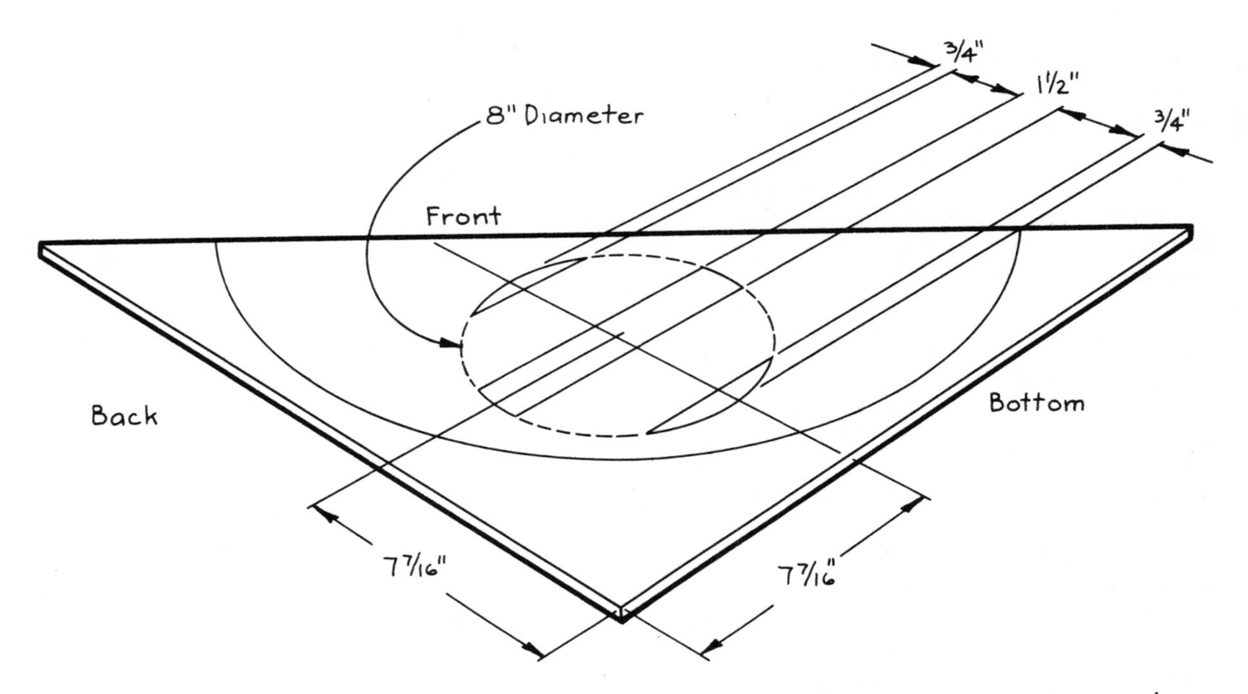

Illustration 7–3—Layout details for the vents, tray openings, and stovepipe placement on the sides.

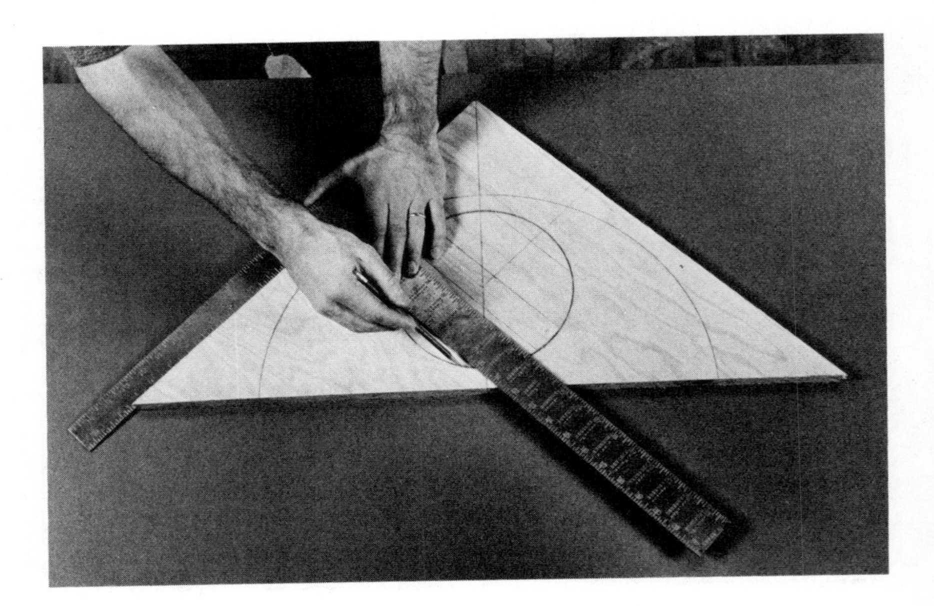

Photo 7–1—Mark the vent opening positions with a carpenter's square within the 8-inch circle.

reassemble them in their original position. Clamp one of the half-circles to your workbench before cutting the vent and tray openings. To start the saw cut, first drill a hole large enough to receive the saber saw blade in the area to be removed and near the line you will cut. Cut out the vent and tray openings on both sidepieces.

On the triangular sides, mark the cuts to fit the top and front cross members, as shown in illustration 7–4. First mark off a strip ¾ inch wide along the front corner, from the half-circle to the bottom edge. Next, measure 18⅟16 inches from the 90-degree corner toward the top corner, and square a line across to meet the end of your half-circle line. Cut off these two pieces on both sides.

Cut the top cross member, the back cross member, and the front cross member all from the same 23½ × 10½-inch section of ¾-inch plywood. Mark the plywood as shown in illustration 7–5. Cut the 45-degree angle for the top and front cross members at the same time. Set the saw to cut a 45-degree angle, and cut the top and back cross member angles. Reset the saw and cut the 2 × 22-inch strip for the back cross member.

Construct the dryer frame with waterproof wood glue and 4d cement-coated box nails. Fasten the back cross member flat between the sides at the lower back corner. Then nail and glue the top cross member to the sides, keeping the ends flush with the sides and the 45-degree angle in front. Fasten the front cross member across the sides at the lower front corners with the 45-degree angle at the bottom.

On the remaining ¾-inch plywood, lay out two circles, each 10 inches in diameter, to make the doughnut-shaped drying chamber supports. Inside each circle draw a second circle with a diameter of 8 inches. Drill entry holes near the circles and cut out the inside piece with the saber saw. Check the fit of the 8-inch stovepipe in the support rings. If the circles need to be trimmed, use a rasp or sandpaper.

Cut four 2 × 7-inch pieces of fiberglass screen, and staple these

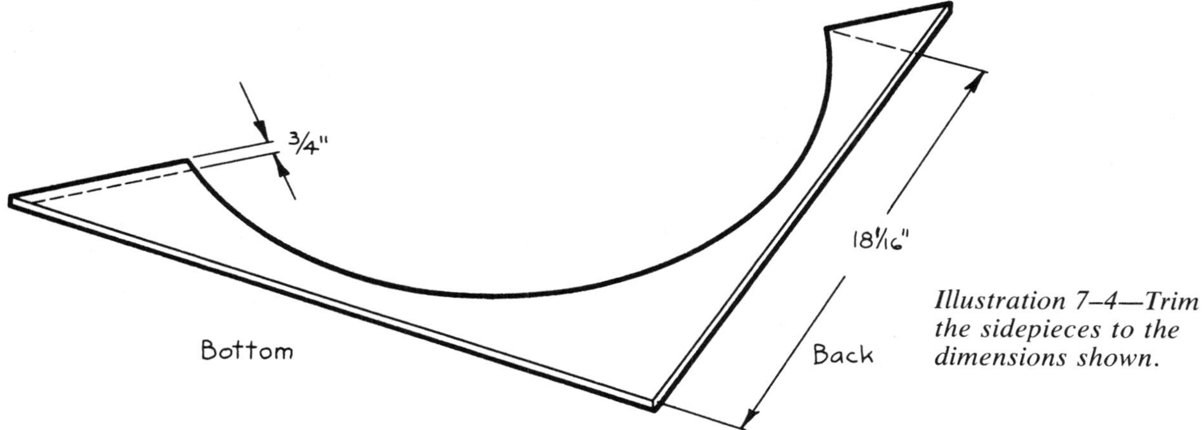

Illustration 7–4—Trim the sidepieces to the dimensions shown.

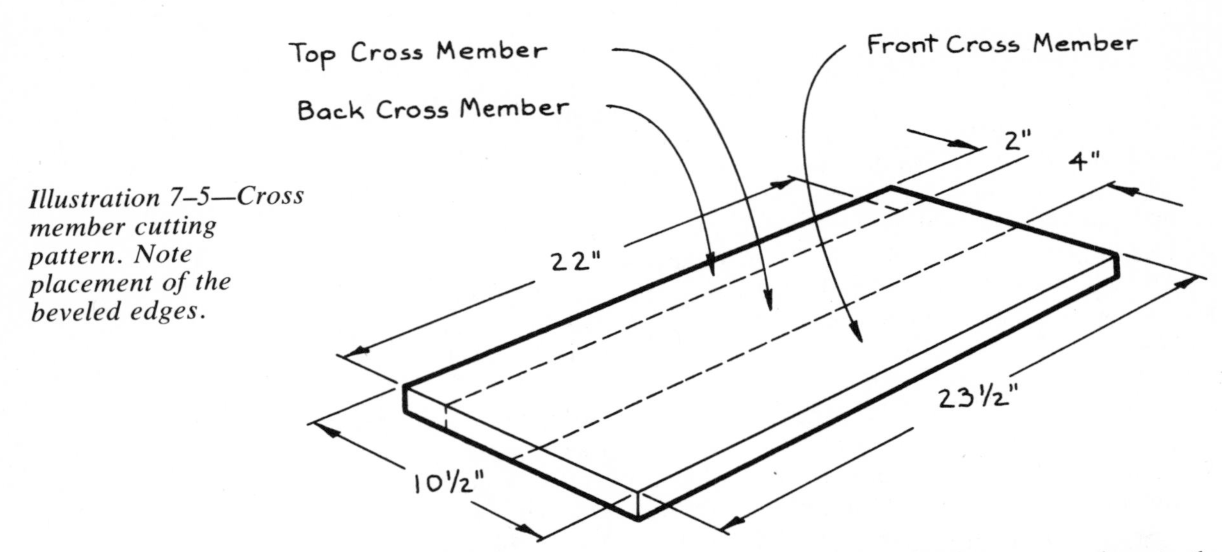

Illustration 7–5—Cross member cutting pattern. Note placement of the beveled edges.

across the inside of each vent opening. Fasten the support rings to the inside faces of the semicircles, over the screening, lining them up with the tray and vent openings. Use waterproof glue and 1¼-inch 16-gauge brads to hold the supports in place.

For the tray fronts, cut two 2½ × 8½-inch pieces of ⅜-inch plywood. Use wood putty to fill cracks in the plywood of the dryer frame, particularly the edges, and between framing pieces. Sand the frame and tray fronts. Use a primer and two coats of exterior enamel to cover the tray fronts, the half-circles, and the dryer frame. The interior of the half-circles should be white to reflect light onto the drying chamber. The outside of the semicircles, the frame, and the tray fronts may be painted the color of your choice.

Photo 7–2—Carefully cut the stovepipe to length with a tin shears. Be very careful of the sharp edges.

The drying chamber is made from a section of 8-inch-diameter stovepipe. The pipe can be unhooked along its seam and laid out flat, then later rehooked into a cylindrical shape. With a tin shears, cut the pipe to a length of 22 inches. Photo 7–2 shows the stovepipe opened up with the tin shears cutting along the side. In this way, the narrow cutoff strip curls up out of the way of the shears. Rehook the stovepipe seams after cutting the length.

Cut a 7 × 23-inch piece of aluminum flashing to form the intake baffle. Draw two lines diagonally from corner to corner to find the center point. As in illustration 7–6, mark out the 2 × 6-inch rectangular opening in the center of the baffle. Use the tin shears to cut out this space, after punching a starting hole with a screwdriver or drill.

Shape the corner notches in the intake baffle, ½ inch from each corner, referring again to illustration 7–6. Use the tin shears to snip out the notches. Draw a line on each side, ½ inch from the edge of the baffle. Lay the intake baffle on a block of wood, with the line at the edge of the block. With a straightedge, bend the flange down (approximately 45 degrees) to conform to the inside diameter of the stovepipe (see photo 7–3). Measure ½ inch from the baffle ends, and use a straightedge again to bend the shorter flanges 90 degrees in the same direction.

A small piece of flashing, the cross baffle, is cut to match the curve of the stovepipe and to fit in the space below the intake baffle opening. When this baffle unit is installed in the drying chamber, the cool air drawn in through the lower vents is heated and forced up through the drying trays and out the upper vents, drying the food. Cut a 2 × 6-inch piece of flashing. Use the stovepipe as a pattern to mark the curve. Cut the baffle pattern, then bend a 90-degree angle at the top ½ inch. Hold the flange of the cross baffle with pliers across the center of the

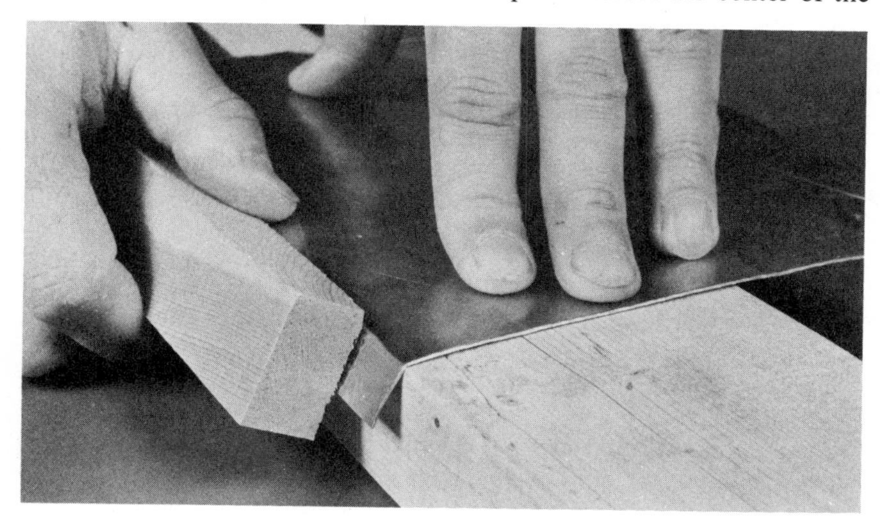

Photo 7–3—Bend the baffle by clamping the flashing material between two boards.

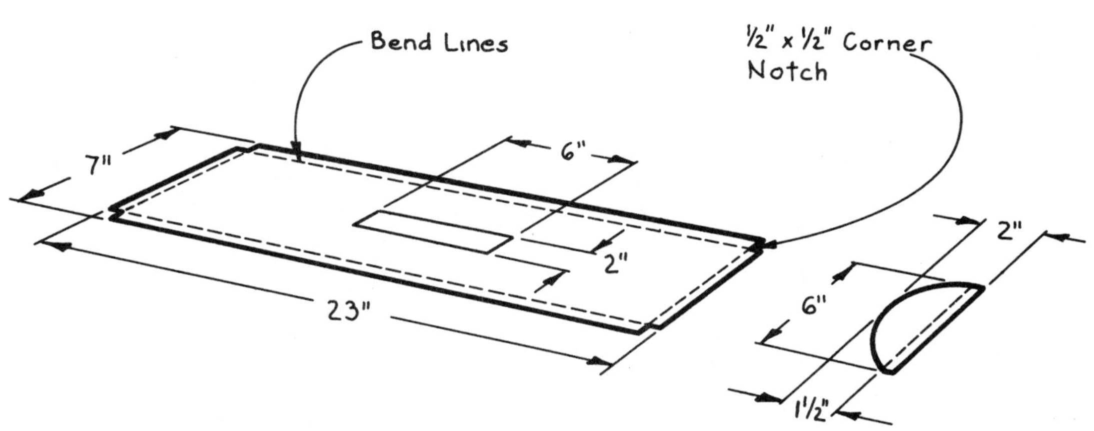

Illustration 7–6—Layout details for the aluminum flashing air baffles.

intake baffle opening. Drill two ⅛-inch holes through both pieces. Rivet the cross baffle to the intake baffle with ⅛ × ⅛-inch pop rivets.

Place the baffle inside the stovepipe, and, holding the two pieces firmly with pliers, drill ⅛-inch holes through the pipe and the flanges of the baffle. Rivet the baffle in place. An interior view of the drying chamber is shown in illustration 7–7. Paint the outside of the stovepipe with flat black absorber paint to make it a more efficient absorber of solar energy.

Cut a 23½ × 32⁷⁄₁₆-inch piece of aluminum flashing for the reflector. Bend one 23½-inch edge to make a 1-inch flange. Place the aluminum sheet into the semicircle of the sides with the flange at the top, as shown in illustration 7–8. The flange overlaps the top cross member. Nail the reflector to the plywood top, sides, and front with ⅝-inch 16-gauge brads.

Spray trim adhesive on the aluminum collector, and press aluminum foil into place, keeping it as smooth as possible to maximize reflection.

Assemble the drying chamber and the semicircular sides by slipping the stovepipe into the supports. The intake baffle must be parallel with the bottom of the dryer. Fasten the drying chamber unit into the triangular frame with four 2-inch #8 flathead wood screws, as shown in illustration 7–9. Drill ³⁄₁₆-inch holes in the frame and ¹⁄₁₆-inch pilot holes in the semicircles. Place the screws 11 inches from the lower back corner of the frame, at the points where the frame is most narrow.

Cut the acrylic glazing to 23½ × 22 inches, and position the sheet over the front of the dryer to check the fit. File the edges of the sheet to remove any sharp edges or burrs. Tape the glazing onto the dryer temporarily so that the mounting holes can be drilled. Drill ¹⁄₁₆-inch holes through the glazing and into the plywood edges. Space the holes evenly, with three in each side. Remove the tape and drill the holes in the glazing larger, to ³⁄₁₆ inch, to allow for expansion of the glazing material.

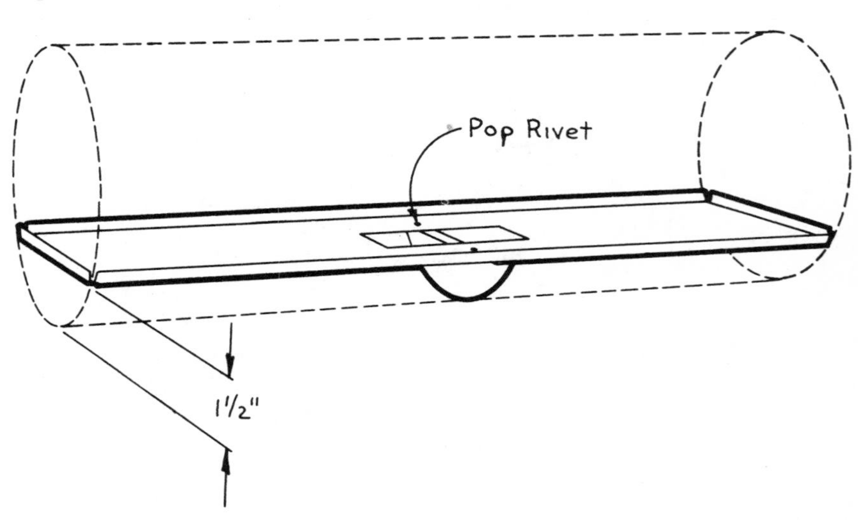

Illustration 7–7—Assembled baffle arrangement inside the stovepipe.

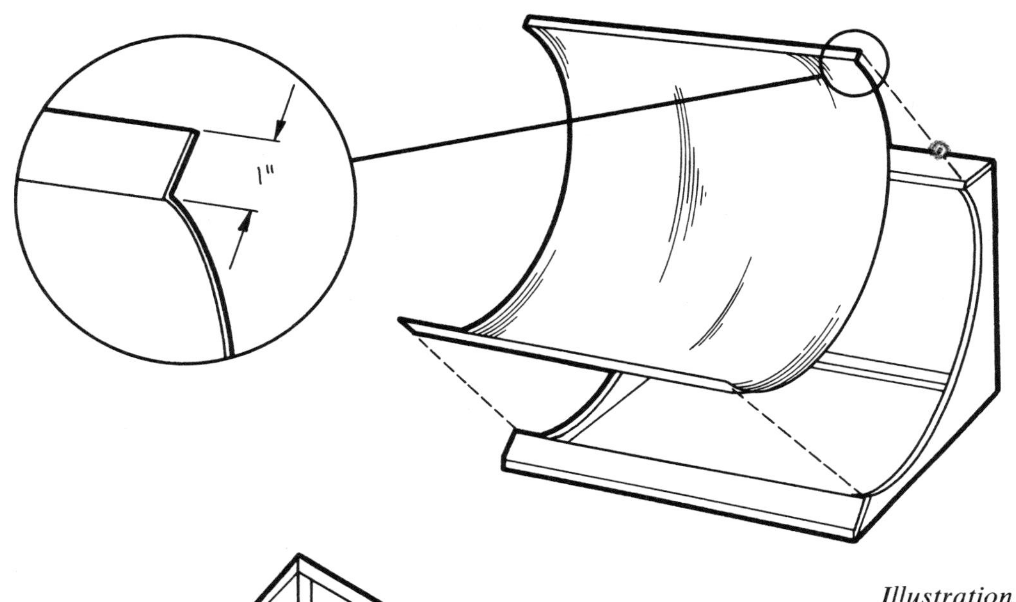

Illustration 7–8—Reflector attachment details; note placement of bend.

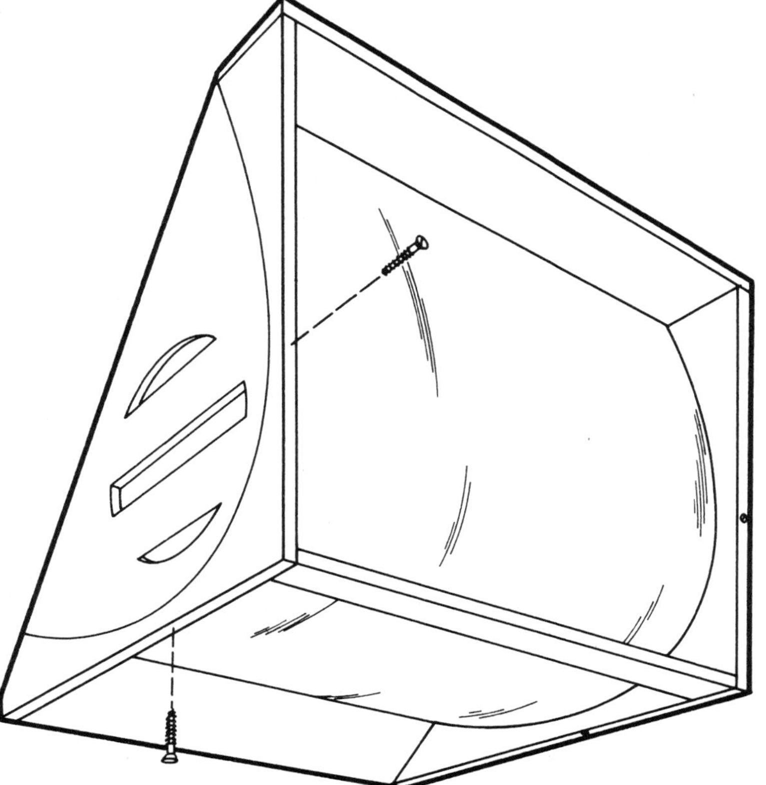

Illustration 7–9—Fasten the side sections together with screws placed as shown.

Apply weather stripping to the front edges of the dryer where the glazing will be mounted. Clean the glazing and set it on the weather strip. Realign the screw holes, then fasten the glazing to the front of the dryer with ¾-inch #6 brass roundhead wood screws. *Note:* Do not overtighten the screws, as this will crack the glazing.

Next, construct the sides and ends of the drying trays from ⅜ × ¾-inch pieces of pine. Cut four 11⅝-inch pieces and four 7⅜-inch pieces. Shape a rabbet on the end of the tray sides to make a stronger joint than butting the pieces together would make. To cut the rabbet, shown in illustration 7–10, set an end piece in position against the end of a sidepiece. Mark the width of the rabbet on the sidepiece. Mark 3/16 inch as the depth of your cut. Use a handsaw to cut just to the marked depth. With a ½-inch wood chisel, remove the excess material from the rabbet, as shown in photo 7–4. Repeat the rabbet on each end of the drying tray sides, a total of eight rabbets.

Screen molding holds the screen to the tray bottoms. Cut trellis stock to 7 feet long. Then rip the 7-foot piece to ⅜ inch wide. Use a rip guide on the saber saw, and clamp the trellis stock to your workbench. From this ripped piece cut four 11⅝-inch pieces and four 7-inch pieces. Cut two 7¾ × 11⅝-inch pieces of fiberglass screen for the bottoms of the trays.

Referring to illustration 7–10, assemble the trays by first fastening the tray fronts to two of the ends with waterproof glue and one ⅝-inch brad at each corner. Staple the screen to the bottoms of the trays, keeping the frames square. Tack the screen molding strips over the staples, using ⅝-inch brads. Fill and sand all the voids in the tray joints with wood putty. Fasten a 1-inch wooden knob in the side-to-side center of each tray front but keep it above the end piece.

Make tray guides from two 1 × 23½-inch pieces of aluminum flashing. Clamp a strip lengthwise between two straightedges with ½ inch

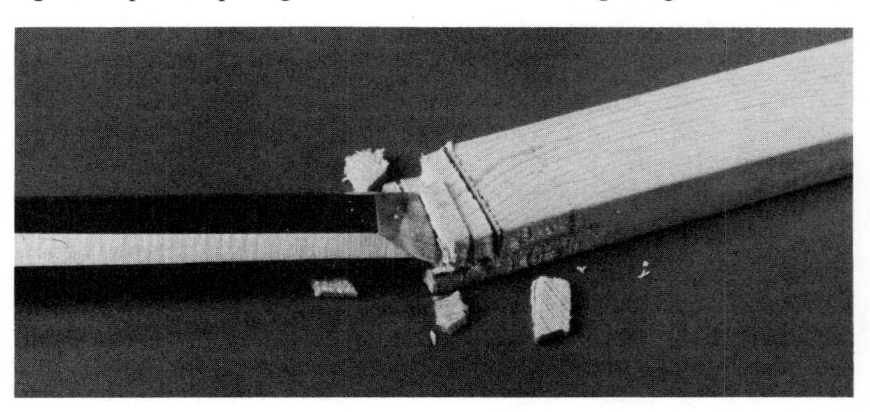

Photo 7–4—Clean the rabbet areas for the tray piece with a sharp wood chisel.

extending the length of the strip. Bend the aluminum into a right angle. Slide the tray guides into the dryer chamber, and fasten them to the sides at the edge of the tray opening with ⅝-inch nails. Nail through the end of the guide into the plywood of the openings in the dryer sides, as shown in illustration 7–11.

To use the food dryer, load the trays with thinly sliced fruits or vegetables. Leave space between the pieces so air can circulate. Set the oven in bright sun with the glazing facing the sun. You might need to reorient the dryer as the sun moves overhead. Enjoy those tasty pieces of dried food!

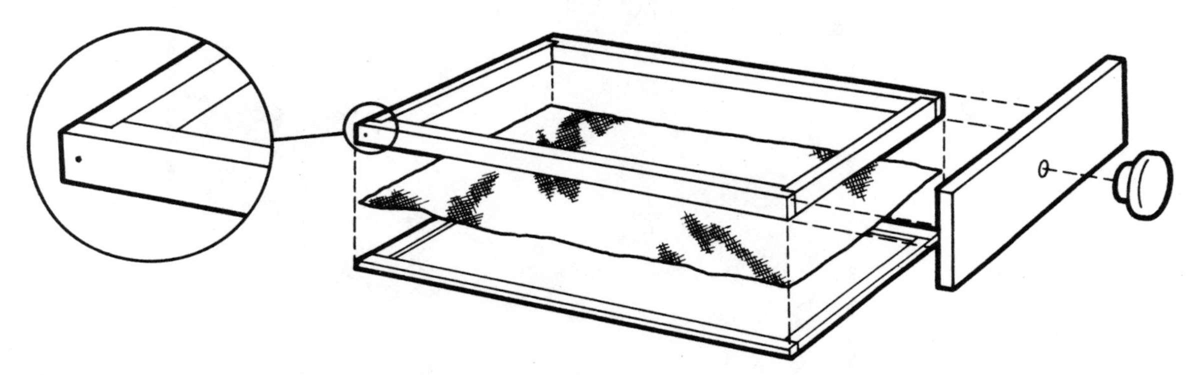

Illustration 7–10—Tray assembly details; note the rabbet detail at the corners.

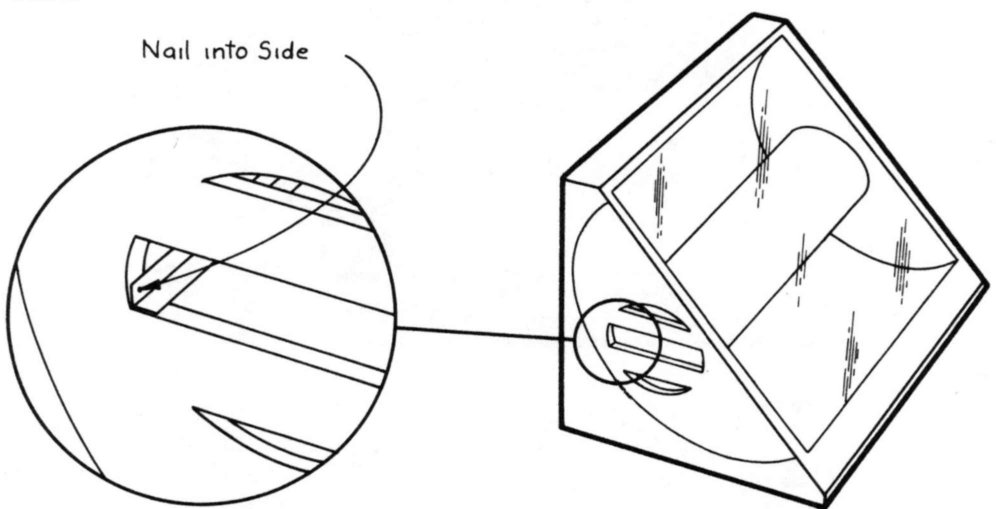

Illustration 7–11—The trays slide on small angles installed inside the side openings.

8 Solar Window Herb Dryer

To retain their fragrance and flavor, herbs need to be dried in a warm, dry, dark, well-ventilated place. Although it may be more picturesque to hang bunches of herbs from your kitchen beams, our herb dryer provides near-ideal drying conditions for you herbs.

This herb dryer is a simple plywood box. It mounts on the inside sill of a south-facing window. A metal back on the dryer faces out the

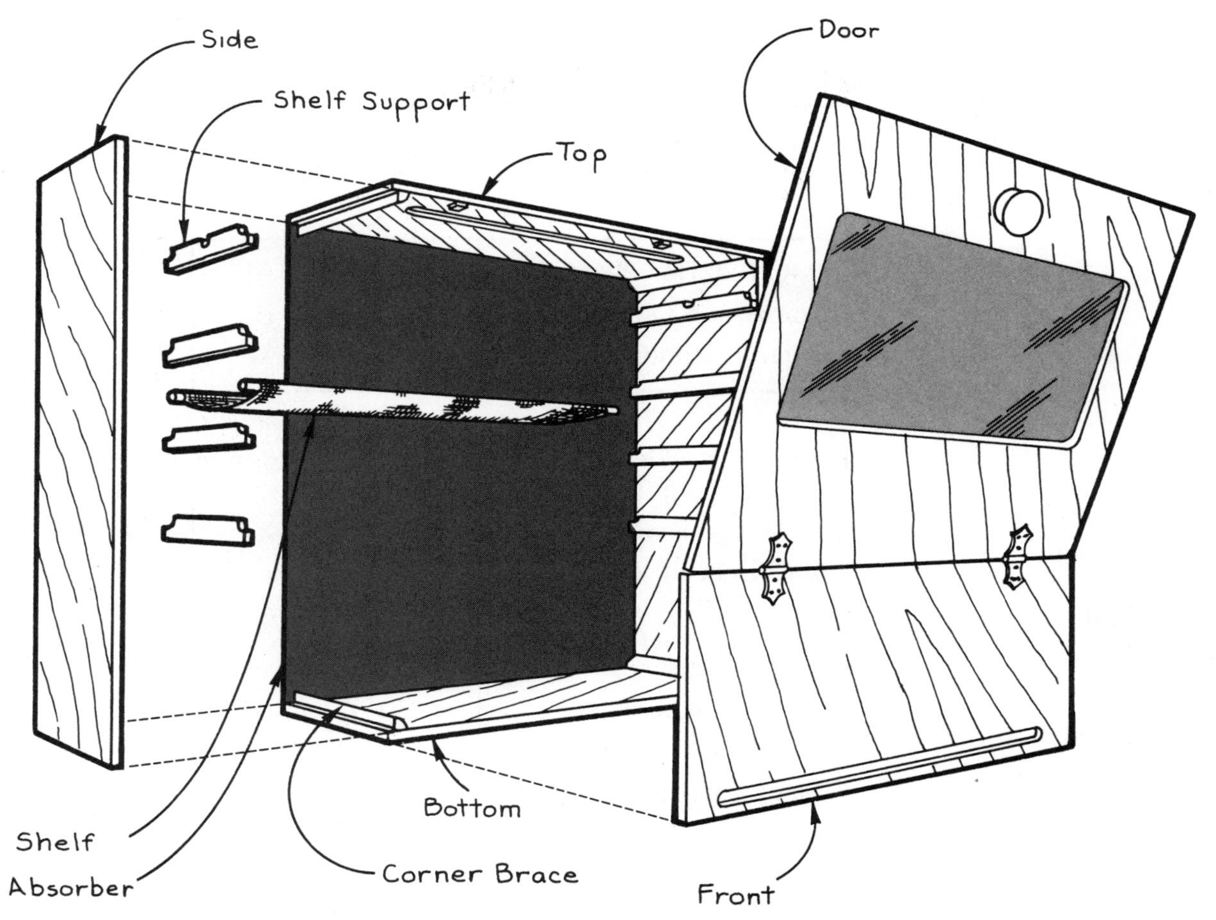

Illustration 8–1—Exploded view of the window herb dryer.

window and collects solar heat. Toward your room, a door with a window in it allows convenient loading and observation. Air circulates freely, assisted by the "warm air rises" principle. Room air enters a vent low in the front of the dryer, passes through the screen shelves (gaining heat from the metal absorber as it goes), and flows out a top vent.

Herbs for drying should be picked just before the plants bloom, when their flavor is highest. The early morning of a clear day is the best time, before the sun's heat diminishes the oils in the herbs.

Dill, caraway, and other seed heads should be harvested in the early stages of ripening. The seeds are ready to be picked when they have just turned from green to brown or gray. The screen shelves work well to hold all the small seeds as they dry.

CHART 8–1—
Materials

DESCRIPTION	SIZE	AMOUNT
	Lumber	
A-C Exterior Plywood	½″ × 4′ × 4′	1
Hardwood Dowel	5⁄16″ × A − 1½″	6
Quarter-Round Molding	¾″	36″
	Hardware	
Finishing Nails	4d	1 lb.
16-Gauge Brads	¾″	1 box
#4 Flathead Wood Screws	½″	10
Flush Cabinet Hinges with Screws	¾″ × 2″	2
Wooden Knob	1½″ dia.	1
Magnetic Catches	1″	2
Snap Catches	...	2
Aluminum Flashing	A − ¼″ × B	1
Fiberglass Screen	2′ × 3′ or to fit	1
	Miscellaneous	
Wood Glue	...	1 pt.
Primer	...	1 qt.
Interior Enamel	...	1 qt.
Flat Black Absorber Paint	...	1 can
Wood Putty	...	trace
Acrylic Sheet	⅛″ × 9″ × A − 4″	1
Thread	...	1 spool

CHART 8–2—
Tools

- Saber Saw
- Circular Saw (optional)
- Drill and Bits
- Tin Shears
- Acrylic-Cutting Knife
- Countersink
- Sewing Machine (optional)

Store dried herbs in tightly sealed glass jars. Leaves should be stored whole and in the dark, to preserve their fragrance. After filling the jars, check them in a few hours and then in several days to be sure that no moisture has collected inside. If there is moisture, redry the herbs until they are crisp. Label your jars as you fill them.

We show an exploded view of the herb dryer in illustration 8–1. Chart 8–1 is a materials list; chart 8–2 lists the tools you will need in addition to the usual hand tools.

Before you purchase materials for your dryer, measure the window in which it will be used, as shown in illustration 8–2. In the formula, measurement A is the distance from side to side and measurement B is the distance from the sill to the top edge of the lower window sash. Put the figures for measurements A and B into the materials list formulas to help determine your quantities of materials.

Begin construction of the window herb dryer by laying out the main pieces on a 4 × 4-foot sheet of ½-inch A-C plywood. (Unless you have a very large window and want a large dryer, you should be able to cut all the pieces from one 4 × 4-foot sheet.) One side of the A-C plywood is smooth and the other side is rough. Use the better side for the exterior of the pieces of your herb dryer. On half of the 4 × 4-foot sheet lay out the two sides, 8 inches by B minus 1 inch. Also lay out the top and the bottom, each 8 inches by A minus ¼ inch. On the second half of the plywood sheet, lay out the front, 8½ inches by A minus ¼ inch, and the door, A minus ¼ inch by B minus 8½ inches.

Cut a vent opening in the top piece, as shown in illustration 8–3. Lay out the ¾-inch-wide vent 1½ inches from the edge, and allow 2 inches at the ends of the piece. At each end draw a curve with a radius of ⅜ inch. Drill a starting hole within the vent area. Cut out the slot with a saber saw, then trim where necessary with a wood file.

Cut four 8-inch lengths of corner bracing from ¾-inch quarter-round molding. Next, cut out eight 1 × 8-inch shelf supports from the ½-inch plywood. Cut ½-inch-radius notches in the ends of the shelf supports

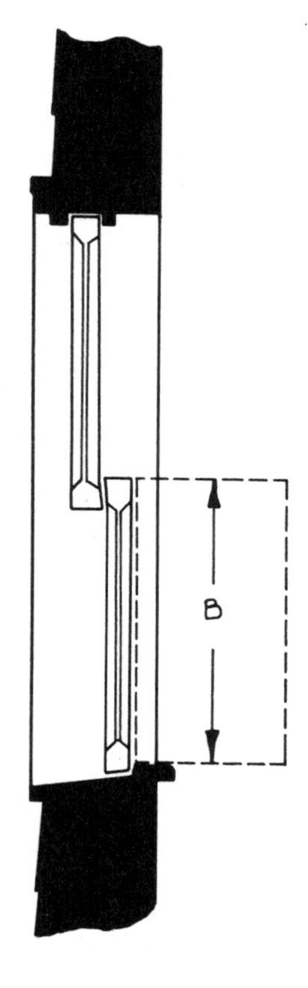

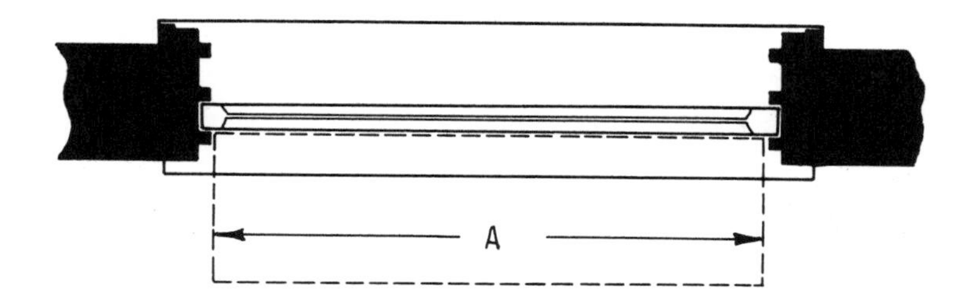

Illustration 8–2—The two measurements needed to build the herb dryer are A, the width of the window, and B, the height.

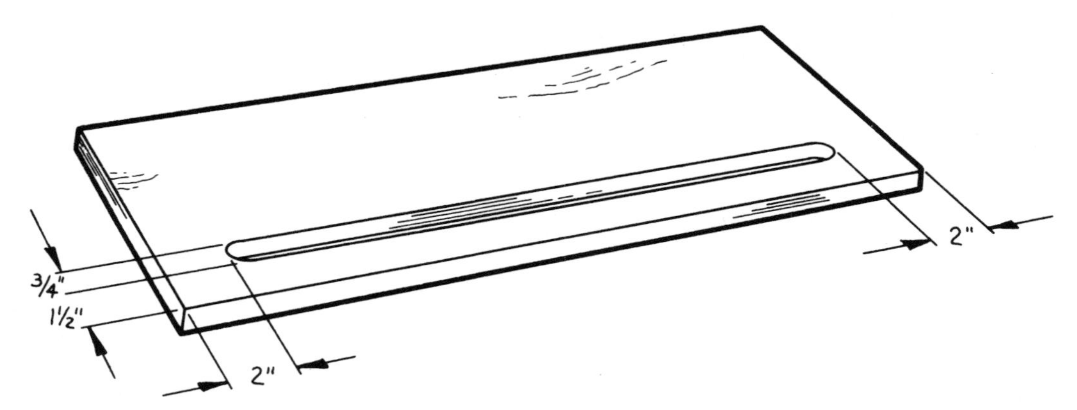

Illustration 8–3—The top of the herb dryer is slotted for ventilation.

to hold the shelves, as shown in illustration 8–4. In two of the supports, also cut out a ¼-inch-radius center notch.

Position the shelf supports on the sides, as shown in illustration 8–5. Start by placing the upper edge of the highest support 2¾ inches from the top edge, then space the supports 4 inches apart. Note that the supports with the center notch go at the top. Fasten the supports to the sides with wood glue and ¾-inch 16-gauge brads.

Next, assemble the dryer frame. Using wood glue and 4d finishing nails, fasten the top and bottom to the sides of the dryer. Note that the top and bottom overlap the sides. Fasten a brace into each corner with glue and 16-gauge brads to add strength.

Cut a vent slot on the front identical to the one on the top, but 1 inch from the lower edge of the piece. Fasten the dryer front to the bottom and sides with wood glue and 4d finishing nails. Remember to have the vent slot at the bottom.

Cut the door window opening, as shown in illustration 8–6. The dimensions are 8 inches high by measurement A minus 5¼ inches wide. If your dryer is considerably higher than 24 inches, you might also want

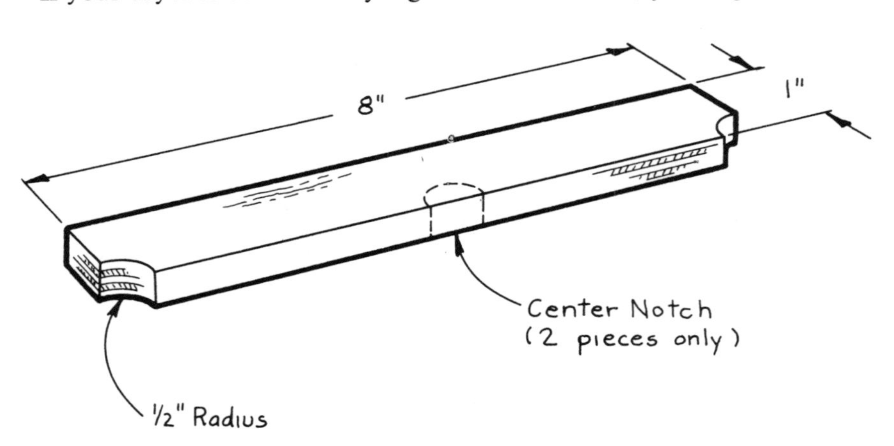

Illustration 8–4—Layout details for the shelf brackets.

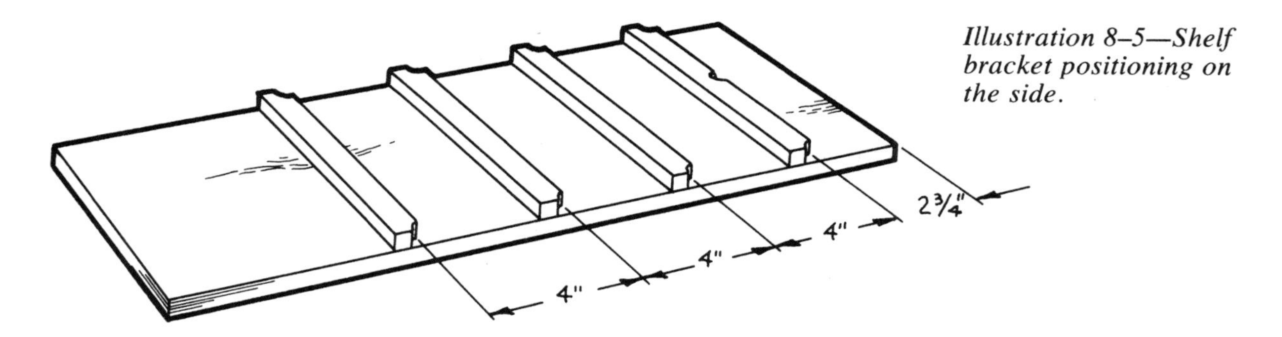

Illustration 8–5—Shelf bracket positioning on the side.

to increase the window height. In that case, be sure to cut the glazing larger as well.

Cut a ⅛-inch sheet of acrylic to fit the window opening area with ½-inch overlap on all four sides. The acrylic can be cut with a acrylic-cutting knife against a straightedge or with a saber saw with a fine blade, as shown in photo 8–1. Hold the acrylic and the saw firmly to minimize vibration.

Use wood putty to fill in any cracks and gaps in the wood frame. Sand the dryer as necessary, and paint it to match your interior decor.

Center the glazing inside the door and mark ten holes, evenly spaced, around the edges. Remove the glazing. Drill and countersink ⅛-inch holes in the acrylic sheet. Be careful to hold the acrylic sheet down, as

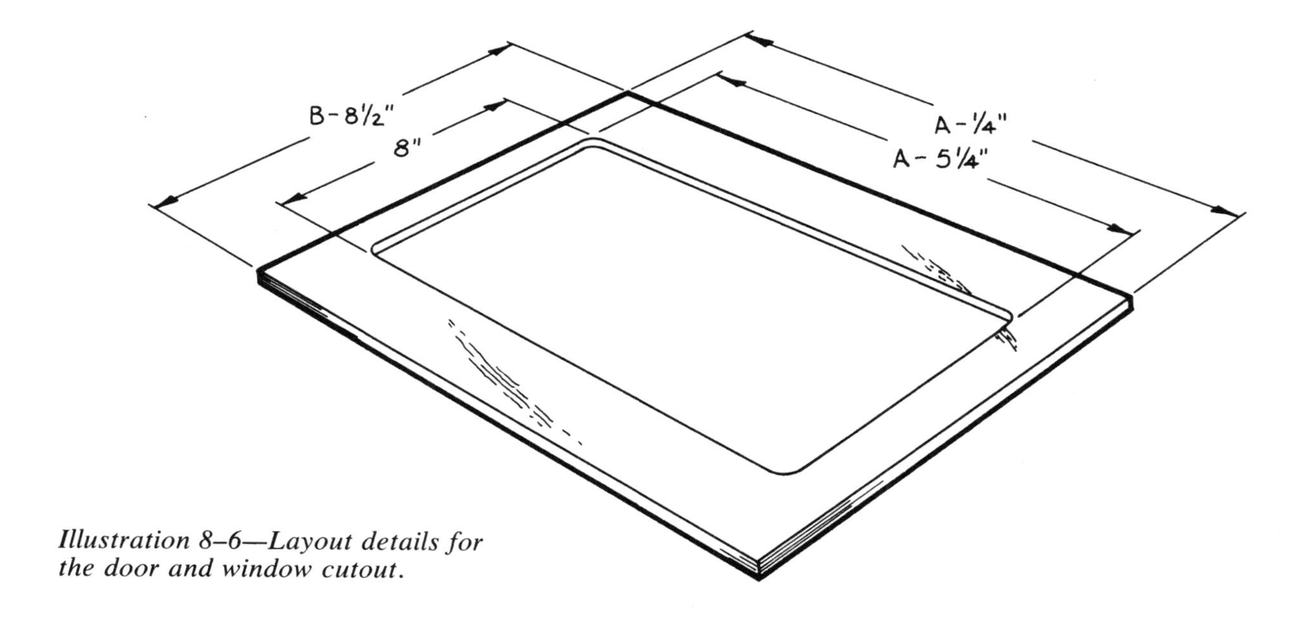

Illustration 8–6—Layout details for the door and window cutout.

Photo 8–1—Clamp the acrylic sheet and straightedge guide firmly to the workbench before attempting to cut with the saber saw. A slow speed and medium/coarse blade work best.

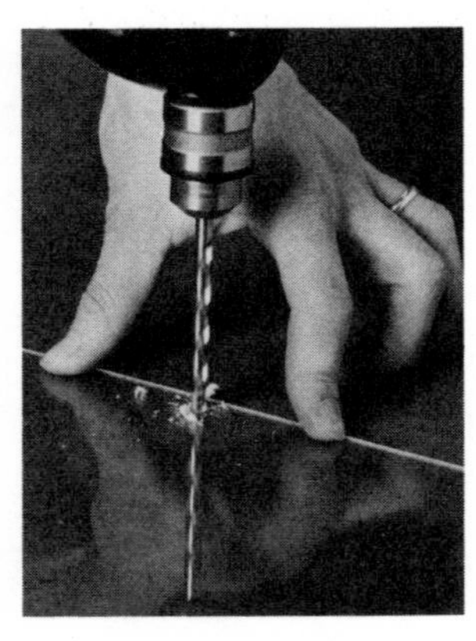

Photo 8–2—Hold the acrylic sheet firmly as it is drilled to avoid having it catch on the drill bit.

shown in photo 8–2, as it tends to catch on the drill bit. Replace the glazing and mark the location of the holes in the door. Drill 1/16-inch pilot holes in the door to a depth of 1/4 inch. Be careful not to drill all the way through. Fasten the acrylic glazing to the door with ten 1/2-inch #4 flathead wood screws.

Mount the wooden knob on the outside of the door. Drill through the door toward the top and screw the knob onto the door.

The back of the dryer, which faces the existing window glass, acts as the heat absorber. It is cut from one piece of aluminum flashing. Use measurement A minus 1/4 inch by measurement B to mark out the back. Cut the flashing with tin shears, being careful of the sharp edges.

Paint one side of this absorber panel with flat black absorber paint. Paint the inside of the cabinet whatever color you choose. When the paint is dry, fasten the back to the dryer cabinet with 3/4-inch 16-gauge brads so that the black side faces out.

Fasten the door to the dryer front with two cabinet hinges. Attach two magnetic catches just under the top and match their position with the other pieces on the door. Place the herb dryer in its location on your windowsill, and line up two snap catches, as shown in photo 8–3. When properly installed, the dryer may easily be removed when necessary by simply undoing the catches and lifting it out of the window.

Make the lightweight shelves, as shown in photo 8–4, from six pieces of 5/16-inch-diameter dowels cut to measurement A minus 1 1/2 inches. These will form the shelf sides and fit into the notches in the shelf supports. Cut fiberglass screen to fit the shelves, 10 1/2 inches wide and measurement A minus 1 3/4 inches long. Sew a 1/4-inch hem across the ends of the shelf screens. Stitch a hem 1 1/4 inches wide to hold the dowels.

Slide the shelf sides into the hems and install the shelves on the supports. For hanging herbs, simply tie them to a dowel and put the dowel in the center notch of the top set of shelf supports.

You are now set to start drying herbs in your new window herb dryer! As you experiment with the dryer, use a thermometer to help monitor temperatures. Herbs retain their delicate flavoring oils best if drying temperatures are kept between 90°F and 100°F.

Photo 8–3—The hardware to hold the herb dryer to the window sash can be decorative snap catches or inexpensive hook-and-eye catches, as shown here.

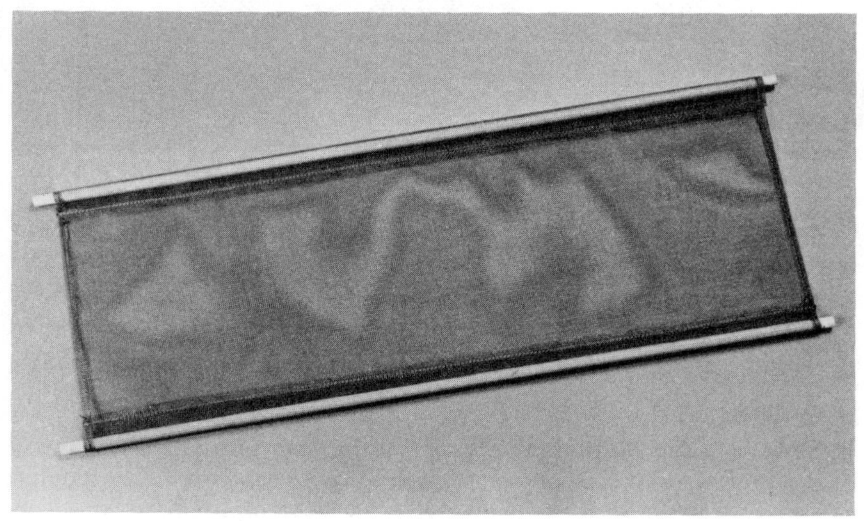

Photo 8–4—The shelves for the herb dryer are made of dowels and fiberglass screen.

9 Solar Doghouse

The solar doghouse is a snug and well-insulated shelter for your special canine. Utilizing the same sound principles as solar home construction, the doghouse benefits from the heat of the sun, draft-free construction, and substantial insulation. Animals generate more body heat than people do, and their living quarters will stay comfortable when the dog's body heat is the only source of heat, if it is retained by insulation.

With a south-facing window of acrylic glazing, sunlight fills and warms the enclosed sleeping area. There is also a sun porch that doubles as a windbreak, sheltering the door into the sleeping compartment. The

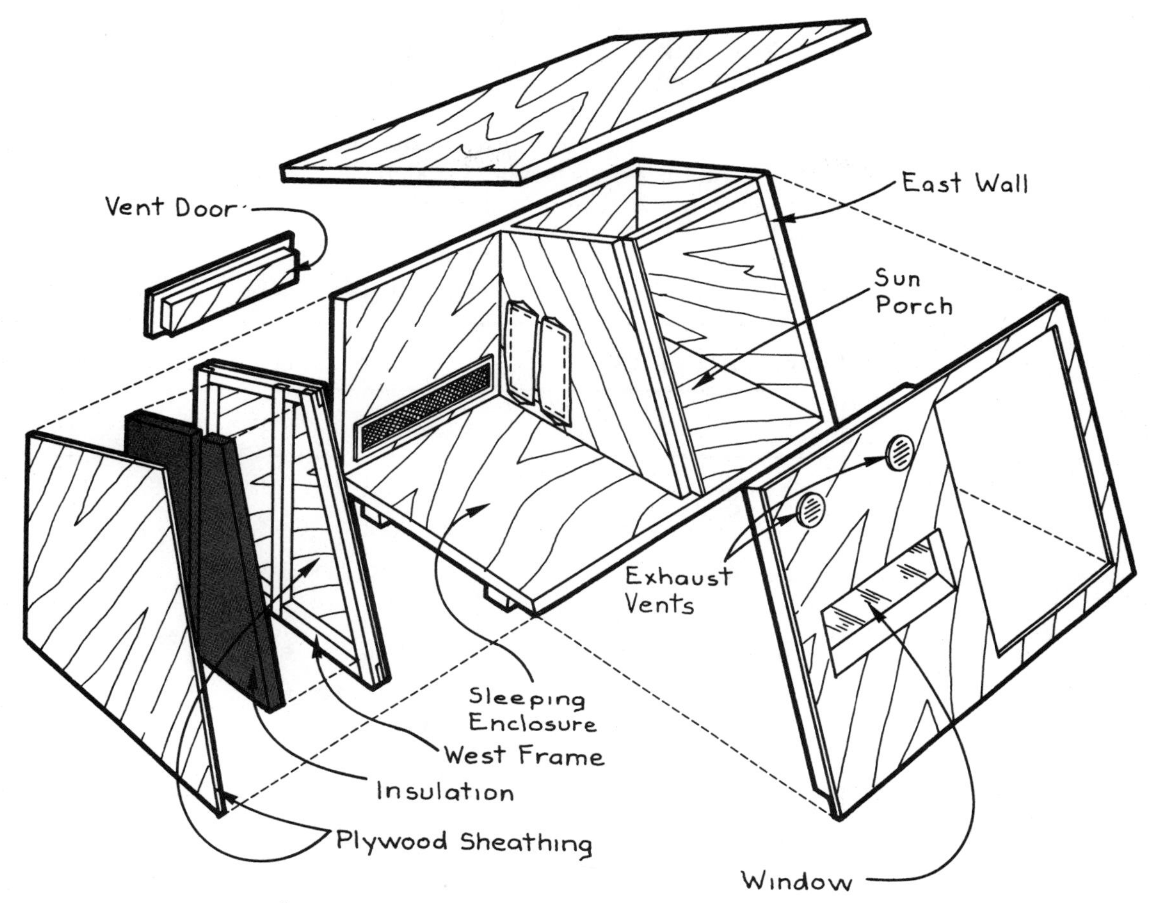

Illustration 9–1—Exploded view of the solar doghouse.

porch is a good outdoor, wind-protected sunning area for year-round use.

An enclosed area, about 6 square feet, occupies half the floor space, and the porch takes up the remaining space. Dogs prefer a cavelike den, and even quite a large dog of 30 to 40 pounds will have plenty of space in a house of these dimensions. Loose straw in winter will make the house more comfortable.

The doghouse roof overhangs the south wall to shade the window from the high summer sun, preventing overheating. The hinged roof makes it easy to keep the house clean. A low vent in the north side is opened in summer so that a cooling breeze flows through the enclosed area and out through high vents on the south wall.

Construction of the doghouse consists mainly of "sandwiches" of rigid foam insulation between plywood sheathing. With tight construction, drafts are very effectively eliminated. The house is finished with primer and two coats of high-quality exterior enamel for weather resistance.

Illustration 9–1 shows an exploded view of the solar doghouse. The materials you will need to buy are detailed in chart 9–1. Chart 9–2 is a tools list.

CHART 9–1—
Materials

DESCRIPTION	SIZE	AMOUNT
	Lumber	
#2 Pine	2 × 4 × 8′	2
#2 Pine	2 × 2 × 8′	20
#2 Pine	1 × 2 × 10′	1
A-C Exterior Plywood	¼″ × 4′ × 8′	7
Hardwood Dowel	5/16″ × 3′	1
Screen Molding	¼″ × ¾″ × 5′	1
	Hardware	
Finishing Nails	6d	6
Aluminum Nails	1¼″	2 lb.
Rustproof Nails	10d	1 lb.
#8 Flathead Wood Screws	1″	34
#8 Flathead Wood Screws	¾″	4
#6 Flathead Wood Screws	¾″	10
Staples	⅜″	1 box
Wire Clips	½″ × 1″	4
Strap Hinges with Screws	6″ × 1″	5
Utility Handle	...	1

The solar doghouse is built by constructing a frame for each wall, covering it on one side with plywood, and inserting 1½-inch-thick rigid foam insulation. An exterior layer of plywood covers the insulation and the frame. The frame uses 2 × 2s, which may be 1½ × 1⅝ inches or the dimensions obtained by ripping a standard 2 × 4 in half. Have ten 8-foot 2 × 4s ripped at the lumberyard (if you have the equipment, rip them yourself), or buy 20 precut 2 × 2s.

CHART 9–1—*Continued*

DESCRIPTION	SIZE	AMOUNT
Hardware—*continued*		
Screen Door Hooks and Eyes	4″	3
Louvered Vent	2″ dia.	4
Window Screen	5″ × 2′	1
Miscellaneous		
Rigid Foam Insulation	1½″ × 2′ × 8′	1
Vinyl Tube Weather Stripping	¼″ × ½″	106″
Wood and Vinyl Weather Stripping	¼″ × ¾″ × 8′	2
Construction Adhesive	...	4 tubes
Primer	...	1 gal.
Exterior Enamel	...	1 gal.
Silicone Caulk	...	1 tube
Wood Putty	...	4 oz.
Acrylic Sheet	¼″ × 8″ × 2′	1
Neoprene Fabric or Other Heavy-Duty Canvas	6″ × 18″	2
Elastic Strip or Heavy Rubber Strip	½″ × 14″	1

CHART 9–2—
Tools

Tools
Saber Saw
Circular Saw
Hole Saw: 2-inch-dia.
Drill and Bits
Heavy-Duty Staple Gun

Begin construction with the floor framing. From the 2 × 2s, cut two 53½-inch-long floor sidepieces and five 35½-inch-long floor crosspieces. The frame, shown in illustration 9–2, is built with half-lap joints. The half-laps were designed to give you a more durable and rigid frame, but if you wish to simplify the construction you can modify the dimensions and use butt joints.

Begin the half-lap joints by positioning the crosspieces over the sidepieces, with all the boards resting on the 1⅝-inch sides. One crosspiece is flush and square with each end of the side boards, and the other three boards are evenly spaced and squared off. Referring again to illustration 9–2, from one end of the sidepieces, measure 13 inches, 26 inches, and 39 inches, and square lines across. Place the inner crosspieces against these lines, on the side away from the starting point. Making sure that each joint is square, mark the overlapping position on both the sidepieces and the crosspieces, as shown in photo 9–1.

A circular saw can do most of the cutting for half-laps. Set the blade depth to ¾ inch. Clamp the board firmly to the workbench and make the first cut just to the lap side of the marked line. Then make several cuts over the lap area, and finish clearing out the excess wood with a chisel, as shown in photo 9–2. Repeat these steps to make a half-lap at each joint in the floor frame.

Assemble the floor frame, using two 1¼-inch aluminum nails at each joint. Use a square to be sure the crosspieces are at right angles to the sidepieces. Label one side of the frame *front,* and keep the nails 1 inch

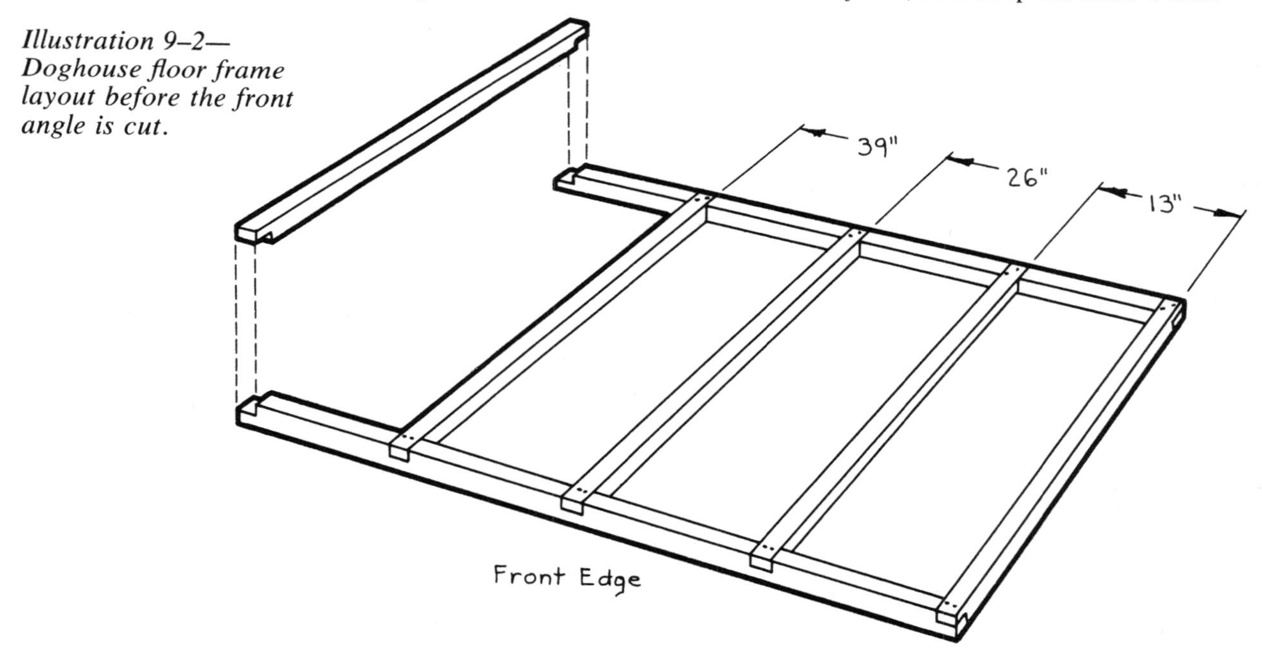

Illustration 9–2—Doghouse floor frame layout before the front angle is cut.

back from this edge. It will be ripped at an angle after the floor is built.

Square up the frame before attaching the sheathing by measuring diagonally from corner to corner in each direction. When the two diagonal measurements are equal, the frame is square. For the floor sheathing, cut two 35½ × 53½-inch pieces of ¼-inch A-C exterior plywood. Choose the better side of the plywood to face out. Use construction adhesive and 1¼-inch nails to fasten one of the pieces of sheathing to the frame. Again, be careful not to nail within 1 inch of the front edge.

Insulate the floor area with pieces of 1½-inch rigid foam insulation. Measure the four cavities within the floor frame, and cut insulation to fit tightly into each one. Place the insulation in the spaces between the frame members, cover it and the frame with a second piece of plywood

Photo 9–1—Lay out the pieces for the doghouse floor, then carefully square and mark the half-lap joints.

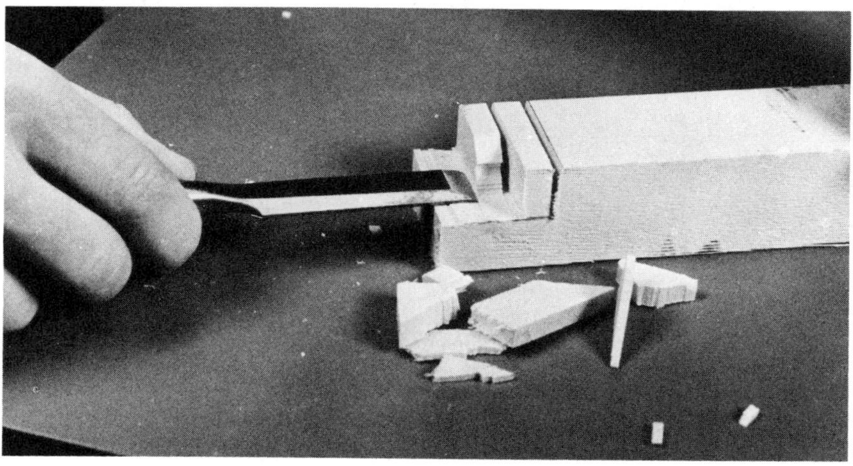

Photo 9–2—Clean the excess wood from the half-lap joints with a sharp wood chisel.

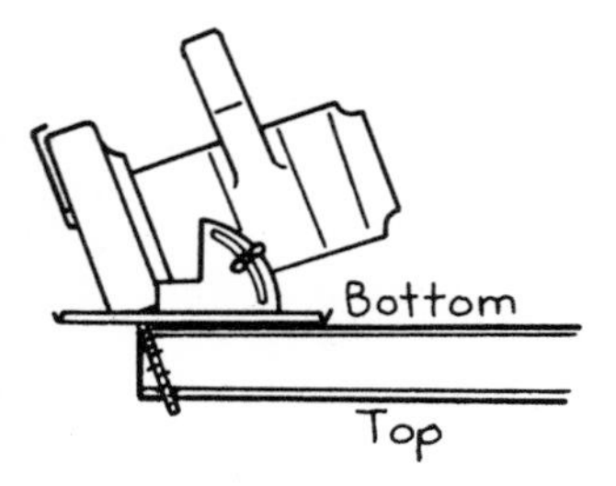

Illustration 9–3—Floor angle cut.

sheathing (with the better side out), and fasten the plywood in place with construction adhesive and 1¼-inch aluminum nails.

With your circular saw unplugged, set it to cut a 20-degree angle. Use a protractor to check the accuracy of the setting. Rip the angle along the front edge of the floor, as shown in photo 9–3. Cut from the bottom of the panel, keeping the weight of the saw on the panel, and follow the front edge, as shown in illustration 9–3.

Fasten skids under the doghouse both to facilitate moving it and to keep it dry and help prevent deterioration. Cut two pieces of 2 × 4, each 54 inches long, for the skids. With the floor unit upside down, place the skids parallel to the front edge. They are set in 4 inches from the front and back edges and are fastened flat. Use construction adhesive and 10d rustproof nails to hold the skids in place. You may want to use treated lumber and galvanized nails for the skids to prolong their life.

Now you are ready to construct the back, or north, wall of your doghouse. Cut the following lengths of 2 × 2 stock: two horizontal pieces, 53½ inches; four vertical pieces, 26¼ inches; one vent cross-piece, 27⅝ inches; and one vent vertical piece, 20½ inches.

Illustration 9–4 shows the plan for assembling the back wall frame. In the same way you handled the floor frame, lay the vertical pieces over the horizontal pieces with the corners square. The vertical pieces are set 13 inches, 26 inches, and 39 inches from one end. Use a square at each corner and mark the joints carefully on all the boards. Cut out the half-laps as you did in the floor frame. Use 1¼-inch aluminum nails

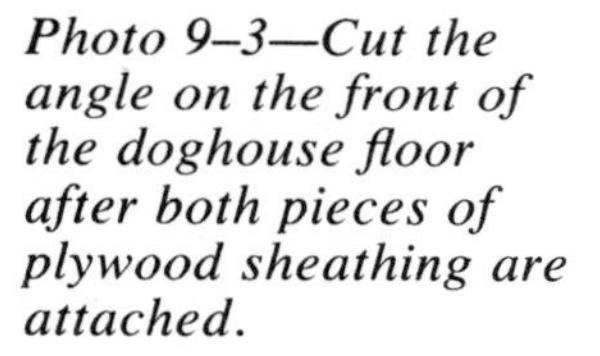

Photo 9–3—Cut the angle on the front of the doghouse floor after both pieces of plywood sheathing are attached.

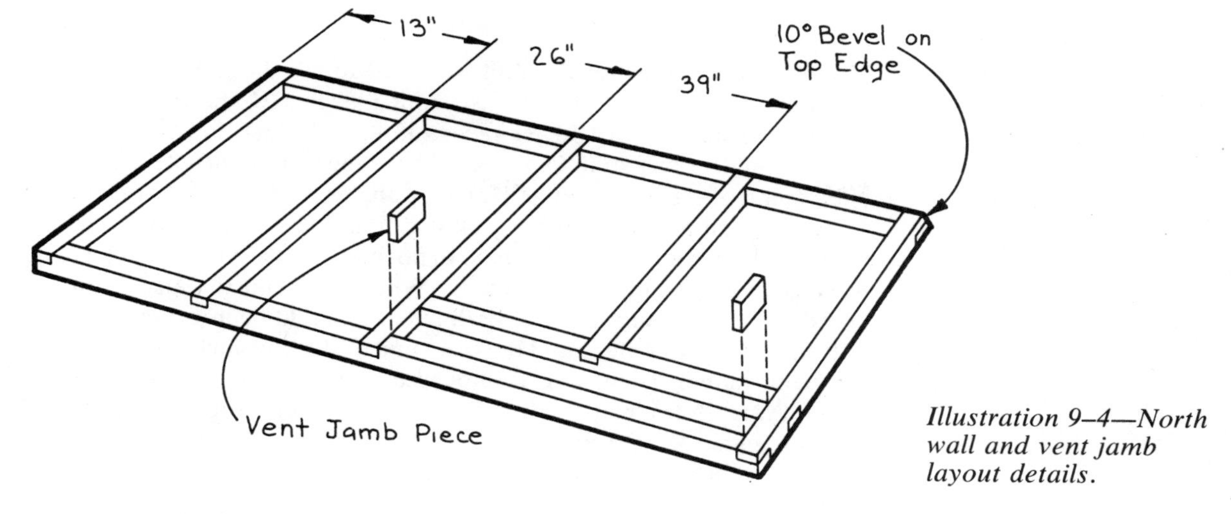

Illustration 9–4—North wall and vent jamb layout details.

to construct the frame, keeping all nails 1 inch away from the top back edge.

From a 1 × 2 cut two 4⅛-inch vent jamb pieces, and fasten them into the ends of the vent opening with 1¼-inch nails.

Cut inside sheathing for the north wall from ¼-inch plywood, 26¼ × 53½ inches. With the better side out, place the frame over the sheathing with the vent on the west side, and mark the vent opening. Cut the vent opening into the plywood after first drilling a hole near the line to accept a saber saw blade. Fasten the sheathing to the frame with construction adhesive and 1¼-inch nails. Keep the nails 1 inch back from the top outer edge of this wall, as it will be ripped to match the roof's slope.

Lay the wall on the bench, with the plywood sheathing facing up. Set your saw for a 10-degree cut and trim off the top edge of the frame, leaving the plywood edge. This slight trimming of the frame will enable the roof to fit tightly to the rear wall.

The north wall frame is now ready to be fastened to the floor. Lay a bead of construction adhesive on the back edge of the floor. Stand the back frame on the doghouse floor, making sure the two units are at right angles to each other and the back is flush along the edge of the doghouse floor. With 10d rustproof nails, nail through the lower horizontal back piece into the floor frame. Then nail a temporary diagonal brace across the west end of the frame to maintain the right angle, as shown in illustration 9–5.

In the same way that you insulated the floor frame, cut pieces of insulation to fill the spaces in the north wall frame.

The next unit to build is the east end wall. As you face the doghouse from the front, or south, the east wall is on your right. The construction is similar to the north wall. You will need to cut the following framing members from 2 × 2 stock: east front piece, 32¼ inches; east back piece, 26½ inches; east lower piece, 33 inches; east upper piece, 22¾ inches; and the east stud, 29 inches. These measurements are for square cuts; some of the pieces must be cut at angles to match the slope of the front wall and the roof. Refer to illustration 9–6, which shows the east wall. Lay out your frame members in position so that you can determine where the angle cuts will be made. When you trim the angles on the boards, do not change their overall length. Cut 20-degree angles on the bottom of the front piece and the front end of the bottom board, to match the front slope. Then, cut a 10-degree angle on the top of the front piece. *Note:* Be sure to cut in the correct direction in relation to the bottom angle. Next, repeat the 10-degree angle on the back of the top piece. Trim a matching 10-degree angle from the top of the middle and back studs.

Lay out the east wall frame with the back corner square, the middle stud at a right angle to the lower piece, and the remaining pieces matching at the corners. Mark and cut the half-lap joints where each board is fastened to the other, checking angles for accuracy.

Build the east wall frame with 1¼-inch aluminum nails as you did the floor and north wall. To mark the interior sheathing for this wall, lay the frame over the rough side of your plywood and use it as a pattern. It must be square at the back corner. Mark and then cut out the sheathing. Fasten the sheathing to the inside of the frame with construction adhesive and 1¼-inch nails.

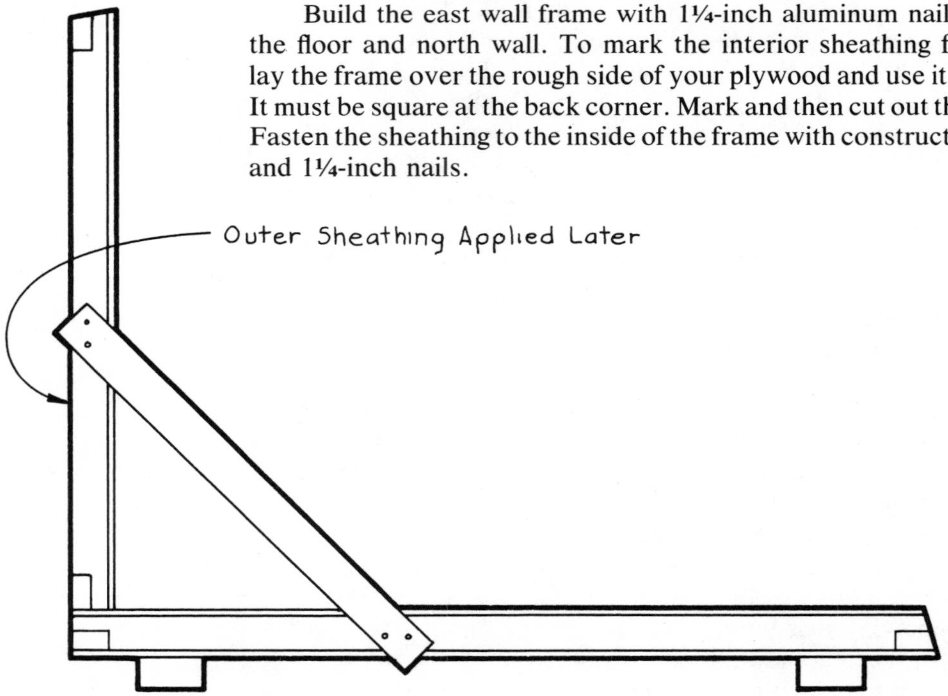

Illustration 9–5—The north wall and floor are held in position with a temporary brace.

Following the same procedure by which you joined the north wall to the floor, add the east wall to your assembly. The front edge of the wall should be flush with the front of the floor, and the back edge butts to the north wall. Use construction adhesive and 10d rustproof nails to fasten the east wall to the floor and to the back of the doghouse.

Insulate the east wall the same as the other units. Measure each space and cut the foam to fit; place the insulation in the wall.

The procedure for building the west and center walls is very much like that of framing the east and north walls. However, the dimensions of the two walls are smaller, to accommodate the front, or south, wall.

From the 2 × 2 lumber, cut the following pieces: west front piece, 32⅛ inches; west back piece, 26½ inches; west lower piece, 31¼ inches; west upper piece, 21 inches; and the west stud, 29 inches.

Lay the frame pieces in their approximate position so you can determine where to make the angle cuts. Once you are sure all the pieces are properly positioned, trim the ends to the proper angles where needed, and assemble the wall as you did the others.

With the wall assembled and the inside sheathing attached, you are ready to join the west panel to the doghouse. Remove the temporary brace, and place the frame in position against the back wall and on top of the floor. *Note:* The front edge is set back from the front of the floor approximately 2 inches to accommodate the front wall section, as shown in illustration 9–7. Use 10d rustproof nails and construction adhesive to join the wall to the back and to the floor.

As you did in the previous sections, cut and fit insulation into the west wall cavities.

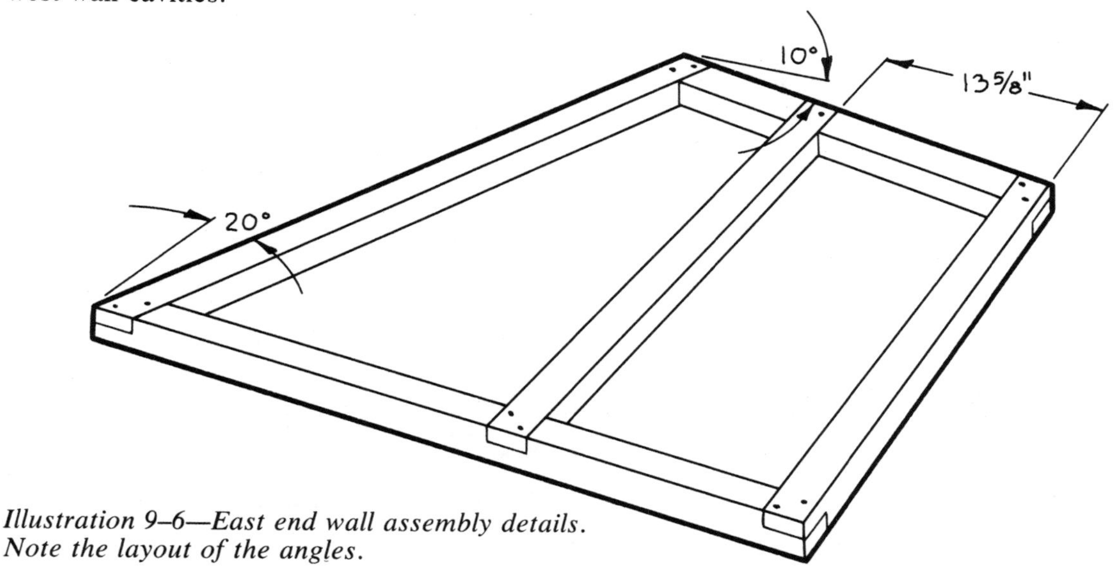

Illustration 9–6—East end wall assembly details. Note the layout of the angles.

Construct the center wall, which is identical to the west wall, except for the addition of a doorway for access into the enclosed sleeping area from the porch. Use the same measurements and procedures as you did for the west wall. Cut out the framing members, cut the appropriate angles on each end, and cut half-laps for the joints. Be sure there is a 12-inch space between the middle stud and the back stud. Nail the frame together.

To build the doorway, cut the sill and header from 2 × 2s, each one 12 inches long. Also cut a right doorjamb, 14 inches long. Following illustration 9–8, fasten the sill into the doorway on the lower frame piece, using 10d rustproof nails. Stand the jamb on the sill and nail it into the back piece. Last, place the header on the jamb and, keeping it level, nail through the middle stud and through the back stud into the header. Be careful to keep both faces flush.

The sheathing is laid out using the frame as a pattern. Set the plywood against the frame, and mark the door opening. In order to have continuous sheathing at the doorway, drill a hole inside the opening for the saber saw blade, then cut out the doorway and the piece of sheathing. Before installing the sheathing, use it as a pattern for a second piece. Use construction adhesive and 1¼-inch aluminum nails to fasten the sheathing to the center frame.

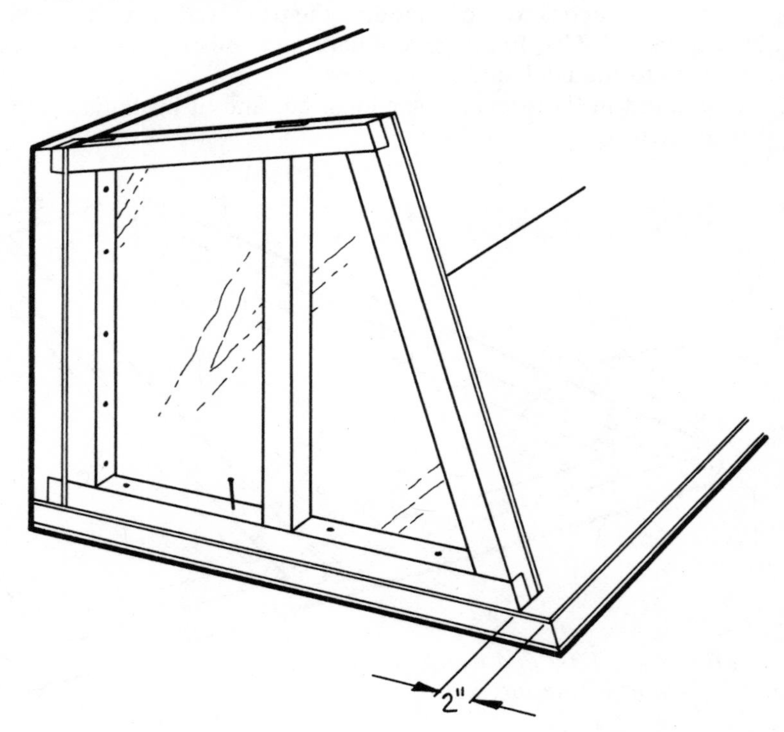

Illustration 9–7—Set the west wall back 2 inches to allow for the south wall.

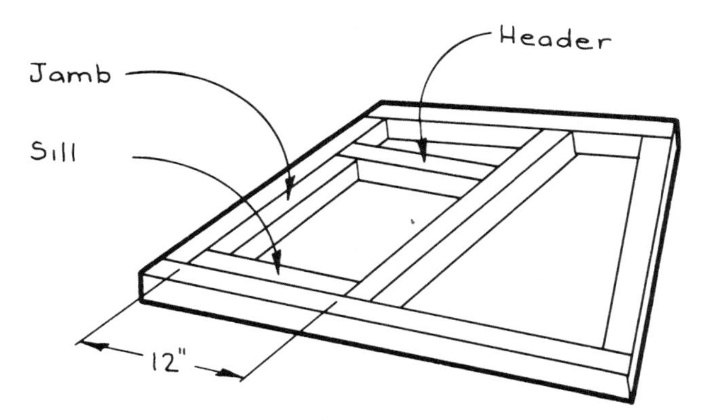

Illustration 9–8—Center wall and header, sill, and jamb layout details.

Next, attach the center wall to the doghouse frame. Place the back stud against the back wall, directly over the center floor crosspiece, or 24 inches from the inside of the west wall to the inside (or west side) of the center wall. The front edge is set back 2 inches for the south wall as was the west wall. As before, glue and nail the center wall to the back wall and to the floor. To insulate this section, again cut and fit rigid foam into the wall cavities. With the insulation and wall in place, cut and attach the other plywood face.

The last wall section of the doghouse is the front wall, which contains a window to let sunshine in. This wall is set on a 20-degree angle, which will capture more heat in winter than a vertical wall. Using the 2 × 2s, cut the following pieces: two front sidepieces, 32¼ inches each; four front crosspieces, 27½ inches each; and a front windowsill, 24¼ inches.

Place the two sidepieces together and, measuring from the bottom, mark them both at 10¼ inches and 20⅜ inches. Following illustration 9–9, lay the four crosspieces at right angles to the sidepieces. Place the

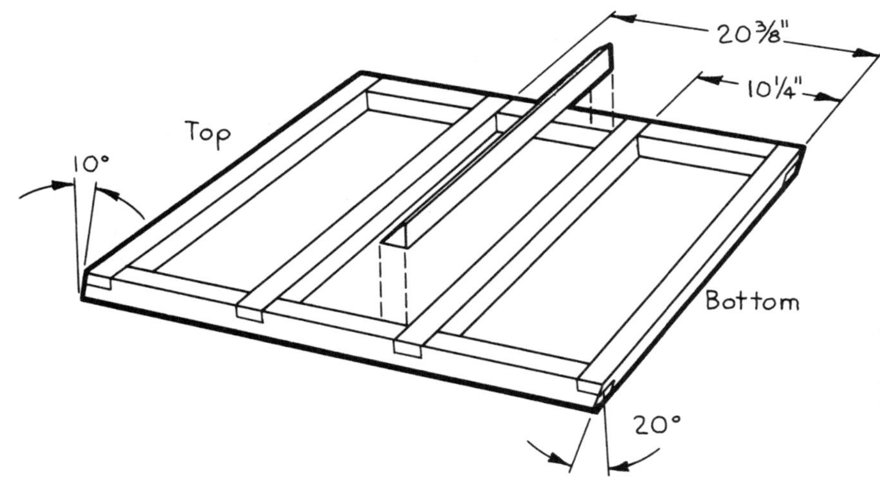

Illustration 9–9—Windowsill positioning details in the south wall.

lower edge of the middle crosspieces at each mark. Draw lines for the half-lap joints on each piece, cut out the half-laps, then nail the frame together. Be careful to keep nails 1 inch back from the top and bottom edges, as these will both be ripped on an angle later.

The inside sheathing of the front wall contains an opening 8¼ × 24¼ inches for the window. First cut the piece of sheathing 27½ × 32¼ inches. Mark 1⅝ inches from each side and square lines down the plywood. Measure down from the top edge 11¼ inches and 19½ inches, and square lines across the sheathing to meet the vertical lines, as shown in illustration 9–10. Cut out the rectangular window opening. Fasten the sheathing to the interior of the front wall frame.

To rip the front wall, first set the saw to cut at 20 degrees. Set the panel with the plywood facing down. Rest the weight of the saw on the

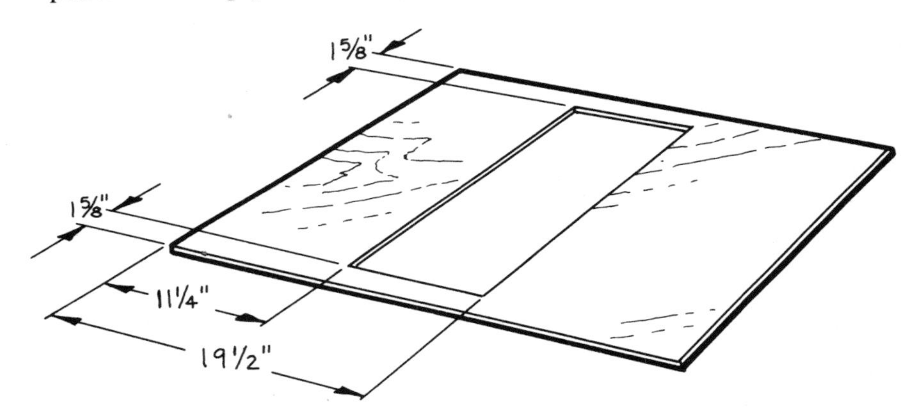

Illustration 9–10—South wall sheathing with window layout details.

Photo 9–4—Rip-cut the 45-degree angle on the windowsill with a circular saw.

frame. Following the edge of the framing member, rip the bottom edge. Reset the saw to cut at 10 degrees. Turn the panel over and rip the top edge.

Next, set the saw to cut at 45 degrees and rip the windowsill, as shown in photo 9–4. Then fasten the sill in place on the lower crosspiece by nailing through the sidepieces into the sill. Accurately measure and cut two pieces of 1 × 2 for the window jambs. On one end of each board cut a 45-degree angle to match the angle of the sill. Use 6d finishing nails to fasten the jambs to the sides of the frame between the sill and the upper crosspiece, as shown in illustration 9–11. Set the nails and fill the holes with wood putty.

Place the south wall on the front of the doghouse against the west and center walls, and fasten it to the floor, to the west wall, and to the center wall with construction adhesive and 10d rustproof nails. Finally, insulate this wall as you did the previous ones. There you have the basic structure of the doghouse.

Before attaching the plywood, cut a 2 × 4 for the header across the sun porch area. It is 23¾ inches long to fit between the east and center walls. Rip a 10-degree angle along the top edge of this board, then fasten it between the walls.

Measure and lay out two pieces of plywood, one each for the east and west end walls. Note that these exterior pieces are larger than the interior sheathing so they entirely overlap the joints between the end

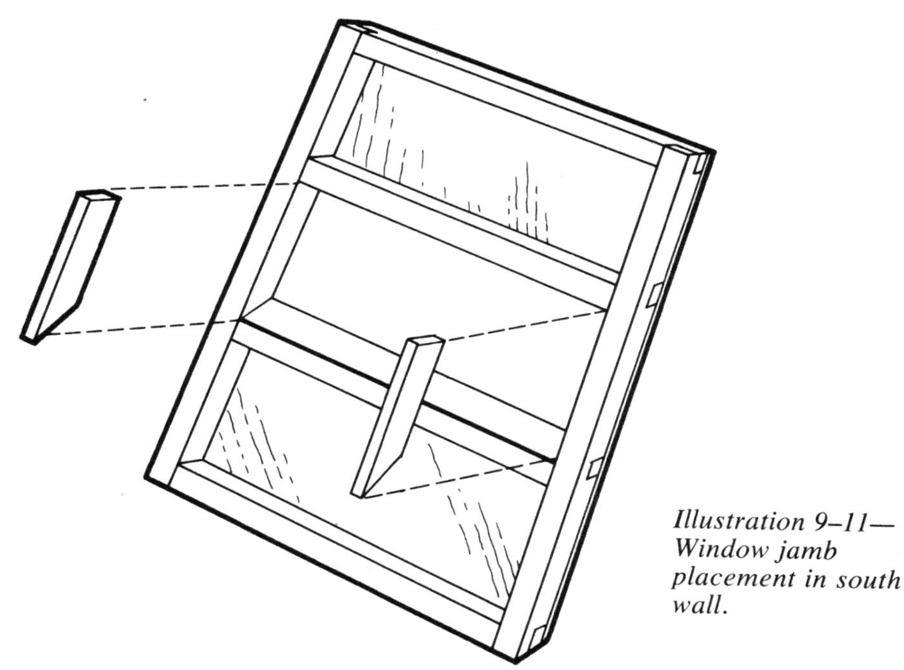

Illustration 9–11—Window jamb placement in south wall.

walls and the back, floor, and front. Illustration 9–12 shows a cutaway view of a north wall corner. Unwanted air infiltration is reduced to a minimum with this construction technique. Cut out the exterior sheathing, and fasten it to the frame with construction adhesive and 1¼-inch nails.

Cut a piece of plywood to fit the south wall, approximately 34¼ × 54 inches. Measure first to fit the pieces accurately. Set the piece in place against the front of the doghouse, and mark the window and porch openings. The sheathing is continued around the porch, covering the edge of the floor and the header. Cut out the openings and fasten the plywood to the front of the doghouse.

The north wall sheathing is handled in the same manner as that on the south wall. First measure to recheck the dimensions, and then cut a piece of plywood to fit. It should cover the entire back wall. Also, mark and cut out the vent opening in the back wall. Fasten the back wall sheathing to the frame with construction adhesive and 1¼-inch aluminum nails.

The vent door is constructed in "sandwich" fashion, just like the house itself. In winter you can keep the door in place to block cold air, and in summer remove it entirely so a breeze is pulled through the enclosed area. Referring to illustration 9–13, cut two horizontal pieces from 1 × 2 pine, 22⅝ inches each. Then cut two vertical pieces from 1 × 2 pine, 2⅜ inches each. Next, from ¼-inch plywood, cut two sheathing pieces, 3⅞ × 22⅝ inches, and one front piece, 5½ × 24½ inches. Assemble the vent by nailing the four framing members together, the short vertical pieces between the longer horizontal ones, as shown in illustration 9–13. Check the size of this frame by slipping it into the vent

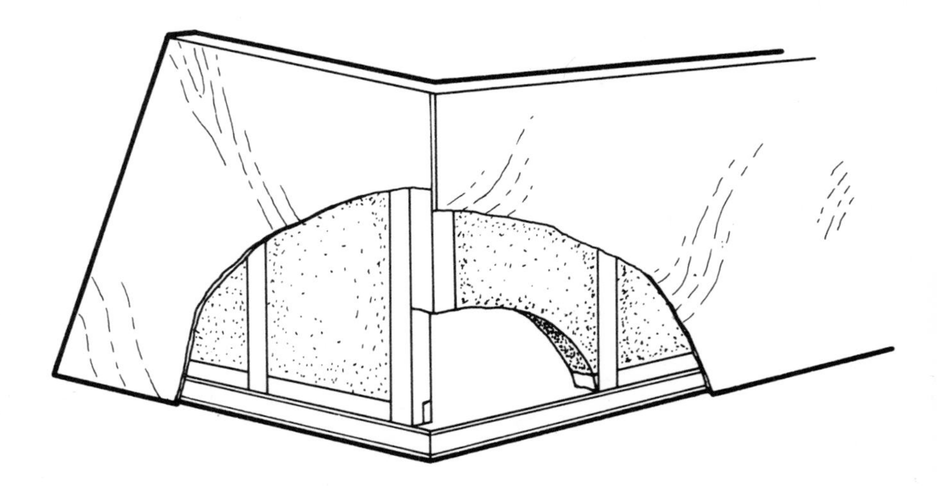

Illustration 9–12—
North and east walls.

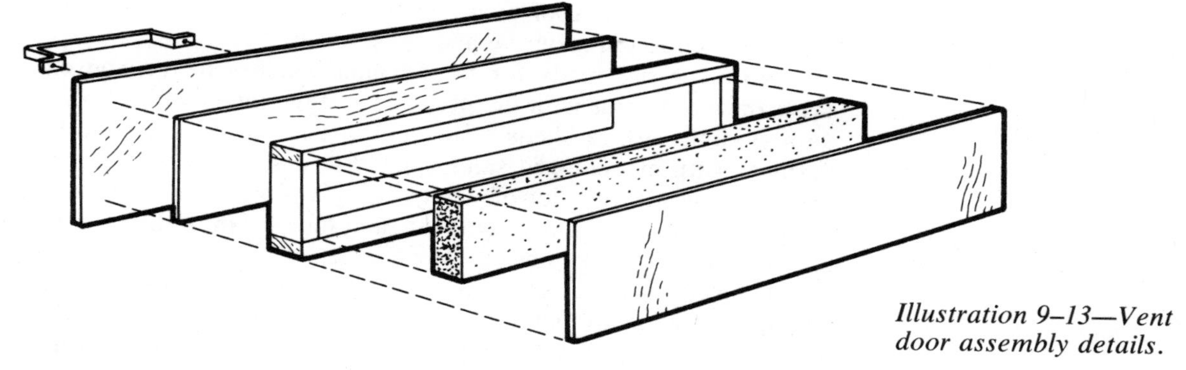

Illustration 9–13—Vent door assembly details.

opening. Glue and nail one of the sheathing pieces to the frame. Measure the cavity and cut a piece of insulation to fill it. Cover the insulation and frame with the second sheathing piece. Center the door front over the door frame, and fasten the larger piece of sheathing to the door with construction adhesive and nails. Be sure to leave equal overhangs.

The doghouse roof is your next construction area. From 2 × 2 stock, cut two sidepieces, 60 inches each; five crosspieces, 48 inches each; and two end pieces, approximately 44¾ inches each. Lay out the roof boards, as shown in illustration 9–14. The end crosspieces are not half-lapped and are shorter. Position the five crosspieces, starting 3¼ inches from the ends; the inner crosspieces are 13 inches apart on center. Mark and cut the half-lap joints, then measure for the two end crosspieces. Nail the roof frame together, with the end pieces filling out the area.

Cut two 48 × 60-inch pieces of plywood for the roof. With the better side out, glue and nail one piece of sheathing to the roof frame. Fill the four large cavities with insulation. You do not need to insulate the end

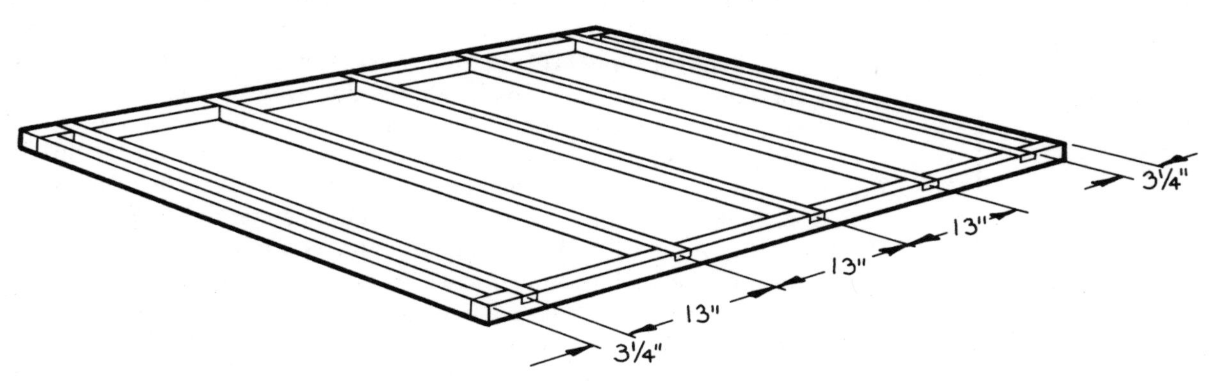

Illustration 9–14—Layout details for the roof.

spaces. Close in the roof with the second sheet of plywood, fastening it with construction adhesive and nails.

Now you are ready for the finishing touches to the house. It is important to fill cracks in the plywood and the joints with wood putty, as a smooth surface sheds water and dirt more readily. Sand all the surfaces with medium-grit sandpaper. Use a good-quality exterior enamel; you have put a lot of work into this project, and it should last many years. Use a primer coat first, and then finish off by covering all surfaces with two coats of exterior enamel.

To make a tight seal around the vent door, staple a vinyl tube weather stripping around the sides, toward the edge, where it will press against the inside of the vent opening. Also mount a utility handle in the center of the door. The door is designed as a snug press fit, but you may want to add latches to the inside for a tighter fit.

Cut a piece of ¼-inch acrylic glazing 8 × 24 inches for the south wall window. Place it in the window opening from the inside of the house to check the fit. Mark the placement of ten holes around the edge. Removing the glazing to your bench and drill and countersink ten 5⁄32-inch-diameter holes. Hold down the glazing as you drill, since it tends to catch on the drill bit. Set the glazing in position, and mark the frame so that you can drill 1⁄16-inch pilot holes. Fasten the glazing to the inside of the window frame with ten ¾-inch #6 flathead wood screws. Caulk the outside perimeter of the window where it joins the wood frame with a paintable silicone caulk.

The door flaps are flexible to let your dog push into his den easily, yet serve to block the wind. They are made of durable neoprene fabric. Cut two 6 × 18-inch pieces of fabric. Fold the ends over and sew them to form loops. The loops should be big enough so you can insert a 5⁄16-inch-diameter dowel. The finished length of the flaps should be approximately 15½ inches. Cut four flap hangers, each 3½ inches long, from ½-inch elastic or heavy rubber strip, and sew these elastics to the flaps, as shown in illustration 9–15. Cut four 6-inch-long dowels and slide them into the flaps. Mount the flaps inside the doorway with four wire clips and ¾-inch #8 flathead wood screws, as shown in illustration 9–16.

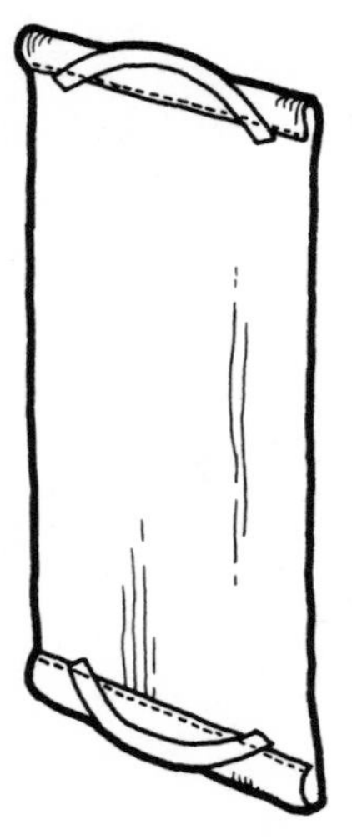

Illustration 9–15—Door flap.

Measure and cut window screen to cover the opening on the inside of the vent opening. Staple the screen in place and cover the edges with ¼ × ¾-inch screen molding. Cut two 2-inch-diameter holes in the upper edge of the front wall for venting. Be sure you don't cut through the framing member. Measure 3 inches from the outside top edge of the front, and 5 inches in from each side of the front wall in the den area to find the center of the vent holes. Drill through the wall at the two center points. Cut out the holes with a 2-inch-diameter hole saw, cutting the sheathing from outside and from inside to avoid splintering. Place

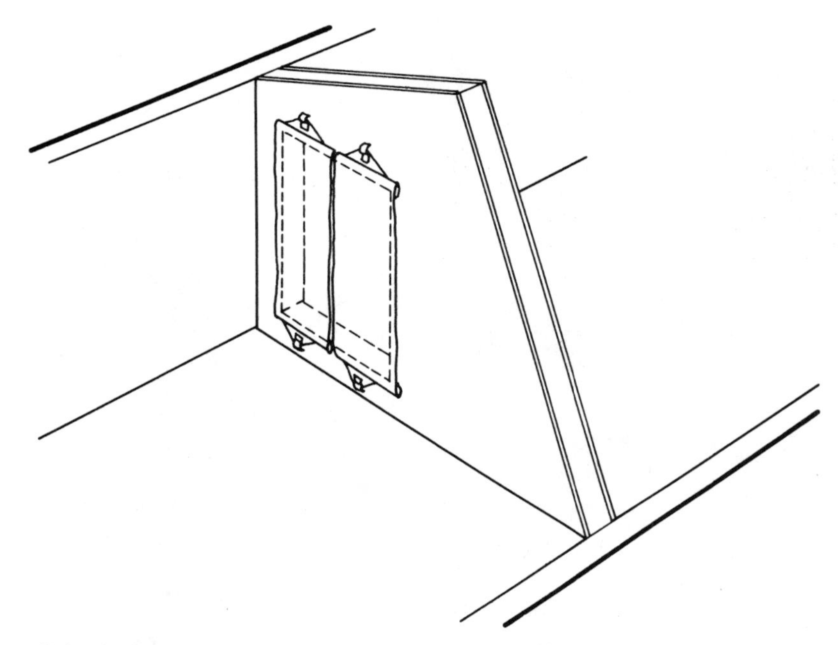

Illustration 9–16—Installed door flaps allow access but block wind.

four 2-inch-diameter soffit vents into the holes, one on the inside and one on the outside of each hole.

The roof is positioned with a 15-inch overhang in front to provide shade in warm weather. It is attached to the back wall with five 6-inch strap hinges. Place the hinges so that they can be screwed into the back vertical frame members, which will also bring them in line with the roof crosspieces. Set the roof on the doghouse, with even side overhangs and with the necessary overhang in front. Mark the hinge locations on the roof and back. Drill pilot holes, and fasten the hinges to the roof and the back wall. In front, mark the points at which you will attach three 4-inch screen door hooks to latch the roof closed. Drill ⅛-inch pilot holes and screw in the hooks. Add screw eyes to the front wall to match the hooks.

We recommend that you add white selvage roofing to the doghouse roof if it will get a lot of wear from being your dog's favorite lookout point. A white roof will be cooler during the summer.

The final trim job is to nail pieces of wood and vinyl weather stripping around the top outside edges of the entire den and sun porch areas, so that the crack between the roof and walls is sealed. Also fasten these strips between the hinges on the back wall.

Step back and admire your handiwork! Some coaxing might be required to get your dog to use the inner door. After a few tries, however, he should use it easily.

10 Firewood Dryer

The best firewood is cut and stacked a year in advance to give it plenty of time to age, but all too often, logs need faster drying. Rodale's solar firewood dryer is a good-looking and sensible enclosure to hold your wood and protect it from rain and snow. The dryer is designed to speed the drying process through both heat and air movement. The clear fiberglass-reinforced roofing and the polyethylene siding allow the sun's rays to heat the stacked wood as well as the air in the dryer enclosure. This warm air rises and is vented out the ends, and cool air is drawn in through spaces at the lower edges. The wood's moisture is removed with the air.

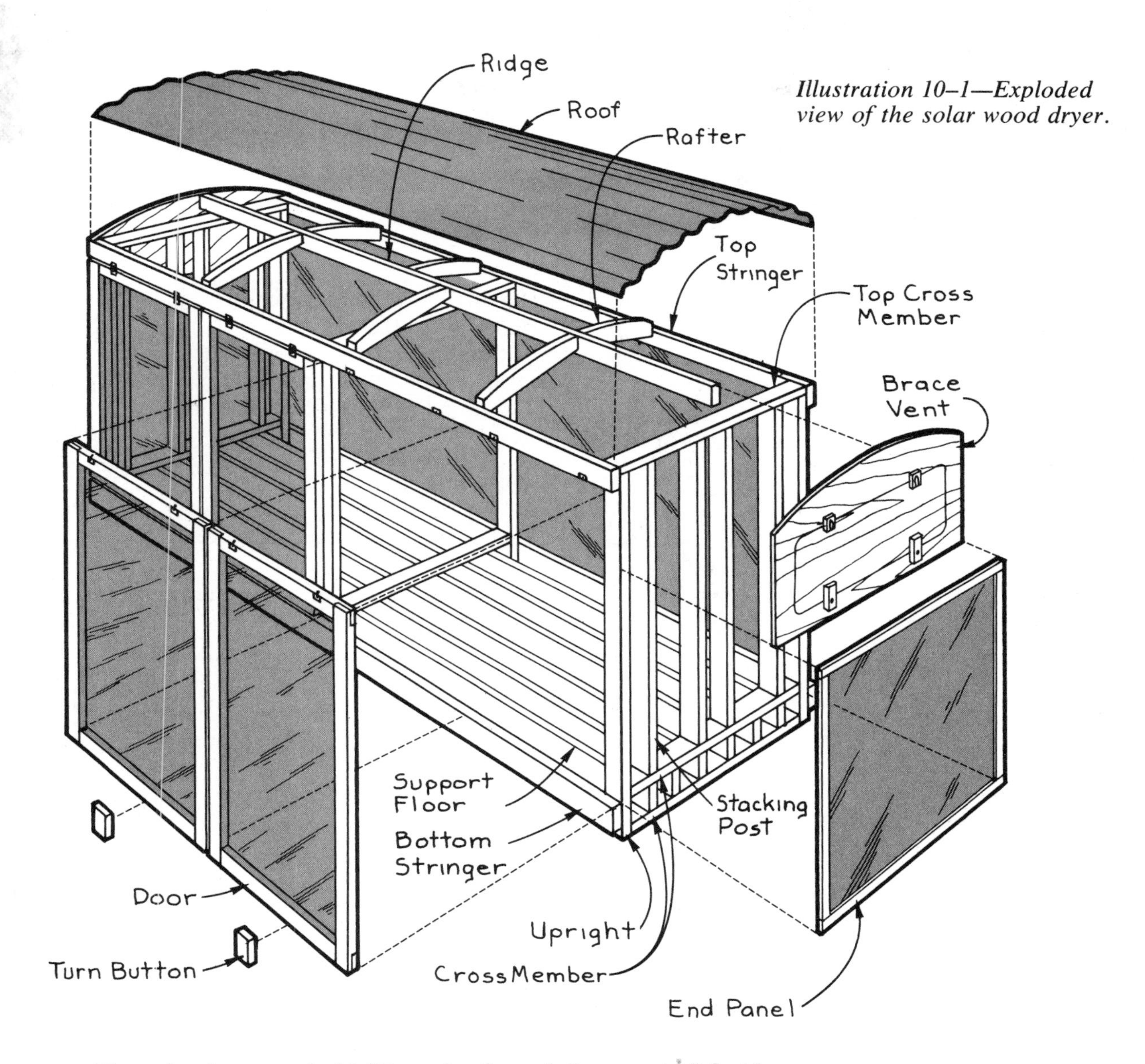

Illustration 10–1—Exploded view of the solar wood dryer.

The solar dryer can hold 1½ cords of wood. Logs up to 3 feet long are stacked on a 2 × 4 deck which serves as flooring and allows air to circulate throughout. When stacking logs, keep them in from the outside walls 3 to 4 inches to facilitate airflow. In the winter months, keep snow away from the air passages along the ground.

The polyethylene walls and door covers are reasonable in cost even if the film has to be replaced every couple of years. Stacking posts are built into the ends of the dryer to protect the polyethylene when the wood is loaded. A sheet of polyethylene under the dryer will help to prevent the migration of moisture from the ground up into the wood and will prevent weed growth.

The exploded view, illustration 10–1, shows the parts of the dryer and their relationship to each other. A materials list, chart 10–1, and a tools list, chart 10–2, follow.

The firewood dryer is a straightforward project that goes together quite easily. It will be useful to label the pieces as you cut them. Start with the 12-foot 2 × 4s. These form the six support floor stringers, the two top and two bottom stringers, and the single ridge piece. For each one, measure 142 inches, square a line across the board, and cut it with a circular saw.

CHART 10–1—
Materials

DESCRIPTION	SIZE	AMOUNT
	Lumber	
#2 Pine	2 × 4 × 12′	12
#2 Pine	2 × 4 × 10′	12
#2 Pine	2 × 3 × 8′	8
#2 Pine	1 × 3 × 1′	1
Wood Lath	⅜″ × 1½″ × 3′	21
A-C Exterior Plywood	⅜″ × 4′ × 4′	1
	Hardware	
Galvanized Common Nails	16d	3 lb.
Galvanized Common Nails	4d	½ lb.
Aluminum Roof Nails with Rubber Washers	1½″	¼ lb.
#12 Flathead Wood Screws	3″	4

Using four 10-foot 2 × 4s, cut eight stacking posts, each 53½ inches long. Then, from three 10-foot 2 × 4s, cut three 60-inch uprights. From one board, cut three top cross members, each 42 inches long. Next, use three additional 10-foot 2 × 4s to cut the three remaining uprights, each 60 inches long, and three cross members, each 39 inches long. Cut the last three 39-inch-long cross members from the last 10-foot 2 × 4.

It is important that, before you assemble your solar firewood dryer, you coat all sides of each piece with a wood preservative. We recommend copper naphthenate because it is not toxic as are the preservatives containing pentachlorophenol. Dip the ends first, then brush the wood preservative on all the parts you have cut so far. An alternative is to buy treated lumber.

In order to maintain an even spacing of the stringers, mark all six of the cross members at the same time. With the six cross members on edge and the ends aligned, mark the points where they will cross the

CHART 10–1—*Continued*

DESCRIPTION	SIZE	AMOUNT
Hardware—*continued*		
#12 Flathead Wood Screws	1½″	4
#8 Flathead Wood Screws	1¼″	32
#8 Flathead Wood Screws	½″	32
Staples	5⁄16″	1 box
Screen Hangers	...	6 pr.
Miscellaneous		
Waterproof Wood Glue	...	4 oz.
Copper Naphthenate Wood Preserver	...	½ gal.
Clear Corrugated Fiberglass Roofing	26″ × 12′	2
6-Mil Polyethylene	15′ × 25′	1

CHART 10–2—
Tools

Tools
Saber Saw
Circular Saw
Drill and Bits
Wood Chisel

Photo 10–1—Mark the position of the stringers on the cross members by squaring across all six pieces simultaneously.

stringers. Measuring from one end, mark on the edge of the cross members at the center points of the support floor stringers, 4½ inches, 10½ inches, 16½ inches, 22½ inches, 28½ inches, and 34½ inches. Mark on the first cross member, then square across the rest, as shown in photo 10–1. As you assemble the flooring, be sure the center of each stringer matches a mark on the cross members.

Start construction of your wood dryer by laying out the six support floor stringers. They are 6 inches apart on center. Position a cross member at each end and in the center of the stringers. The ends of the three cross members extend beyond the stringers 3¾ inches on each side, as shown in illustration 10–2.

Use 16d galvanized nails to fasten the first cross member to the stringers, two nails to each joint. Use a square to be sure the stringers and cross member form right angles. Keeping the stringers parallel to each other, nail a second cross member to the other end of the stringers, again with an extension of 3¾ inches at each end. Then fasten a cross

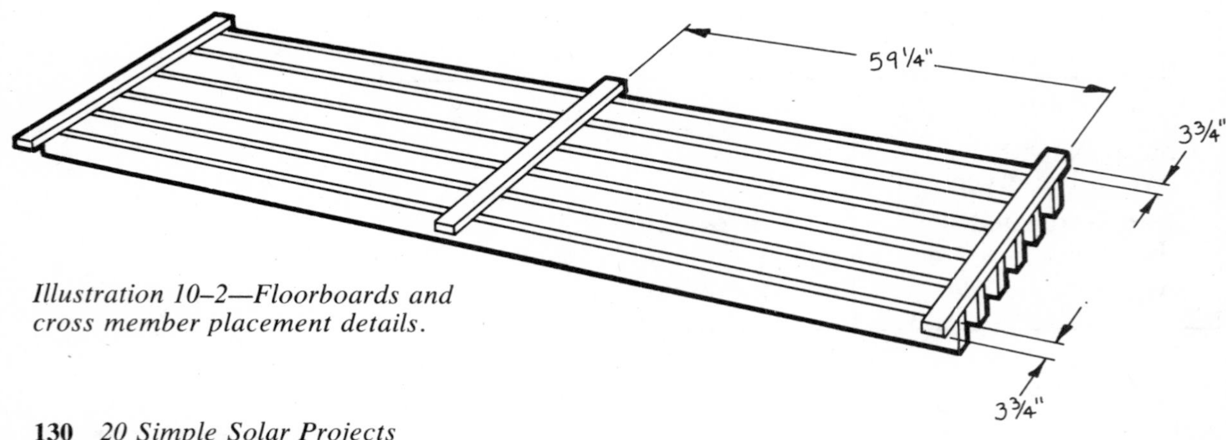

Illustration 10–2—Floorboards and cross member placement details.

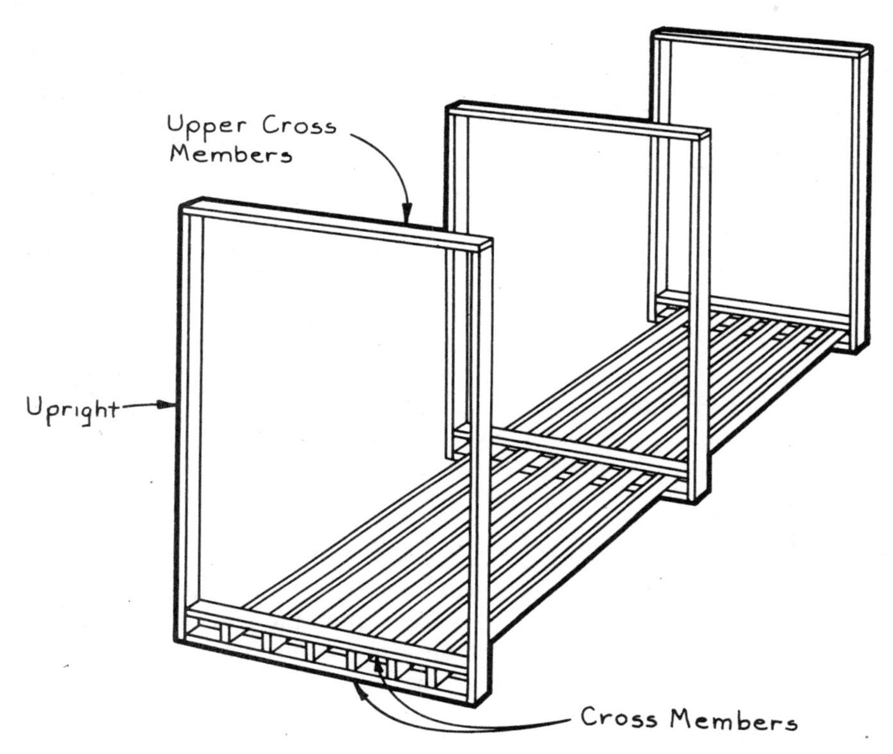

Illustration 10–3—Cross members, uprights, and upper cross members in position on the floor assembly.

member in the center of the stringers. Turn the floor unit over and nail the three remaining cross members directly over the ones already in place.

Begin to construct the wood dryer walls, as shown in illustration 10–3. Nail the six uprights to the ends of the cross members, using 16d nails. Position the lower end of each upright flush with the bottom edge of the cross member. Next, the top cross members are nailed to the tops of the uprights with 16d nails. *Note:* They are not between the uprights, as the lower cross members are, but on top of them.

Position the bottom stringers on the outside of the uprights. They are 1½ inches from the ground, at the same level as the floor supports, as shown in illustration 10–4. Using 16d nails, fasten the bottom stringers to the uprights on both sides. Position and nail the two top stringers flush with the upper edge of the top cross members.

Next assemble the stacking posts. Four posts are fastened vertically at each end of the dryer, as shown in illustration 10–4. The posts are directly above the first, third, fourth, and last support floor stringers. Nail the first post in place at the bottom, measure the distance between the stacking post and the upright, and position the top of the post at the

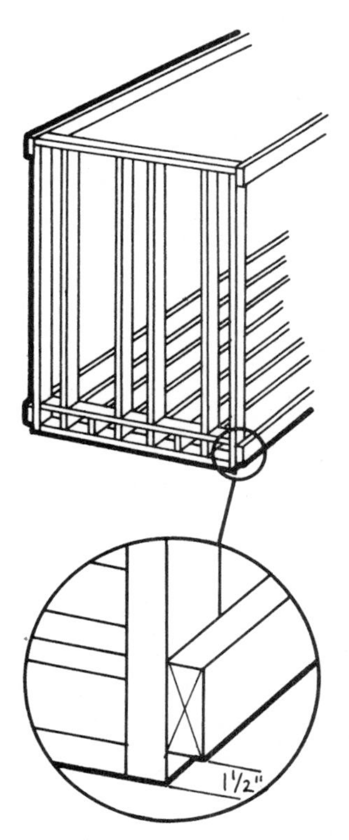

Illustration 10–4—Bottom stringer position is important; note spacing.

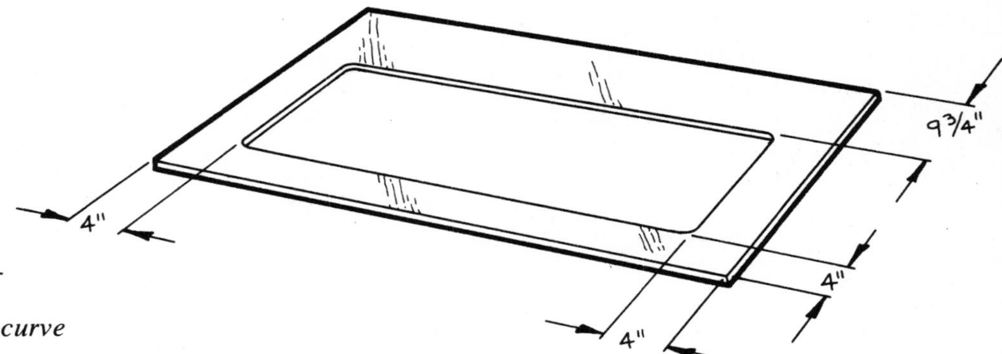

Illustration 10–5—Brace-vent layout details before the curve is cut in the top.

same distance from the upright. Nail it in place to the top cross member. Repeat this procedure for the remaining seven stacking posts.

The basic dryer frame is now assembled, and the plywood brace-vents can be constructed. As their name implies, these pieces serve to give some rigidity to the frame and to allow ventilation through the removable doors.

From ⅜-inch plywood, cut out two 24 × 42-inch rectangles for the brace-vents. Within each brace-vent, mark out the vent door, following illustration 10–5. A 4-inch space is left around the door on three sides and a 9¾-inch space at the top. Drill a small entry hole on the cutting line for the saw blade, or make a plunge cut, and cut out the vent doors with a saber saw. When cutting, stay on the line, because the piece you remove will become the vent door. Round the corners when cutting, for an easier job.

Refer to illustration 10–6 to mark the upper edge of the brace-vent. Mark the center of the longer sides (21 inches from a short side). Square a line from this point, and continue the line off the board until you reach a point 33 inches below the upper edge. Using this point as the center,

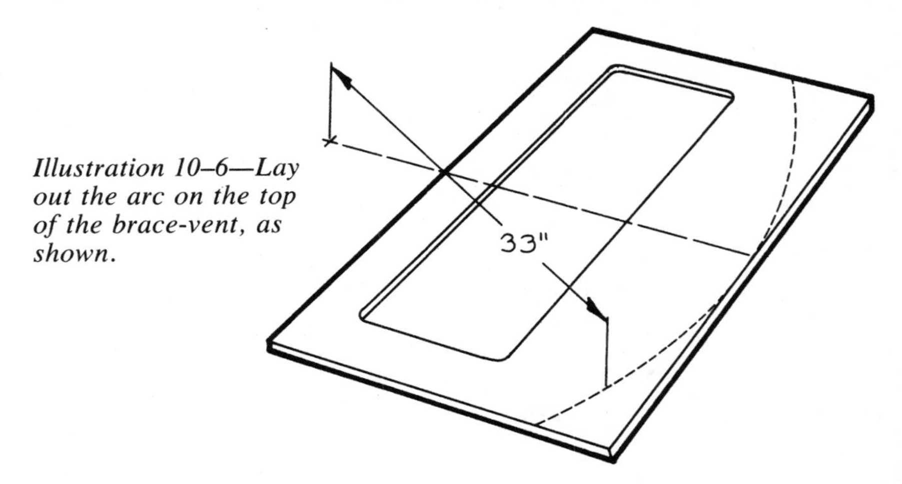

Illustration 10–6—Lay out the arc on the top of the brace-vent, as shown.

draw an arc from one side of the brace-vent to the other. Cut out the curved top edge with a saber saw. Trace the first one onto the second one, and cut it out. Use two screen hangers to fasten each door in place, screwing the hangers into the door and the brace-vent with ½-inch #8 flathead wood screws.

The rafters have the same curve as the brace-vents, so you can use the brace-vents as a pattern. First, from the remaining 12-foot 2 × 4, cut six pieces, each 22½ inches long. Referring to illustration 10–7, set one brace-vent on your workbench. At the center of the top edge, pencil in the position of the ridge board. Lay a rafter on top of the brace-vent, overlapping the vertical edge and the ridge board marking, with the upper edge of the rafter touching the curve of the brace-vent. Mark the cutoff angles at each end. Lay the second brace-vent over the rafter with another 2 × 4 for a spacer, and use it as a pattern to draw the curved top edge of the rafter. Repeat these steps for the five remaining rafters, then cut the curves and angles.

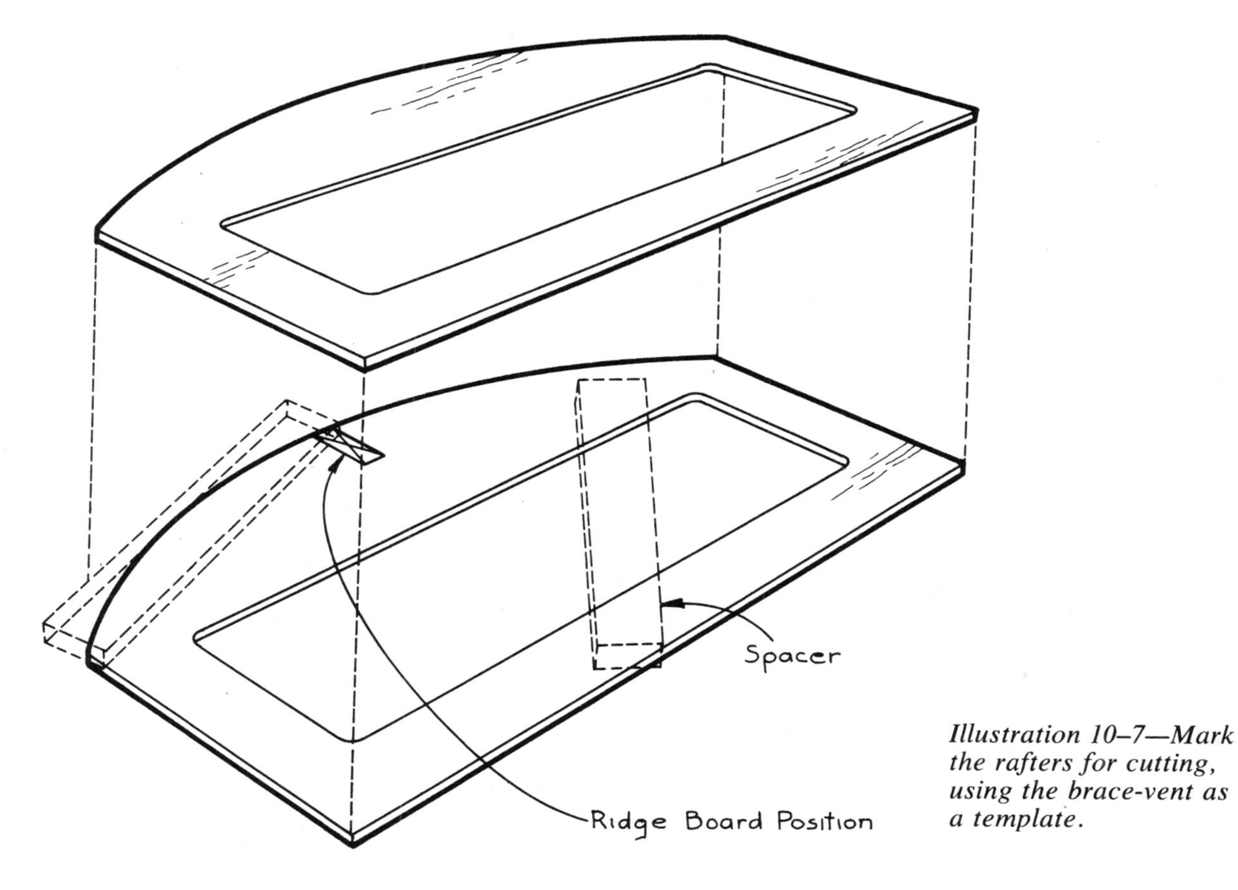

Illustration 10–7—Mark the rafters for cutting, using the brace-vent as a template.

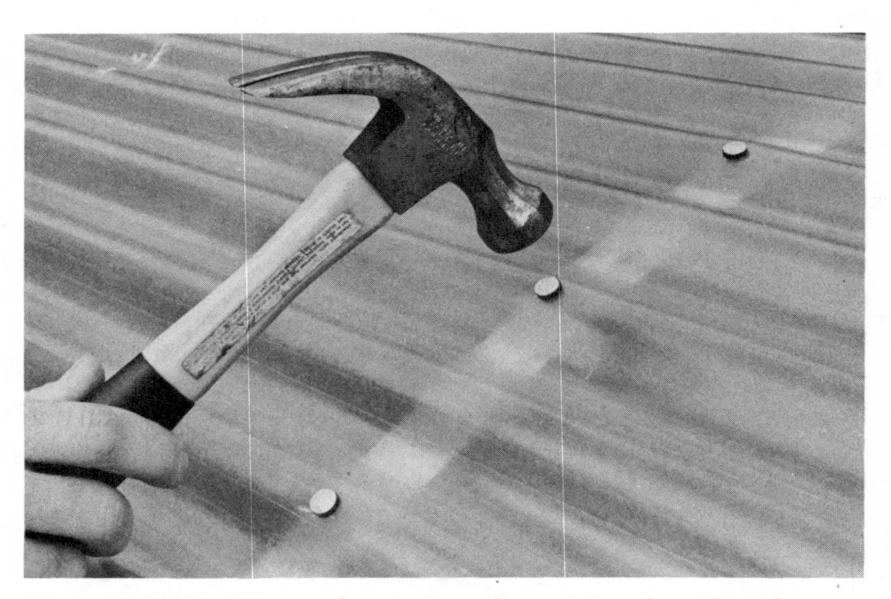

Photo 10–2—Nail the roofing to the rafters with nails with rubber washers. Nail into the ribs, not the valleys.

Now you are ready to assemble the upper parts of the firewood dryer. Set the brace-vents in position at the outside ends of the dryer against the stacking posts and uprights. Be sure the lower edge of the brace is at a right angle to the vertical pieces. The end of the curved upper edge should be about ½ inch above the edge of the top stringer, to let the fiberglass roofing curve down smoothly. Nail the brace-vent in position with 4d galvanized nails. Attach the second brace-vent on the opposite end. At the top center of each brace-vent, nail in the roof ridge board with 16d galvanized nails. Measure 33¾ inches from each end of the top stringer and square a line across. Set a pair of rafters at these lines. Toenail them into the ridge board, and nail through the top stringer. In the center, stagger the rafters one on either side of the cross member.

Use roofing nails with rubber washers to fasten the corrugated fiberglass roofing to the rafters, ridge, and top stringers. Place the nails through the ribs, not the valleys, of the corrugated material, as shown in photo 10–2. At the seams, overlap the roofing at least two ribs.

The firewood dryer is designed with four removable doors of 2 × 3 framing and plastic film to allow you to stack your wood easily. Cut out out eight 2 × 3 side rails, each 52½ inches long. Then cut eight cross rails, each 35¼ inches long. Cut a half-lap at both ends of each framing member to make sturdy corners. Cut the half-laps by laying one board

Photo 10–3—Square and fasten the doorframes together with waterproof glue and wood screws.

across the other at the ends, making sure they are square. Mark the joint on each board. Set your circular saw to cut ¾ inch deep, and clamp the board securely. Cut first at the marked line, and then make successive cuts across the lap area to remove wood from the joint. Clear away the excess material from the joint with a wood chisel. Repeat this for the second piece and check the fit.

Using a square at the corners, fasten the doorframes together with a waterproof wood glue and 1¼-inch #8 flathead wood screws. See photo 10–3. Coat the doorframes, rafters, and brace-vents with copper naphthenate wood preserver.

To make the turn buttons that hold the doors closed at the bottom, cut four pieces of scrap 2 × 3 pine, each 6 inches long. For the vent-door buttons, cut four pieces of 1 × 3 pine, each 3 inches long. Drill a hole ¼ inch in diameter in the center of each button. Paint the eight buttons with copper naphthenate wood preserver. Using the 3-inch #12 wood screws, fasten the four 6-inch turn buttons to the bottom stringer, one for each door. Use 1½-inch #12 wood screws to hold the 3-inch turn buttons in place, two for each vent door.

Next, the dryer must be covered with clear polyethylene, which will help keep your wood dry and increase heat buildup. Starting with

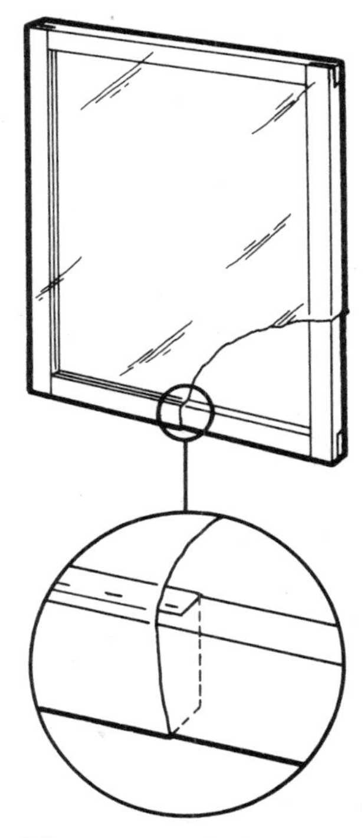

Illustration 10–8—Polyethylene wrapping detail.

Photo 10–4—Fold and staple the polyethylene to the doorframes so that the folds are on the inside of the doors.

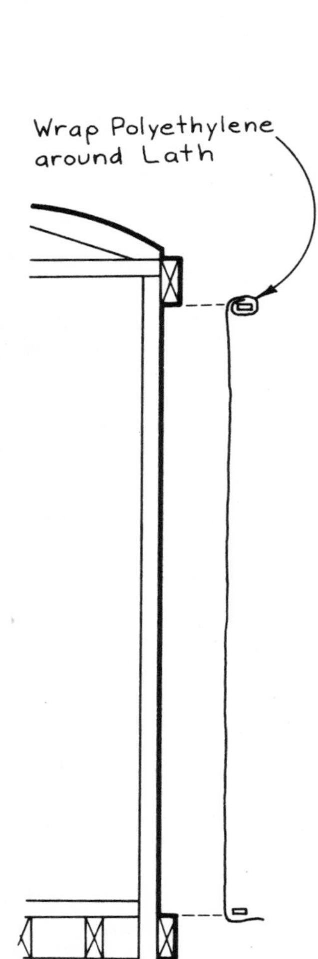

Illustration 10–9—Polyethylene and lath details.

each doorframe, wrap the polyethylene around and to the inside of the 2 × 3s, as shown in illustration 10–8. Fold the plastic film at the corners and staple it in place with 5⁄16-inch staples. See photo 10–4. Hang the doors below the top stringers with two screen hangers on each door. Drill pilot holes and screw the hangers to the doors and to the top stringers.

Cover the ends and the back of the dryer with the polyethylene film. Roll out the film the length of the dryer back and cut it off, allowing 6 to 8 inches extra on each end. Use pieces of lath to fit under the top stringer. Staple the polyethylene to this lath and roll the lath at least one full turn. Then tack the lath to the lower edge of the top stringer with 4d galvanized nails. Also, use pieces of lath for the lower edge. Bring the polyethylene down smoothly, and staple it to the top edge of the bottom stringer. Simply nail the lath over the staples instead of wrapping it in the film, so that moisture can run off under the lath. Trim the polyethylene. Illustration 10–9 shows the complete procedure.

For the dryer back, use pieces of lath to fit the space between the top and bottom stringers. Wrap the edges of the plastic film over the lath and staple it in place. Nail the lath to the uprights, and nail a third lath to the center upright.

Use the same procedure to cover the end panels of the dryer with polyethylene. Start by wrapping the plastic over a strip of lath that is cut to fit just under the brace-vent. Staple the plastic to the lath, roll it, and nail the lath in place to the uprights and stacking posts, just below the brace-vent. Carry the plastic down only to the upper floor cross member, leaving space between the stringers for air to enter the dryer. Staple the polyethylene to the cross member. Cut lath to fit, and nail it over the polyethylene. Finally add the side strips of lath, wrapping them in the polyethylene and nailing them to the uprights. There you have it—a handsome and sturdy enclosure to keep your firewood neat and drying rapidly.

The solar firewood dryer works most effectively when the vent doors are open to facilitate airflow. During rain or snowstorms they can easily be closed, but it is important not to trap moisture in the dryer.

11 Solar Shower

With this attractive, energy-free shower, you bathe outside in a private room with water heated entirely by the sun. The shower consists of a durable and spacious enclosure topped with a sun-heated water tank.

You can tuck the solar shower into a landscaped corner near a pool; it is large enough to be used as a changing room. The shower needs a

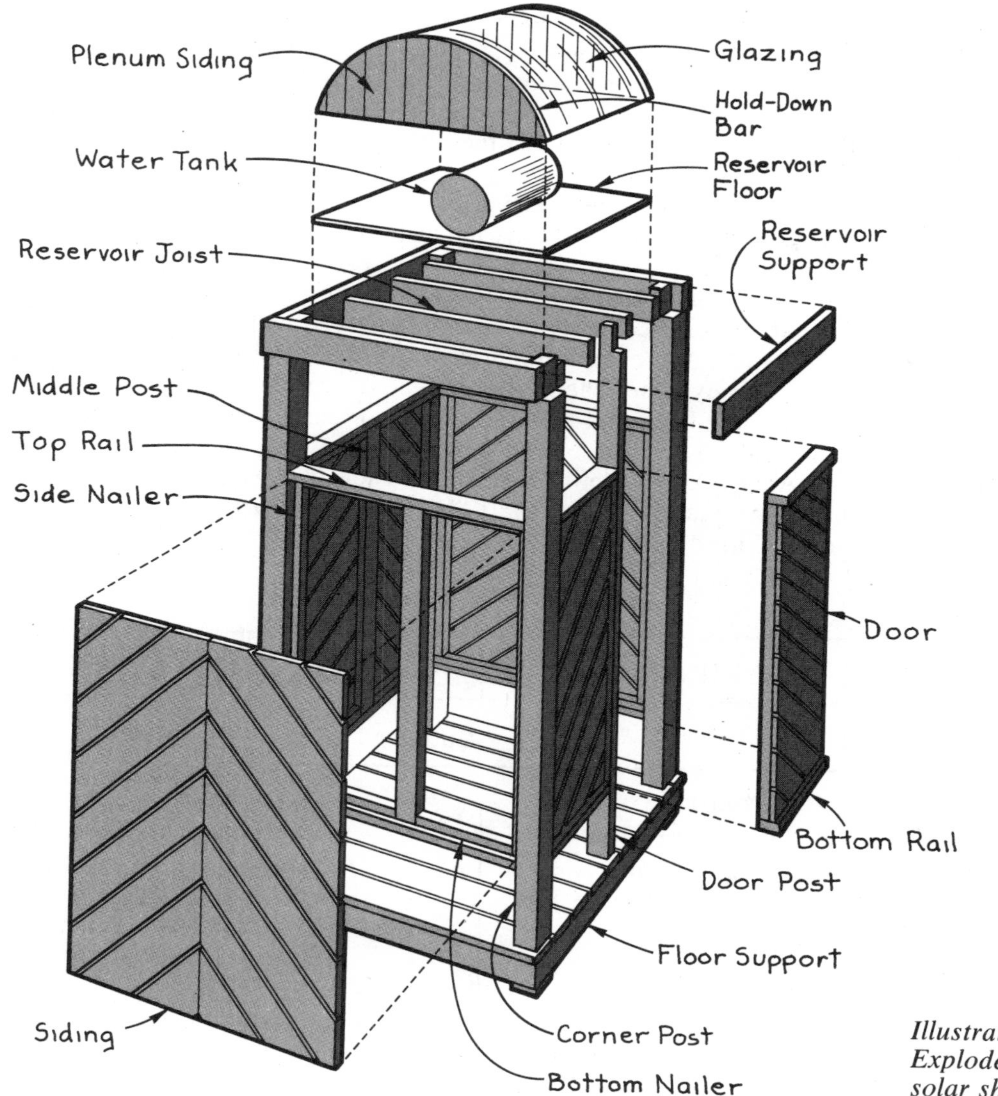

Illustration 11–1—Exploded view of the solar shower.

location with lots of sun, good drainage, and access to an outdoor hose connection. You can extend the outdoor showering season by planting shrubs against prevailing winds to help make your solar shower comfortable even in spring and fall.

For maximum solar exposure, the wooden ends of the shower dome should face east and west. Position the shower unit as close to this orientation as possible. Your hose feeds fresh water to a 30-gallon storage tank mounted on the shower's roof, where it collects the sun's heat. Depending on the variables of air temperatures, hours of direct sunlight, and initial water temperature, you may find the tank's temperature near boiling. Provision is made in the plumbing to mix in cold water when needed.

Construction should take place close to where you plan to have the solar shower, since it is heavy and awkward to move. We suggest using wood that is pressure-treated with preservative to last outdoors or a naturally rot-resistant wood, such as redwood or cedar.

An optional concrete pad can be included with the construction of the shower, and even a drain can be added. The ends of the upright posts, in any case, should be protected from resting directly on the ground. Use large, flat rocks, concrete, or post risers to keep the posts from soaking up water.

An exploded view of the solar shower is shown in illustration 11–1, and a materials list, chart 11–1, follows. Chart 11–2 lists the tools you will need.

CHART 11–1—
Materials

DESCRIPTION	SIZE	AMOUNT
Lumber		
Redwood, Cedar, or Pressure-Treated Pine	4 × 4 × 8′	4
Redwood, Cedar, or Pressure-Treated Pine	2 × 6 × 10′	12
Redwood, Cedar, or Pressure-Treated Pine	2 × 4 × 8′	13
#2 Pine	1 × 6 × 8′	28
A-C Exterior Plywood	¾″ × 4′ × 4′	1
Hardware		
Galvanized Common Nails	16d	1 lb.
Galvanized Common Nails	6d	3 lb.

CHART 11-1—*Continued*

DESCRIPTION	SIZE	AMOUNT
Hardware—*continued*		
#14 Flathead Wood Screws	1½″	6
#6 Roundhead Aluminum Screws	1″	36
Lag Screws and Washers	¼″ × 2½″	4
Carriage Bolts	¼″ × 6″	8
Carriage Bolts	¼″ × 4½″	28
Washers	½″	28
Self-Closing Hinge with Screws	3½″ × 3½″	1
Butt Hinge with Screws	3″ × 3″	1
Barrel Bolt	5″	1
Aluminum Strip	⅛″ × ¾″ × 10′	2
Plumbing Supplies		
30-Gallon Galvanized Tank	18″ dia. × 39″	1
Copper Tubing	¼″ I.D. × 4′	1
CPVC Pipe	½″ × 8′	1
CPVC Compression-Fit Stop Valve	½″	1
CPVC Elbow	½″	1
FPT Galvanized Tee	½″	1
FPT Self-Closing Shower Valve	½″ × ½″	1
FPT Garden Hose Adapter	½″	1
MPT-CPVC Adapters	½″	2

[*Continued on next page*]

CHART 11-2—
Tools

- Hacksaw
- Saber Saw
- Circular Saw
- Drill and Bits
- C-Clamps
- Caulking Gun
- Wood Chisel: 1-inch
- Adjustable Wrenches

CHART 11-1—*Continued*

DESCRIPTION	SIZE	AMOUNT
Plumbing Supplies—*continued*		
MPT-FPT Bushing*	1¼″ × ⅜″	1
MPT-FPT Bushing*	¾″ × ½″	1
MPT-FPT Galvanized Reducer*	¾″ to ½″	1
MPT Plug*	1″	1
MPT Brass Compression Fitting	⅜″	1
MPT Shower Head	½″	1
Galvanized Union	½″	1
Galvanized 90-Degree Service Elbow	½″	1
Galvanized Nipple	½″ × 3″	2
Galvanized Nipple	½″ × 2″	1
Pipe Brackets	½″	2
CPVC Adhesive	...	4 oz.
Miscellaneous		
Waterproof Wood Glue	...	½ pt.
Semitransparent Stain	...	3 qt.
Flat Black Absorber Paint	...	1 qt.
White Exterior Enamel	...	1 pt.
Silicone Caulk	...	1 tube
Fiberglass-Reinforced Plastic Sheet	4′ × 69″	1
Teflon Tape	...	1 roll

NOTE: *MPT = male pipe thread; FPT = female pipe thread; I.D. = inside diameter.*

**These four pieces are sized to fit the tank. Check to see that your tank's openings are the same as ours.*

Begin construction of the shower with the framing. Using the 4 × 4s, trim four corner posts, each to exactly 96 inches. From a 2 × 4, cut the 95-inch door post. For efficient assembly, lightly label the pieces and keep them in separate piles, rechecking the designated position before you drill or fasten them.

To make a strong assembly, the corner posts have two rabbets cut into each end to fit the 2 × 6 supports. Illustration 11–2 shows the rabbets at the top of the post, and illustration 11–3 shows the rabbets at the bottom of the post. Mark 5½ inches from one end of each post and square off lines on two adjacent sides. These are the top of the posts. At the lower end, measure up 6½ inches, and square both lines onto the same two adjacent sides that are marked on the top of the post. The extra inch at the bottom will raise your shower slightly off the ground to improve drainage. On the door post, mark a line 5½ inches from each end on one 1½-inch edge.

The rabbets are cut with a circular saw. With the saw unplugged, set the blade to cut ¾ inch deep. You may want to clamp a straightedge to the post to guide your saw on the first cut. Cut on the rabbet side of the line, then make approximately eight more passes with the saw over the area to be cleared, as shown in photo 11–1.

Remove the excess wood with a sharp wood chisel. Test the rabbet with a piece of 2 × 6, which should fit in halfway. Trim the rabbet if necessary. Follow this procedure to cut two rabbets at each end of the corner posts and one at each end of the door post.

Cut eight pieces of 2 × 6, each 55 inches long, for the floor and reservoir supports. Use a square to mark each piece accurately before

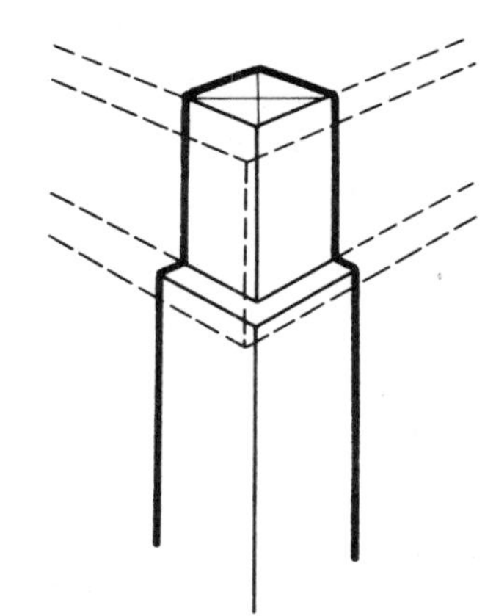

Illustration 11–2—Corner post top.

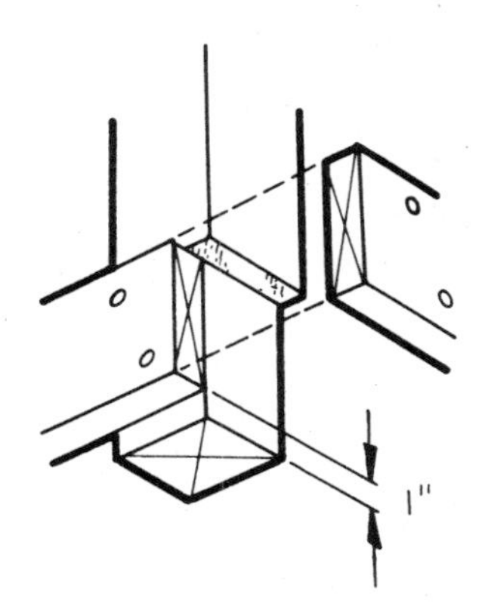

Illustration 11–3—Corner post bottom.

Photo 11–1—Cut the rabbets in the corner posts with a circular saw, making a number of passes.

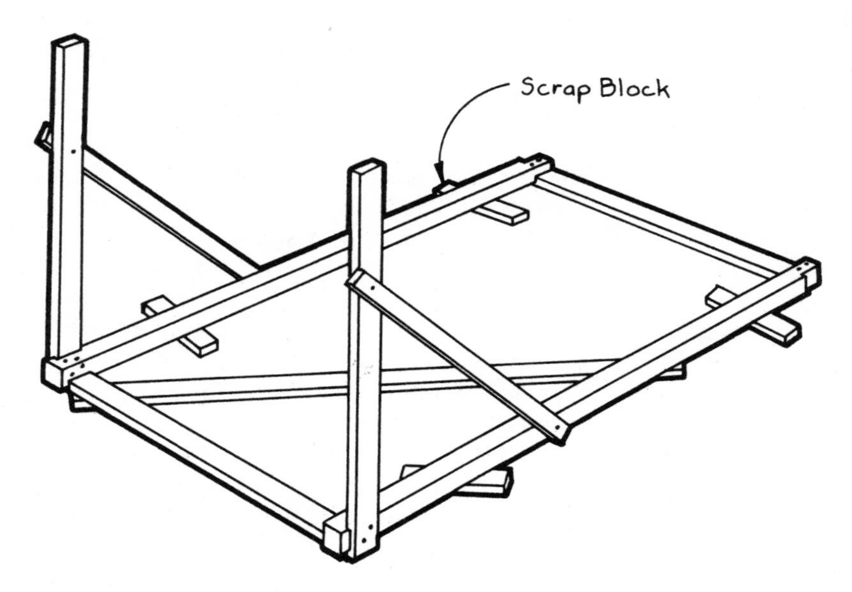

Illustration 11–4—Corner posts and floor supports with temporary braces.

cutting. Cut one 2 × 6 for the middle floor support, 53½ inches long. Cut ten pieces of 2 × 6 for the floorboards, each 59 inches long. Set aside the ten floorboards and the support for now.

The reservoir joists, which hold the plywood reservoir floor, are 2 × 6s ripped to 4¾ inches to accommodate the plywood thickness. First, cut four pieces of 2 × 6, each 53½ inches long. Clamp one to your workbench so that a long edge extends about 12 inches off the bench. Mark 4¾ inches from the clamped edge. Use a straightedge clamped to the piece to guide your circular saw as you trim off the excess wood, or use a rip guide on the saw. Repeat with the three other joists.

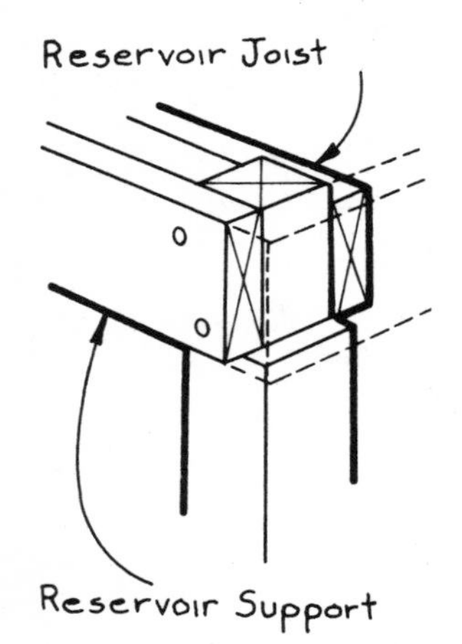

Illustration 11–5—Reservoir joist detail.

Another person to help you assemble the basic framework of your solar shower will be useful, if not essential. Remember to build the shower on or near its permanent location. First, lay out two adjacent corner posts on scrap blocks, 48 inches apart. The blocks will help you to drill holes without damaging your drill bit. Set one of the floor supports into the bottom rabbets. (Remember, the rabbet is 1 inch wider than the board.) One end of the support is flush with the edge of the rabbet in the post's side, while the other extends 1½ inches to overlap the next support board. In this way, all the floor and reservoir supports can be the same length. Use a square to be sure the post and support form a right angle. Two C-clamps, one on each joint, will hold them together while the holes are drilled.

As shown in illustration 11–3, drill two ¼-inch-diameter holes through the support and post. These two holes should not be directly above one another, but at a slight angle, as the bolt stress may split the post along the grain. Also, plan to stagger the bolts that will go through the posts at right angles, so they don't interfere with each other. Fasten the support to the posts with ¼ × 4½-inch carriage bolts, keeping the nuts and washers on the inside face of the post. Fasten the other end of the floor support in the same way. Then align, drill, and bolt a reservoir support to the top of these two posts. A temporary diagonal brace should be nailed across a post and a support to help hold the unit in square.

Assemble the two remaining posts, a floor support board and a reservoir support the same way. Next, turn the assembly over and fasten the two remaining floor supports to this second unit. See illustration 11–4. Again, nail braces between the posts and floor supports. Be careful that you overlap the supports in a way that will be compatible to the other post unit, so the two can be assembled.

Add the remaining two reservoir supports to the other assembly, drilling holes and slipping ¼ × 6-inch carriage bolts through the supports and posts. Thread on nuts temporarily and attach braces at an angle to hold the posts square.

Erect the two units in place, in your chosen sunny location, by raising the second post/support unit to meet the first. Fasten two more ¼ × 6-inch bolts through the reservoir supports and posts, which will make the entire assembly self-supporting. Square everything, and bolt the floor supports in place securely.

Working at the top of your shower frame, place a reservoir joist in position flush with the bottom of the reservoir supports, against the inside face of the posts. Drill through and bolt the support post and joist at both ends with the ¼ × 6-inch carriage bolts installed earlier. Refer to illustration 11–5. Use the same procedure to fasten a joist to the opposite side securely.

To add the door post, mark the 1½-inch side for two bolt holes at the top and two at the bottom. Center the holes in each rabbet at 2 inches and 4 inches from each end. Drill the four holes in the door post. Set the door post in position against the floor support and reservoir support, at exactly 24 inches on center, as shown in illustration 11–6. Mark and drill bolt holes in the support boards. Fasten the door post to the supports with ¼ × 4½-inch carriage bolts.

Now you should be ready to install the flooring. The first step is to install the middle floor support. This piece runs from one floor support to the other. It is nailed parallel to the unit that has the door post attached to it and flush with the top of the two floor support pieces. Nail it in place by nailing through the two supports, into the ends of the piece. Lay out all the floorboards so they run perpendicular to the middle floor support. On the two outer floorboards, measure and cut a notch at both ends to fit around the corner posts. See illustration 11–7. The most

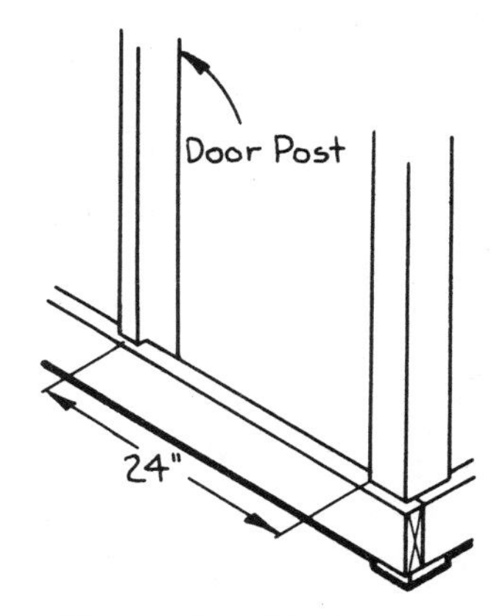

Illustration 11–6—Door post placement.

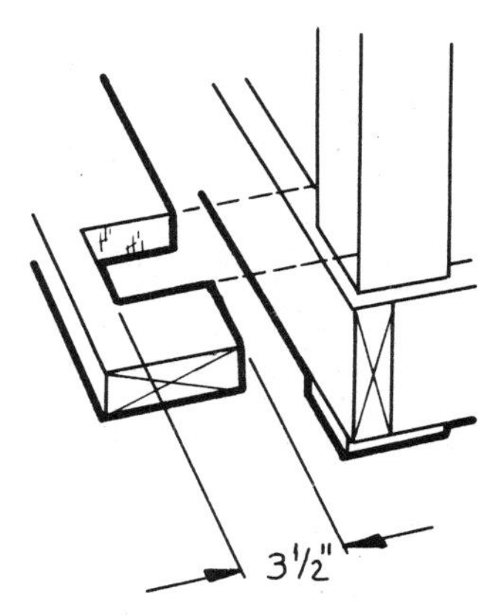

Illustration 11–7—Floorboard notch.

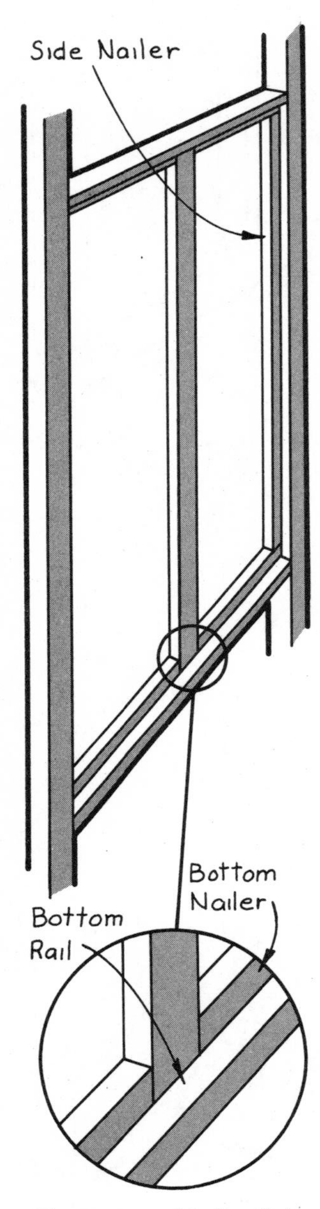

Illustration 11–8—Side panel details.

Photo 11–2—Nail the bottom rails to the corner posts by supporting the rails with 12-inch spacer blocks of scrap 2 × 4.

straightforward method is to hold the boards in position against the post and mark for the notch. Cut the notch 3½ inches deep with a saber or handsaw. Check them against the posts, and trim if necessary.

Maintain a ½-inch space between the floorboards to allow for drainage. It will be necessary to cut additional notches into the center boards to fit around the door post. Nail the boards in place, using 16d galvanized nails.

With a floor to work from, you can complete the reservoir above your shower room. Nail the two remaining reservoir joists in place, spacing them 12 inches on center from the two reservoir joists already bolted in place. Use three 16d galvanized nails through the reservoir supports at each end of the joists to make common butt joints.

The reservoir floor measures 46½ × 48 inches and is cut from ¾-inch exterior plywood sheathing. The floor should be positioned flush with the edges of the two joists and centered in the opening. The plywood will serve as a brace to hold the unit in square, so be sure the framework is square as you fasten down the reservoir floor. Nail the floor to the joists, using 6d galvanized nails.

Now you are ready to close in your shower unit. Again, label all pieces as they are cut to keep them in order. Cut six 2 × 4 pieces, each 48 inches long, for the top and bottom rails of the side panels. Cut three pieces, each 51 inches long, for the middle posts of the side panels. Cut two pieces, each 23¼ inches long, for the top and bottom rails of the panel next to the door. From 2 × 4 or 2 × 6 scrap wood, cut two spacer boards, each 12 inches long.

Drill ⅛-inch-diameter pilot holes, and toenail a 16d galvanized nail into each end of each of the three bottom rails. Use the spacer blocks to support the bottom rail as you toenail it into the post, as shown in photo 11–2. Nail in the remaining bottom rails in the same way. Add nails to the faces, so that each end is toenailed into the post on three sides.

Illustration 11–8 shows the side panel framing. Set a middle post on each end of a bottom rail to position and balance each top rail as you toenail the top rails in place. Nail in the short rails between the door post and a corner post. Then, fasten one middle post between each top and bottom rail, centered between the posts. Make sure that the 3½-inch face is flush with the inner edge of the rails, creating a setback to the outside.

Rip five of the 2 × 4s into 2 × 2s (actually 1½ × 1⅝ inches). Use a table saw or a rip guide attachment for your circular saw.

Working with the 2 × 2s, cut eight pieces, each 48 inches long, for the vertical nailers for the side panels and the panel next to the door. Then cut two pieces, each 51 inches long, for the vertical door nailers. The top and bottom nailers for the panel beside the door, each 23¼ inches long, are cut from the ends of the same boards. Cut two pieces, each 20 inches long, for the top and bottom door nailers. Finally, twelve pieces, each 22¼ inches long, are cut from the remaining 2 × 2s to make top and bottom nailers for the side panels.

Following illustration 11–8, nail the top and bottom nailers in place first and then the vertical nailers on three sides and in the panel next to the door. Like the middle posts, the nailers are flush with the inner edge of the rails and provide a surface on which to attach the siding.

The diagonal 1 × 6 pine boards provide bracing for the shower frame, as well as an attractive finished appearance. Set the siding pieces at 45 degrees with a ⅜-inch space between the boards. Cut the corner pieces first. Start by marking off a 45-degree corner from one board. From the inner end of this line, measure 1½ inches and mark off a second 45-degree line, perpendicular to the first, as shown in illustration 11–9. Cut out this first board. Hold it in place on three sides of the shower unit, reversing it on either side of the middle post. If it fits, use this first piece as a pattern for the second, which will mirror it, and for the four additional siding boards on the other two sides. If necessary, make adjustments for the five similar pieces. Nail the first six pieces of siding in position on the lower edge of the shower sides, using 6d galvanized nails. To maintain even spaces between the 1 × 6s, use a spacer made from a piece of lath or other ⅜-inch-thick material. You will have to

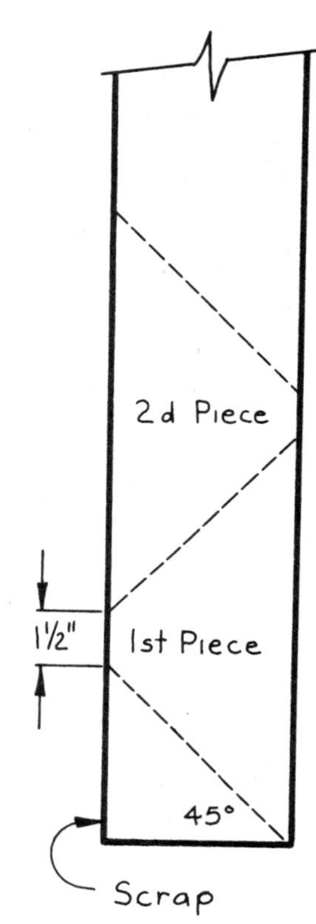

Illustration 11–9—Siding layout.

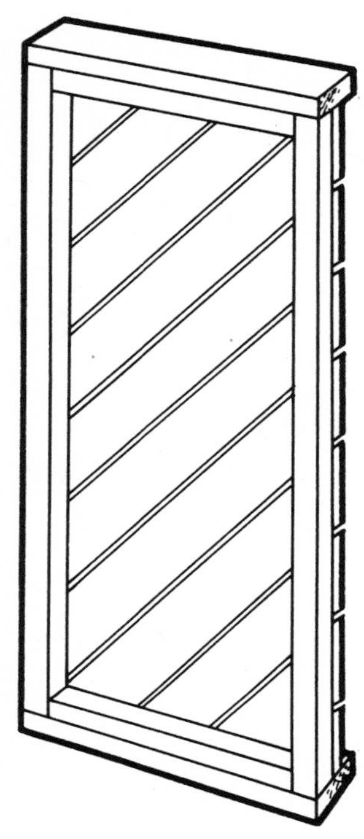

Illustration 11–10—Completed door.

measure individually for all of the shorter pieces. *Note:* Since the dimensions of the panel next to the door are slightly different, work out those measurements separately. With all the diagonal siding in place, you are ready to build the shower door, shown in illustration 11–10. Construct the door with the two 20-inch nailers and two 51-inch nailers you have already cut. Then cut the top and bottom door rails from 2 × 4s, each 23 inches long. Nail the frame together with 16d galvanized nails. Add diagonal siding after you make sure the door is square and fits easily into its opening.

For privacy, we suggest using one self-closing hinge and one butt hinge to hang the door. A barrel bolt may be installed to hold the door shut. So that the door is prevented from swinging too far, you should fabricate a doorstop by nailing several pieces of scrap wood to the outside edge of the post against which the door closes.

Now that you have constructed the shower building, you are ready to assemble the solar-heated water system. The reservoir tank is the heart of the system. It should be as large as possible but not exceed a length of 39 inches and a diameter of 18 inches.

Place your tank up on the platform. If the tank has flanges, screw four ¼ × 2½-inch lag bolts through the flanges and into the platform. If there are no flanges, nail down small blocks of wood on either side of the tank to keep it from rolling.

Following the plumbing diagram, illustration 11–11, install the water system. Use Teflon tape on the galvanized fittings and follow manufacturer's instructions with the CPVC connections. The plastic piping is used for the cold water and the galvanized pipe for the hot.

Start by plugging the unneeded opening at one end of the tank with a 1-inch male pipe thread (MPT) plug. Next, make a gradual curve in the ¼-inch copper overflow tube so that it goes through the port in the end of the tank and curves up to within 1 inch of the top of the tank. Be careful not to make a sharp bend, which would restrict the flow or crack the tube. To hold the tubing in its correct position, slip on a 1¼-inch MPT by ⅜-inch female pipe thread (FPT) bushing and then a ⅜-inch MPT brass compression fitting. With the fitting tightened to the tank and around the tubing, make another gradual bend to allow the tubing to run down through a ⅜-inch-diameter hole in the plywood reservoir floor. Cut it off about an inch below the floor.

On the lower tank opening, fit a ¾-inch MPT by ½-inch FPT bushing, a ½ × 2-inch nipple, and a ½-inch union, which will allow the tank to be removed and replaced if necessary. Continue the assembly with a ½-inch 90-degree service elbow, a ½ × 3-inch nipple through the plywood floor, and a ½-inch tee. If you live in an extremely sunny area, you should replace the tee with a mixing valve, to be sure you don't get burned from the solar-heated water. On one side of the tee, connect

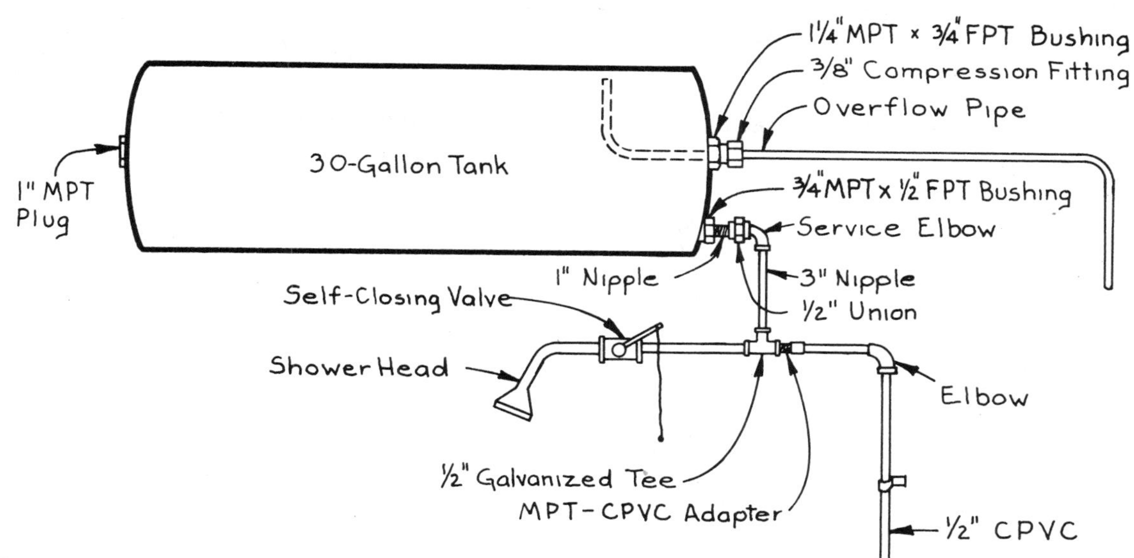

Illustration 11–11—Plumbing diagram for the solar shower.

another ½ × 3-inch nipple to the self-closing valve. A 1-inch-diameter hole drilled in the 2 × 4 reservoir joist provides a good support for the shower assembly. Complete the hot-water line with a water-saver shower head, which will stretch the usefulness of the solar-heated water.

On the cold-water side of the ½-inch tee, connect a ½-inch MPT-CPVC adapter fitting to make the transition to ½-inch CPVC, and continue the CPVC piping down the inside of the shower wall. Cut the CPVC pipe to the necessary lengths with a hacksaw. Insert a ½-inch CPVC compression-fit stop valve at a convenient height for use while showering. At the lower end, use another ½-inch MPT-CPVC adapter, then attach the FPT garden hose coupling. Two pipe brackets are used to fasten the CPVC pipe to the shower walls.

The reservoir tank is enclosed with a dome to help retain the sun's heat. Start by ripping 12 feet of 1 × 6 pine to a 2½-inch width. Use these boards to cut six pieces, each ¾ × 2½ × 48 inches, for the battens on the end panels. Cut 18 pieces of 1 × 6 pine, each 22 inches long, for the siding of the dome.

Referring to illustration 11–12, lay out the siding boards on a flat surface. Place the battens over them, the lowest ¾ inch from one end. The middle and top battens are 7½ inches on center from the first. Find the center point of the semicircle to be drawn for the dome end. The point is centered 3 inches below the ends of the siding boards. Draw the semicircle 24 inches in radius, with a compass made from string and a pencil. Fasten the battens to the siding boards, using waterproof wood glue and 6d galvanized nails, being careful not to nail near the penciled dome line. Cut the end panel with a saber saw, and use the first end panel as a template for the second.

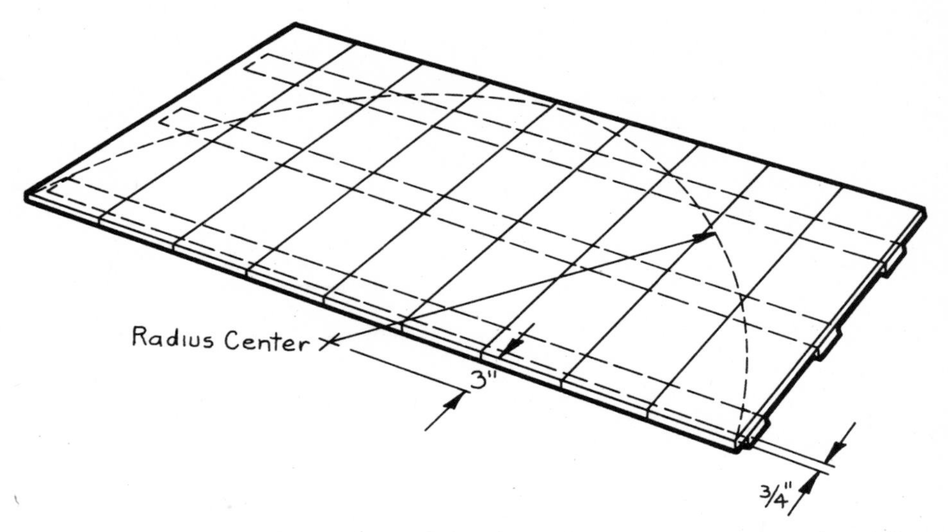

Illustration 11–12—Layout details for the radius on the siding boards.

The next components of the dome frame are two mounting rails, each 45 inches long. Cut these from the 2 × 4 lumber. Cut the top spacer bar from the remaining piece of 2 × 2 stock, 46½ inches long. These pieces are shown in illustration 11–13.

Assemble the dome frame by laying the mounting rails on their 3½-inch face on ¾-inch spacers. Stand an end panel against the ends of the mounting rails with the side with the battens toward the rails. Referring to illustration 11–13, nail through the end panel batten into the mounting rails. This will result in ¾ inch of the end panel extending past the rails. This is so that, when installed, the end panel will overlap the plywood reservoir floor. Add the remaining end panel to the assembly, and nail it together. Finally, nail the top spacer bar in the center between the two end panels.

To make the reservoir area a more effective solar collector, paint the plywood reservoir floor, the inside of the end panels, the mounting rails, and the spacer bar white. Paint the reservoir tank flat black.

The dome frame supports the fiberglass-reinforced plastic glazing. Bands of aluminum hold down the edges of the glazing material. Using ⅛ × ¾-inch aluminum strips, cut two 69-inch pieces and two 46½-inch pieces with your hacksaw. At 12-inch spacing, drill ⅛-inch-diameter holes to accept 1-inch #6 roundhead aluminum screws.

Before cutting the fiberglass-reinforced glazing, lay the glazing over the dome frame and check its fit. Cut the glazing to size, approximately 48 × 69 inches. Lay a bead of silicone caulk on the top edges of the end panels and the outside edges of the mounting rails. Lay the glazing over the dome frame and hold it in place with 1-inch #6 roundhead aluminum screws through the aluminum bars. See photo 11–3.

Mount the dome over the reservoir tank, and fasten it in place with six 1½-inch #14 flathead wood screws through the plywood reservoir

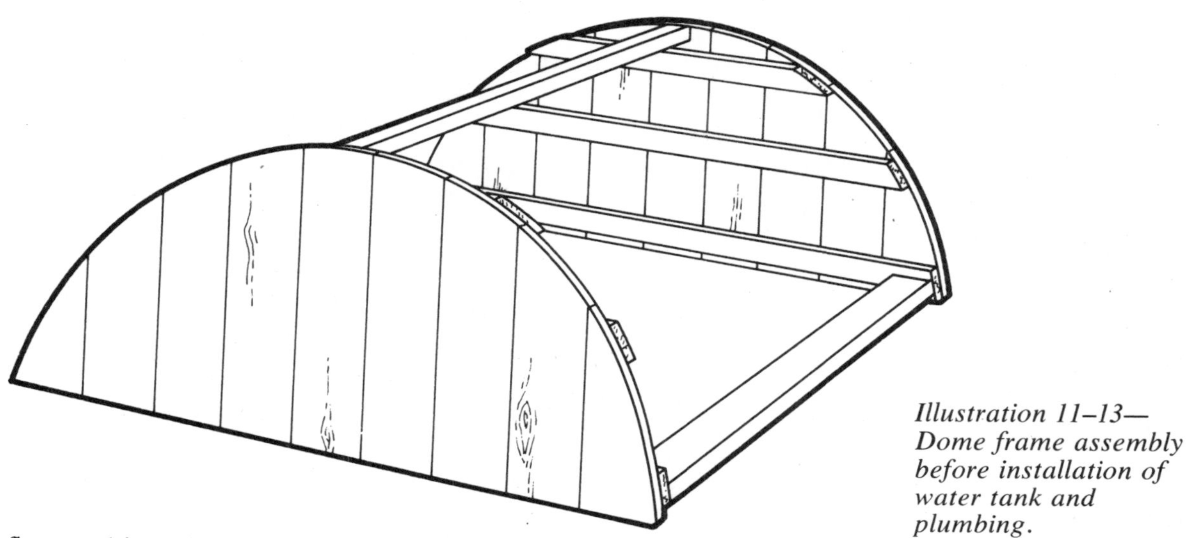

Illustration 11–13—Dome frame assembly before installation of water tank and plumbing.

floor and into the mounting rails. Countersink the screws. To preserve your solar shower, paint all the wood parts with a coat of semitransparent stain.

To operate the shower, fill the reservoir tank from a garden hose with the stop valve open. Close the stop valve when water spurts from the overflow line. Let the sun do its work. To bathe, simply open the self-closing shower valve. Be careful—if the reservoir water runs too hot, add some cold water with the cold-water inlet valve as you shower. Refill the tank when you have finished your shower. Remember to drain the tank and lines after each outdoor shower season.

Photo 11–3—Hold the fiberglass glazing over the dome frame with silicone caulk, aluminum strips, and roundhead aluminum screws.

12 Flat Plate Collector

Did you know that the second largest energy expense in most households is water heating? The expense of buying gas, electricity, or oil to heat water can be substantially reduced by using the sun to heat your water. Install a solar water heater on a sunny portion of your roof or the south side of your house, and you will realize substantial energy savings. Several very effective systems have been developed over the last few years. The collector presented here will work with several types of systems.

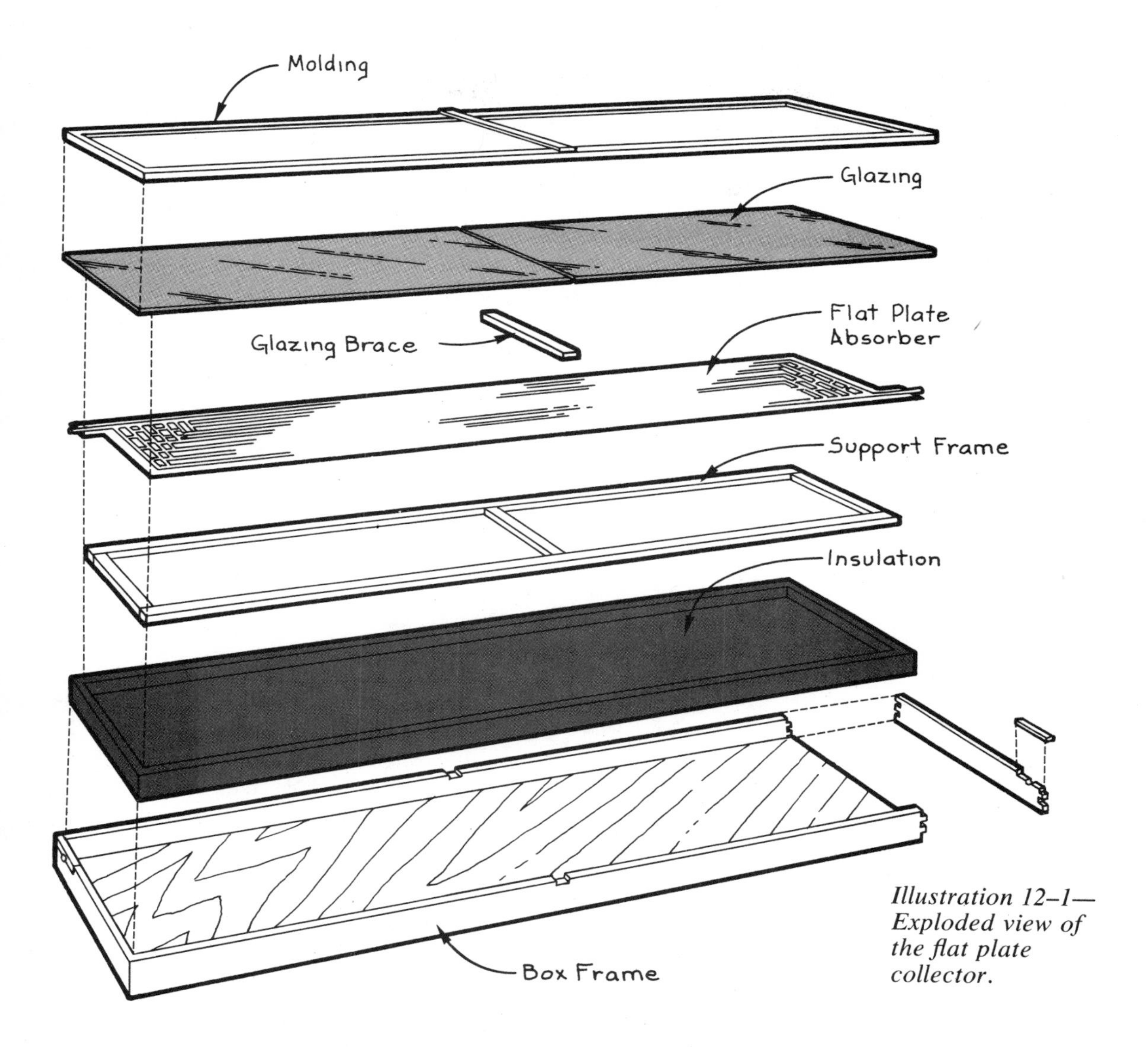

Illustration 12–1—Exploded view of the flat plate collector.

The project described here is only the main component of a solar water-heating system, the collector panel. Illustration 12–1 shows an exploded view of the collector. In order to build a complete system, you will need a storage tank, plumbing lines, possibly a heat exchanger, a circulator pump, and various gauges and valves. Many information sources are available to the do-it-yourselfer who chooses to build such a system. A list of sources for information is at the end of this chapter.

To give you some idea of the systems that are compatible with the collector box, read through the following brief description of each type

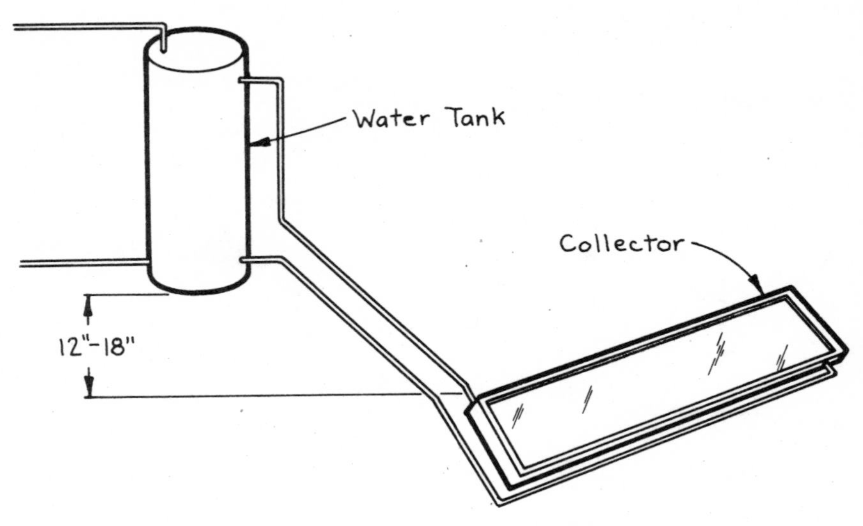

Illustration 12–2—In a thermosiphon system, the collector is below the tank.

of system. They all work well and can be installed by the homeowner. Modifications of the systems can be used to heat swimming pools and hot tubs, but they are usually used for domestic hot water use.

The first of the three systems is the thermosiphon, shown in illustration 12–2. It is based on the principle that warm water rises, to be replaced by cooler water from the supply line. The supply water is heated in the flat plate collector, setting up a flow of water between the collector and a storage tank. The tank must be at least 12 inches higher than the collector, unless special valves are installed, and a direct 1-inch pipe from collector to tank will help the heated water rise. This is an extremely simple, inexpensive, and efficient system. The key to protection against freezing is to locate the flat plate collector, pipes, and tank within the protection of your building: in a greenhouse, under a skylight, or behind a south-facing window.

The drain-back system is the second option, shown in illustration 12–3. The collector plate is mounted above a drain-back tank. Water is pumped through the collector when a differential thermostat determines that solar heat is available. When activated by the thermostat, the water in the collector empties back into the tank. Domestic water is heated through a heat exchanger coil of copper pipe, which is installed in the drain-back tank and is a part of the pressurized house water system. If you include a second pump in the system, the heat gain can be increased by moving hot water through the copper coil even when no one is using water in the house. The drain-back system is reliably freeze-proof and allows the collector to be operated outdoors year round. The heat-exchange feature, however, results in some heat loss and added cost.

A third system, which is extremely efficient in the collection of heat and reliable in cold climates as well as flexible in the collector location,

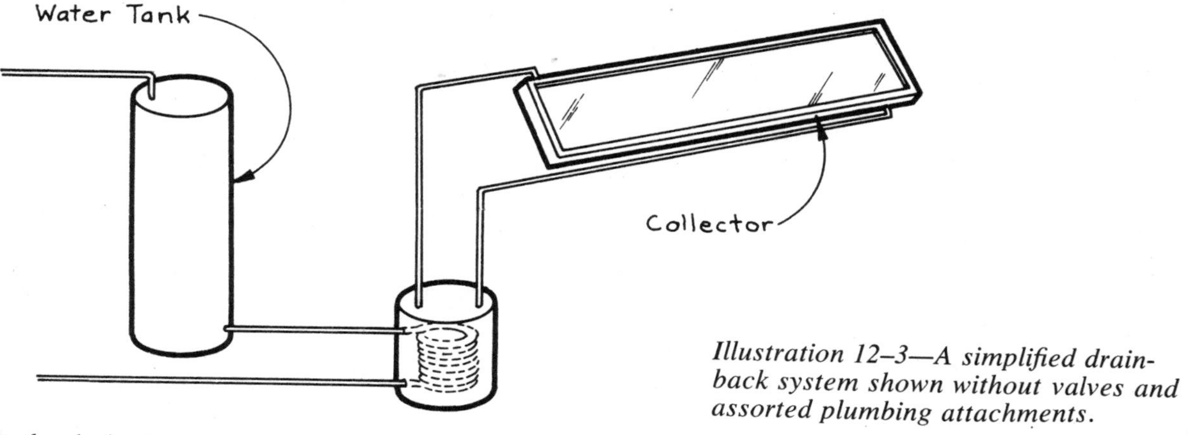

Illustration 12–3—A simplified drain-back system shown without valves and assorted plumbing attachments.

is the drain-down system, shown in illustration 12–4. It consists of one pressurized open loop of hot water from the collector to the storage tank (which may also have a back-up source of heat) and back through the collector. Pressure is provided by the house system. Freeze protection consists of a special three-way differential thermostat that closes the regular line and opens a drain-down valve to release only the water in the collector and exposed pipes. It is a more exacting system to set up initially, but usually worthwhile.

The collector box you are about to build uses glass glazing, which will not discolor or deteriorate in ultraviolet rays (sunlight), as do many plastics. The absorber plate is painted with flat black paint designed for solar use to increase heat absorption. The box must be insulated to maintain its heat gain. The 1-inch, foil-faced foam insulation specified here will be adequate up to 40 degrees latitude. More northerly locations will benefit from the application of 2 inches of insulation. The instructions

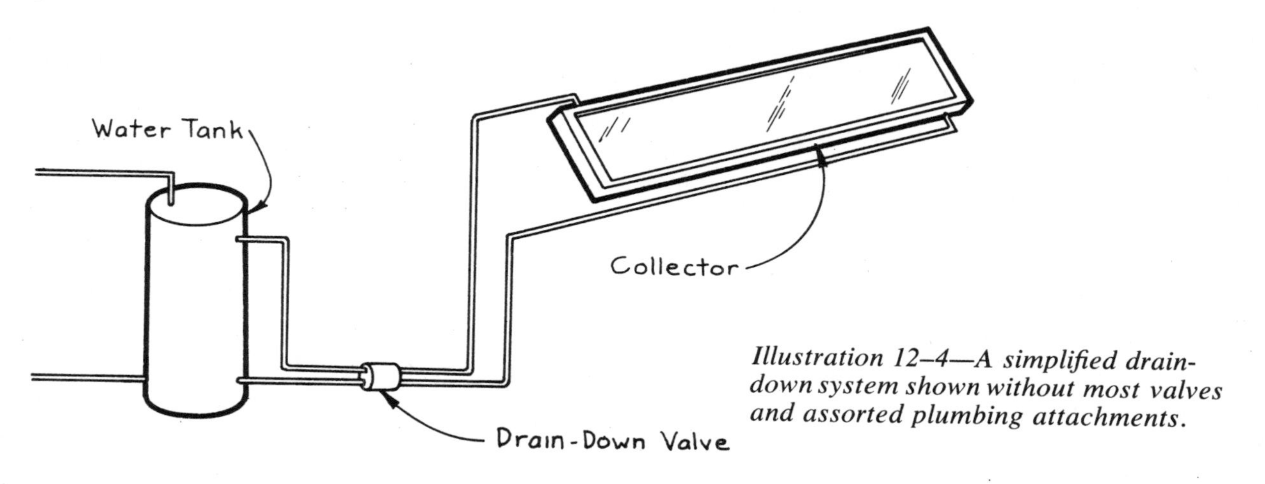

Illustration 12–4—A simplified drain-down system shown without most valves and assorted plumbing attachments.

for fastening the collector box to a support frame, which in turn is mounted on your roof, are included in this chapter.

Chart 12–1 lists the materials needed to build this project, and chart 12–2 lists the tools required.

Since the absorber plates vary in size, you will need to adapt the measurements given to fit the plate you buy. If you use 1-inch insulation, add 4 inches to the length of the sidepieces and endpieces over the dimensions of the absorber plate. Add 6 inches to the absorber plate dimensions if you use 2-inch insulation.

The first step in building the flat plate collector box is to cut the frame pieces. We specify clear pine 1 × 4s, the highest grade of pine. Nearly free of knotholes and cracks, it is used for fine cabinetry. Clear

CHART 12–1—
Materials

DESCRIPTION	SIZE	AMOUNT
Lumber		
#1 Pine	1 × 4 × 12′	2
#2 Pine	1 × 2 × 8′	3
A-C Exterior Plywood	½″ × 4′ × 8′	1
Hardwood Dowel	¼″ × 3′	1
Hardware		
Aluminum Nails	1¼″	30
#9 Aluminum Self-Tapping Screws	¾″	30
#6 Flathead Wood Screws	2″	14
#6 Flathead Wood Screws	1⅝″	8
Tee Nuts	⅜″ × ½″	6
Aluminum Angle	1/64″ × ¾″ × 1¼″ × 12′	2
Aluminum Carpet Strip	1½″ × 26″	1

CHART 12–2—
Tools

Tools
Handsaw
Hacksaw
Saber Saw
Circular Saw
Backsaw
Drill and Bits
Router (optional)
Bar Clamps: 3-foot
Tin Shears
Wood Chisel

pine is used here because the frame pieces must be straight and uniform so that the glazing rests upon an even surface.

Cut two endpieces, each 26 inches long, and two sidepieces, each 96¾ inches long. Each of the four framing boards receives a ½-inch-deep and ⅜-inch-wide rabbet along the bottom edge to fit a ½-inch plywood back, as shown in illustration 12–5.

Cut the rabbets with a circular saw and straightedge guide. Unplug the saw and set the blade to cut at a depth of ⅜ inch. Clamp the frame board to your worktable, and clamp a second board of the same thickness next to the frame board to help support the circular saw. Mark

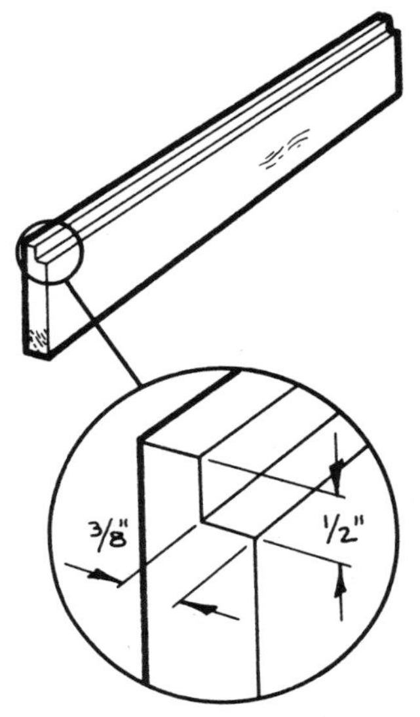

Illustration 12–5—Frame board rabbet.

CHART 12-1—*Continued*

DESCRIPTION	SIZE	AMOUNT
Miscellaneous		
Flat Plate Copper Absorber with Molded Water Tubes*	22″ × 8′	1
Double-Strength Glass	26″ × 4′	2
Urethane or Isocyanurate Foil-Faced Foam Insulation	1″ × 4′ × 8′	1
Waterproof Wood Glue	...	4 oz.
Primer	...	1 pt.
Exterior Enamel	...	1 pt.
Flat Black Absorber Paint	...	1 pt.
Clear Silicone Caulk	...	1 tube
Wood Putty	...	trace
Metal Cleanser or Solvent	...	1 pt.
Vinegar or Muriatic Acid	...	1 pt.
Butyl Glazing Tape	⅛″ × ½″ × 50′	1
Aluminum Iron-On Tape	2″	75′

**Available from Terra-Light, Inc., 30 Manning Road, P.O. Box 493, Billerica, MA 01821.*

Photo 12–1—Cut the rabbets to hold the glazing in the frame with a circular saw guided by a clamped straightedge.

off the ½-inch width of the rabbet. With the circular saw in position to cut to the waste side of the line, clamp a straightedge as a guide for the saw, as shown in photo 12–1. Make the saw cut and then a second cut within the rabbet area. Trim away the excess wood with a wood chisel.

Our design includes a finger joint at each corner of the frame, a detail which gives the frame strength. It is important for the frame to remain weathertight despite the expansion and contraction caused by extreme temperature variations, and the finger joint is the best solution.

Square a line across ¾ inch from each end of each frame board. Referring to illustration 12–6 and starting at the lower edge, mark the following divisions at each end: ½ inch, 1¼ inches, 2 inches, and 2¾ inches. Be careful to cut the end boards differently from the side boards; the ends will have three fingers protruding and the sides will have two fingers. With a pencil, crosshatch the areas from which you will remove the wood. Hold the pieces together before you cut to see how they will fit. Cut on the waste side of the lines with a backsaw, as shown in photo 12–2. Then remove the waste wood with a wood chisel. Or you may use a saber saw, and square up the fingers with a file. Test each joint for a snug fit.

When the frame joints are finished, cut a piece of ½-inch plywood for the collector box back to the dimensions of 25¼ × 96 inches. Test the back in the rabbets of the frame, making any necessary adjustments to the rabbets. Then assemble the box pieces—back, sides, and ends—and clamp or tack them together. Do not fasten them permanently yet.

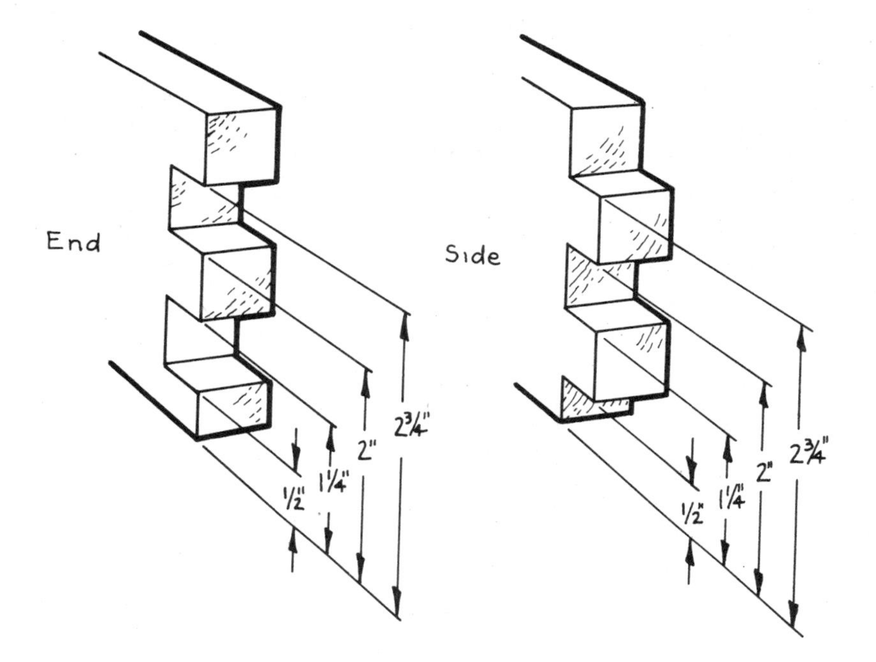

Illustration 12–6—Finger joint layout details for both endpieces and sidepieces.

Insulation is particularly important to prevent loss of heat from the collector during cold days. Cut the back piece from 1-inch foil-faced foam insulation, 24½ × 95¼ inches. Then cut two sidepieces of insulation, each 2 × 95¼ inches. Last, cut two endpieces, each 2 × 22½ inches. In order to prevent the gradual leakage of gases from the insulation due to the high temperatures in solar heaters, carefully cover all edges of the insulation with aluminum iron-on tape, and seal it to the surfaces with a clothes iron. This tape is available from heating and cooling contractors. Install the insulation pieces loosely in the box.

The absorber plate support frame serves to suspend the absorber within the collector box, improving its ability to gain heat. The supports are cut from 1 × 2 pine: two sidepieces, each 93¼ inches long, and

Photo 12–2—Cut the finger joints for the frame with a backsaw or saber saw. Mark the pieces carefully.

three crosspieces, each 19½ inches long. Lay the support pieces on top of the bottom insulation to test their fit.

Trim the absorber plate, if necessary, to fit into the collector box, but be very careful not to cut into a water tube. Use a hacksaw or a metal-cutting blade on a circular saw.

Set the trimmed absorber plate into the box on the supports. Wherever the outlet and intake pipes will pass through the collector box, mark and drill holes to fit the pipes.

Access blocks, as shown in illustration 12–7, make it easy to install and remove the flat plate absorber from the box. Measure 2 inches beyond the drilled hole along the endpiece, and square a line across the board. Next, measure the depth of the center point of the hole, and square a line across. These two lines mark a rectangle (see illustration 12–7) that is to be cut and removed from the end board to release the outlet/intake pipes. Cut out the first access block, then repeat the procedure at the opposite corner of the box, or wherever the second pipe is located. The access blocks are held in position with 1 ⅝-inch #6 flathead wood screws. Drill two ⅛-inch-diameter holes in the access block—one on either side of the outlet hole—replace it, and mark for the pilot holes. Drill 1⁄16-inch-diameter pilot holes in the box ends. Be sure to countersink the holes so that the screws are below the top edge

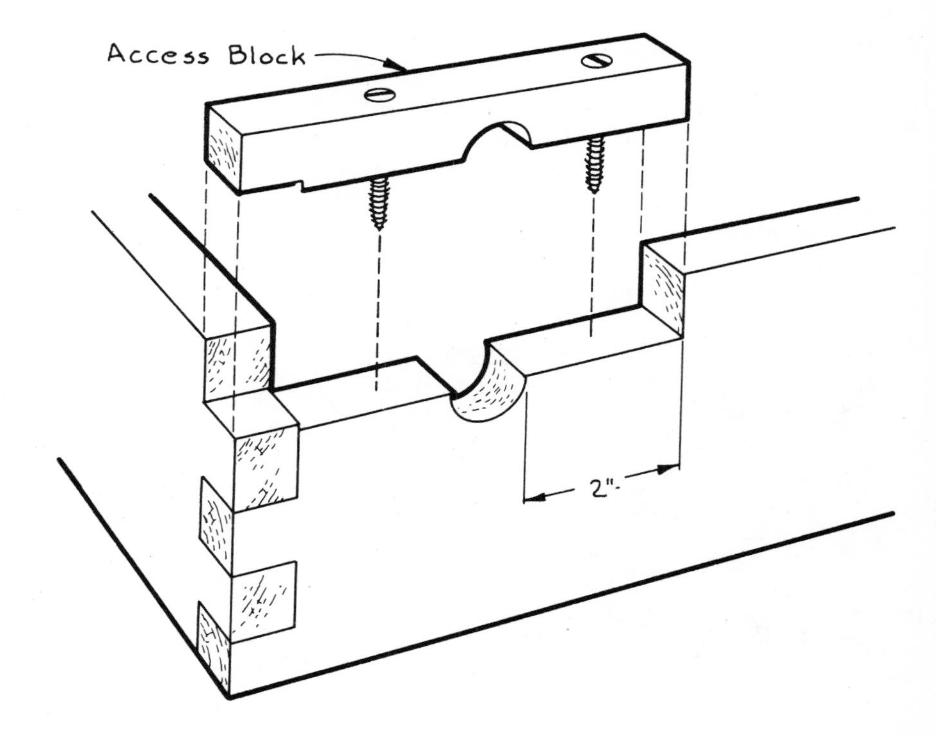

Illustration 12–7—Access blocks go in the endpieces of frame. Their position depends on the absorber you buy.

of the frame. Trim the insulation endpieces to match the shape of the access block. Retape any exposed edges on the insulation.

Now you are ready for the final assembly of the collector box. Remove everything from the box, and clamp the frame and back together again. In the center of each corner joint, drill a ¼-inch-diameter hole vertically through each finger joint from the top, as shown in illustration 12–8. Cut four ¼-inch-diameter dowels, each 4 inches long. Chamfer one end of each dowel by sanding a bevel at the edge. Using waterproof wood glue, fasten the dowels into the four corner joints, tapping them in with a hammer, then trim them off flush to the box.

Fasten the back to the box by first removing the insulation, absorber supports, and back from the frame. Lay a bed of silicone caulk, as shown in photo 12–3, into the rabbet in the frame. Replace the back and fasten it down with 1¼-inch aluminum nails.

The two pieces of glass are supported in the center of the box by a pine 1 × 2 brace, as shown in illustration 12–9. Cut the glazing brace 26 inches long. Lay it across the exact center of the box, and mark the sides of the box with the width of the 1 × 4 to mark the position of the retaining rabbets. On the sides of the box, mark down ¾ inch for the depth of the retaining notches. Cut the notches with a handsaw or saber saw, and clear away the waste wood.

Sand the collector box, and fill any cracks with wood putty. Paint the box first with a primer and then two coats of exterior enamel.

A mounting system of six tee nuts installed in the back of the box will enable you to fasten the box to a support frame on the roof. The tee nuts are positioned 2 inches in from the sides and 8 inches from each end, on the centerline. Drill ½-inch-diameter holes, and tap the ⅜-inch tee nuts into these holes from the inside so they are available for ⅜-inch mounting bolts.

Place the insulation in the box, and lay in the absorber supports. Trim the insulation if needed for the glazing brace. The supports butt

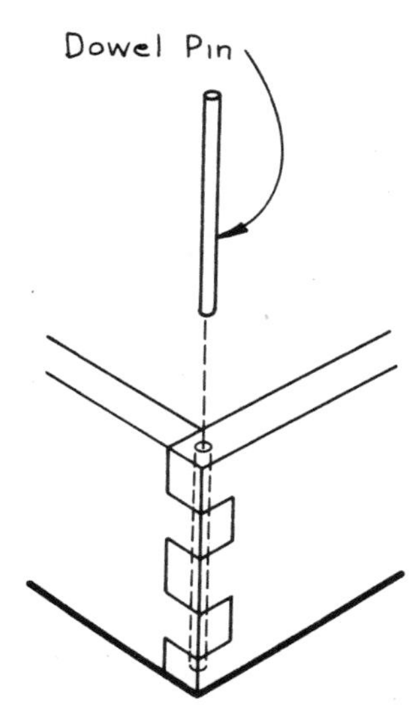

Illustration 12–8— Dowel pin detail.

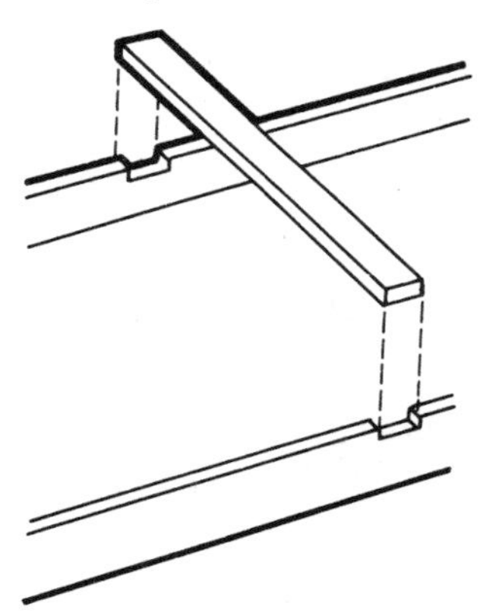

Illustration 12–9— Glazing brace position.

Photo 12–3—Lay a bead of silicone caulk in the rabbets before fastening the back piece in place.

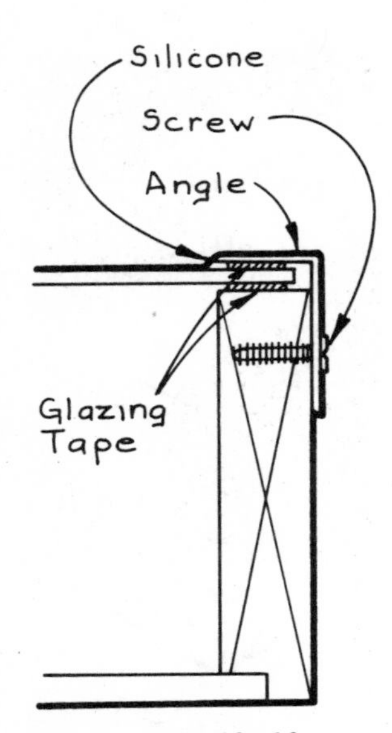

Illustration 12–10—Glazing details.

against the side and end insulation pieces, holding them in place. Drill ⅛-inch-diameter holes evenly spaced in the absorber supports. Drill two holes in each cross support and four holes in each side support. Use fourteen 2-inch #6 flathead wood screws to fasten the absorber supports to the back, screwing through the insulation.

Paint the absorber, absorber supports, and the middle glazing brace with flat black absorber paint. In preparation for painting, clean the metal surfaces with a solvent, such as paint thinner. Then brush on an etching solution, such as vinegar or muriatic acid, which roughens the surface of the metal to give the paint better adherence. Or lightly sand the surfaces.

Install the absorber plate in the collector box. Hold it in place by screwing down the access blocks. Add the middle glazing brace. At each end of the brace, drill two ⅛-inch-diameter holes, and countersink them below the edge. Drill ¹⁄₁₆-inch-diameter pilot holes in the support frame to match. Fasten down the middle glazing brace with four 1⅝-inch #6 flathead wood screws.

Cover the collector box temporarily with a sheet of clear polyethylene, stapled to the sides. Set the box in direct bright sun, and allow it to heat up and naturally evaporate the gases which are found in paint, wood, caulk, and insulation. For best results, this should be done for two days in warm weather. In cold weather, put the collector in a warm area, such as a heated basement, and let it outgas for a week.

The glazing is installed next, as shown in illustration 12–10. Place a strip of butyl glazing tape between the box and the sheets of glass. Add a second strip of butyl tape on top of the glass.

Cover the edges with two ¾ × 1¼-inch pieces of aluminum angle, each 122¾ inches long. Shape the aluminum around the corners by cutting a V of 90 degrees on the ¾-inch side, as shown in illustration 12–11. Cut the V 26 inches from one end of each piece of the molding. Cut a 45-degree miter in each end of each aluminum strip for a neat fit at the remaining corners. Fasten the molding with twenty-six ¾-inch #9 aluminum self-tapping screws with built-in washers after drilling ³⁄₁₆-inch-diameter holes in the metal only. Use three screws in each end and ten in each side. Drill ¹⁄₁₆-inch-diameter pilot holes in the sides and ends of

Illustration 12–11—Aluminum molding covers the glazing and is cut as shown to fit at the corners of the frame.

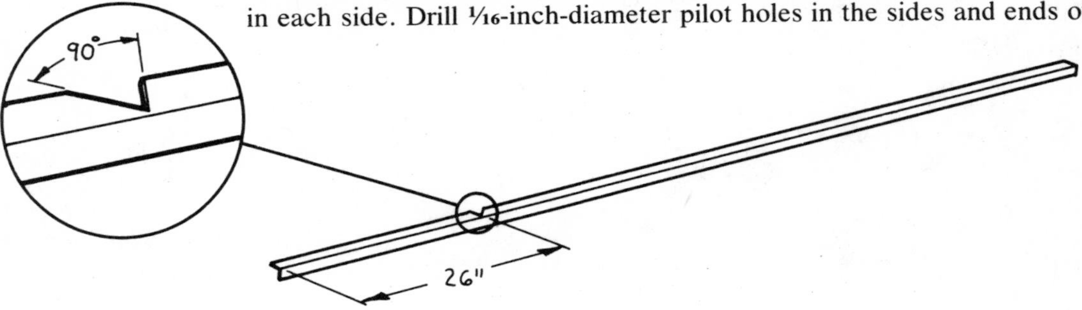

the box. With an assistant to hold the glazing and molding firmly against the frame, fasten the molding, as shown in photo 12–4.

Cover the middle glazing support and the space between the pieces of glass with a 1½ × 26-inch piece of aluminum carpet strip. Apply butyl glazing tape across the seam, then fasten the aluminum strip over the tape. Use four ¾-inch #9 aluminum screws, being careful to place them between the pieces of glass and not touching them.

To prevent water infiltration and air leaks, seal every joint between the glass and aluminum edge guards and middle glazing cover with a narrow bead of silicone caulk. Also seal the access box edges and the outlet/intake holes.

The flat plate collector is now ready to be installed in your water system. Be careful to keep it glazing side up. Otherwise, the absorber may move within the box. The panel must be protected from overheating during installation or storage. The first consideration is to cover the glazing while the unit is being installed until the system's plumbing is complete. Once water is circulating through the absorber, other means, such as pressure relief valves, must be relied upon to protect the collector from overheating. As an extra precaution, the collector box should be covered if it is not being used for a period during the hot summer months. We encourage you to consult an installation expert once you have your system designed and to read further.

Photo 12–4—Use ¾-inch #9 aluminum self-tapping screws with built-in washers to fasten the molding to the frame.

Sources of Additional Information

Build Your Own Solar Water Heater. Stu Campbell and Doug Taff. Garden Way Publishing, Charlotte, VT 05445, 1978.

The Homeowner's Handbook of Solar Water Heating Systems. Bill Keisling. Rodale Press, 33 East Minor Street, Emmaus, PA 18049, 1983.

Hot Water from the Sun: A Consumer Guide to Solar Water Heating. Beth McPherson. Franklin Research Center for the U.S. Department of Housing and Urban Development, Department of Energy; U.S. Government Printing Office, Washington, DC 20402.

Installation Guidelines for Solar Domestic Hot Water Systems in One- and Two-Family Dwellings. U.S. Department of Housing and Urban Development. Superintendent of Documents; U.S. Government Printing Office, Washington, DC 20402.

Solar for Your Present Home. California Energy Commission, 111 Howe Avenue, Sacramento, CA 95825, 1978.

Solarizing Your Present Home. Joe Carter. Rodale Press, 33 East Minor Street, Emmaus, PA 18049, 1981.

13 Window Heater

Hot air rises and cool air falls. This fact of nature helps you heat a small room on a sunny day with our window-mounted solar heater. When the sun strikes the heater, it warms the absorber plate and the air in front of it. The heated air moves up the collector and into the room, pulling cooler room air into the lower vent of the grille. The cool room air is warmed by the absorber plate and reenters

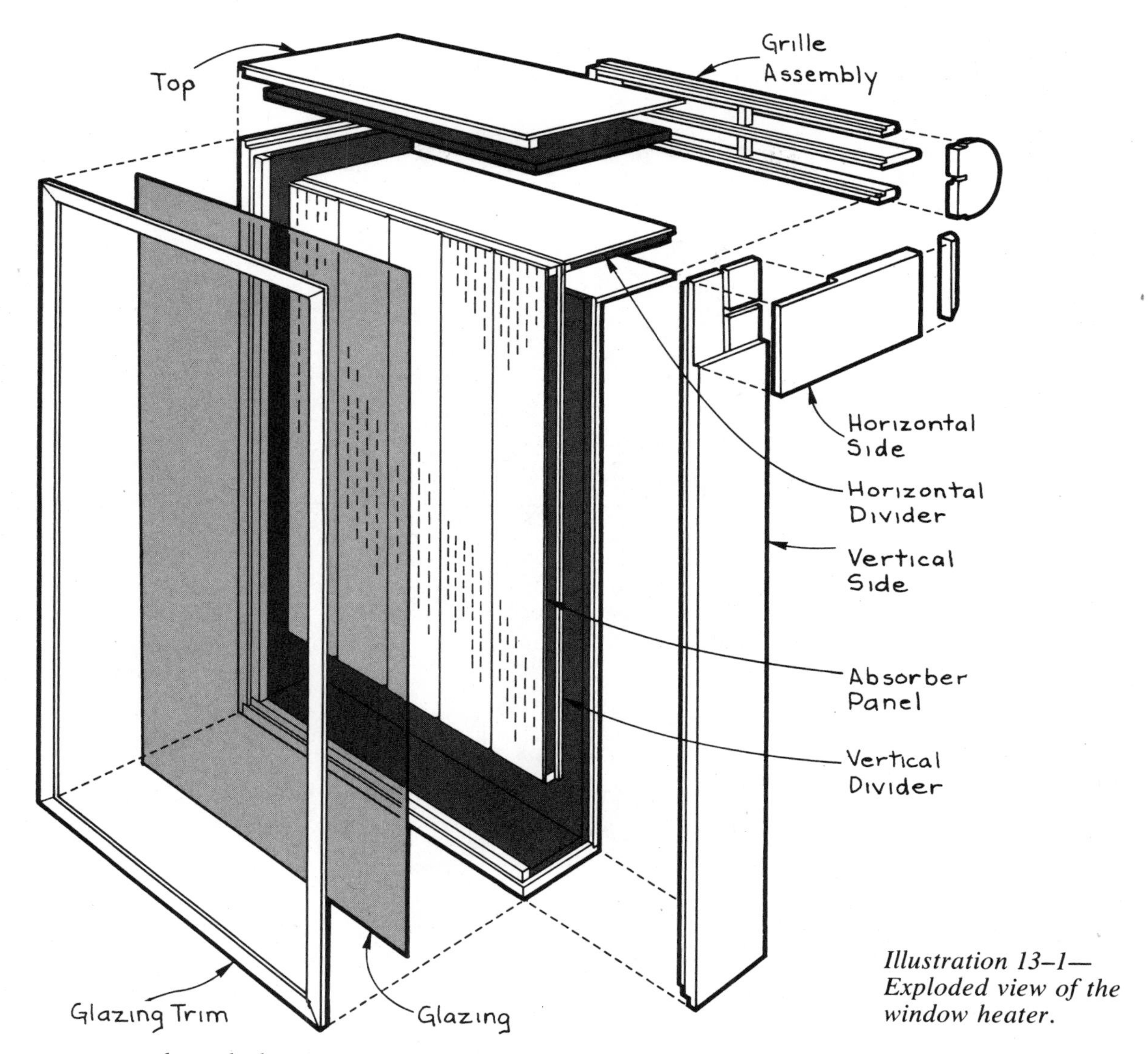

Illustration 13–1—Exploded view of the window heater.

your room through the top vent. The system is called a thermosiphon.

This unit is simple to construct and has no moving parts. The absorber panel is inexpensive vented aluminum soffit material. The vents machined into the surface cause the air to flow unevenly, picking up more heat than a flat surface would.

For the glazing material, we have chosen ⅛-inch acrylic. It is easily cut, lightweight, and durable.

Illustration 13–1 shows an exploded view of the heater, giving the relationships of the parts. Chart 13–1 details the materials needed to build the unit, and chart 13–2 lists the special tools you will need.

It is necessary to size the heater to your windows before buying your materials. The amounts we give in the materials chart are for an average-size heater that will fit a window measuring 27 inches wide and 11 inches deep. Adjust the materials list to fit your window size, using chart 13–3.

Using illustrations 13–2 and 13–3, find measurements A, the width, and B, the depth, for your window. Plug these numbers into the sizing

CHART 13–2—
Tools

- Saber Saw
- Circular Saw
- Backsaw
- Drill and Bits
- Plane
- Clamps
- Caulking Gun
- Tin Snips
- Utility Knife
- Wood Chisel
- Compass
- Nail Set

CHART 13–1—
Materials

DESCRIPTION	SIZE	AMOUNT
Lumber		
#2 Pine	1 × 8 × 8′	2
#2 Pine	1 × 6 × 3′	1
#2 Pine	1 × 2 × 8′	2
Trellis Stock	¼″ × 1⅝″ × 8′	2
Baluster Stock	¾″ × ¾″	20′
A-C Exterior Plywood	¼″ × 4′ × 8′	1
Quarter-Round Molding	¾″	2′
Hardware		
Finishing Nails	6d	6
Finishing Nails	3d	12
Aluminum Nails	1″	22
#8 Flathead Wood Screws	1½″	8
#8 Flathead Wood Screws	⅝″	8
Metal Angle Brackets	⅛″ × 2″ × 2″	4
Vented Aluminum Soffit	16″ × 8′	1

formulas in chart 13–3 to find the correct size of each piece. Measurement A is the distance between the narrowest points of your window, whether it is the storm window frame or the wood mullions. A ⅛-inch clearance on each side of the box is already built into the formula, so record the exact measurements. Measurement B is the depth of the windowsill, including the inside trim and storm window frame.

Our design requires that you have at least 48 inches of clear wall below the window where you will mount the heater. An alternative design should be considered for windows that are less than 21 inches wide, since the limited absorber area will not produce enough heat to justify building the heater.

All of our measurements are given with the assumption that your 1 × 8 lumber measures an exact 7⅛ inches wide. If your wood is wider, trim it to 7⅛ inches. You should be able to cut one vertical side and one horizontal side from each 8-foot length of 1 × 8. First lay out the half-laps, and cut them with your circular saw, then cut the pieces to length. The extra length of the boards can be used to clamp the work

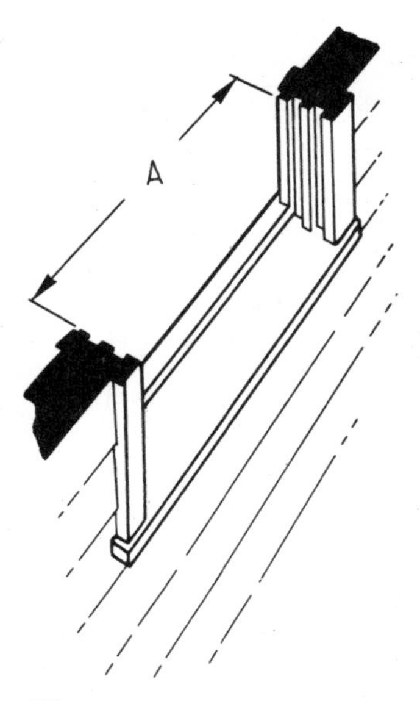

Illustration 13–2—Measurement A.

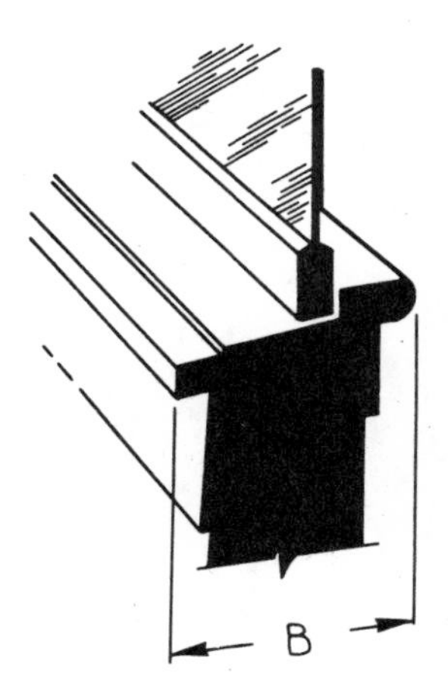

Illustration 13–3—Measurement B.

CHART 13–1—*Continued*

DESCRIPTION	SIZE	AMOUNT
Miscellaneous		
Rigid Foam Insulation	⅝″ × 4′ × 8′	1
Fiberglass Insulation	3½″	scraps
Vinyl Tube Weather Stripping	⅛″ × ½″	8′
Waterproof Wood Glue	...	½ pt.
Panel Adhesive	...	1 tube
Primer	...	1 qt.
White Enamel	...	1 pt.
Exterior Enamel	...	1 qt.
Flat Black Spray Paint	...	1 can
Clear Silicone Caulk	...	1 tube
Wood Putty	...	trace
Acrylic Sheet	⅛″ × A − 1″ × 54⅜″	1
Aluminum Tape	2″	3′

to your workbench when cutting the half-laps. Because of the width of your circular saw blade, you will need to allow at least ⅛ inch between pieces for good measure.

Using an 8-foot length of 1 × 8 pine, measure 55½ inches for the length of one vertical sidepiece. Square a line across the board. Then,

CHART 13-3—

Sizing Chart

DESCRIPTION	SIZE	AMOUNT
Top	A − 1″ × B + 7⅛″	1
Horizontal Divider	A − 1″ × B + 3 7/16″	1
Vent Floor	A − 1″ × B	1
Back	A − 1″ × 48″	1
Vertical Divider	A − 1″ × 45″	1
Horizontal Sides	7⅛″ × B + 7⅛″	2
Vertical Sides	7⅛″ × 55½″	2
Collector Bottom	7⅛″ × A − 1″	1
Grille Sides	7⅛″ × 4¾″	2
Grille Divider	4¾″ × A − 2″	1
Grille Molding	¾″ × 7⅛″	2
Grille Trim	1½″ × A − 2¾″	2
Center Braces	Size to fit	2
Exterior Molding	Size to fit	3
Upper/Lower Glazing Mounts	¾″ × A − 1¾″	2
Side Glazing Mounts	¾″ × 52⅝″	2
Upper/Lower Absorber Mounts	¾″ × A − 3″	2
Horizontal Divider Stiffener	¾″ × A − 3″	1
Side Glazing Trim	1⅝″ × 55½″	2
Upper/Lower Glazing Trim	1⅝″ × A − ¼″	2

allowing a ½-inch space between the pieces, use measurement B plus 7⅛ inches to mark the horizontal sidepiece. Lay out the other two pieces the same way on another 1 × 8 board.

To mark the exact width of the half-lap, lay one of your boards across the other. Be sure you mark squarely on the line. Repeat this for the remaining three half-lap joints. You can also measure the width of your boards and use a square to mark the overlapping area, but using the actual piece is more accurate.

Clamp one board to your workbench. Be sure your circular saw is unplugged while you adjust it. Set the saw to cut ⅜ inch deep, or half the thickness of the board. Then clamp a straightedge across the board so that when the saw is held against the straightedge, the saw blade will cut on the correct side of the line, removing wood only from the joint area, as shown in photo 13–1.

To cut out the half-lap area, keep the saw flat on the board, and make at least 12 passes across the area. Cut close to the end length measurement, but leave the cutoff line intact. When one area is complete, move on to the next board and repeat the same setup and cuts. When all half-laps are cut, cut the pieces to length.

Use a wood chisel to clean out the remaining wood. Match the half-laps to check the fit, then glue and screw the pieces together into two L shapes, as shown in illustration 13–4. Four ⅝-inch #8 flathead wood screws, one near each corner, hold each of the joints in place. Use a

Illustration 13–4—Corner detail.

Photo 13–1—Cut wide half-laps in the vertical and horizontal sidepieces with a circular saw and a clamped guide.

square to make a true 90-degree angle. Drill a pilot hole, and countersink the screw heads.

Cut the 1 × 8 bottom piece to length next, making it measurement A minus 1 inch. Cut a rabbet on one long edge of the bottom to receive the ¼-inch plywood back piece. Adjust your saw to ⅜ inch deep for the rabbet cut. Clamp the straightedge so that the rabbet will be ¼ inch wide, and cut it on one long edge of the piece. You can make one pass with the saw, then chisel the remaining wood away.

The next series of cuts are the ¼-inch dadoes, or grooves, that hold the vertical and horizontal dividers in place. Center a line along the long axis of each assembled side unit, and measure ⅛ inch to each side of it. The grooves must be on the inside of the heater, as shown in illustration 13–4. The vertical groove extends the full length of the side and should be cut first. Extend the horizontal cut past the vertical cut by 1 inch so that the depth of the groove will be a full ⅜ inch all the way. Use a clamped straightedge and make several passes with the saw to get slightly more than a ¼-inch width. Check the grooves with ¼-inch plywood to be sure the joint is a little loose.

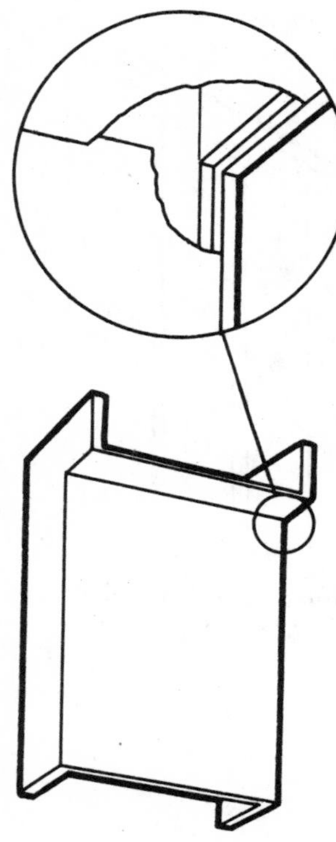

Illustration 13–5—Back and bottom detail.

With the side unit still firmly clamped, use your circular saw to cut out rabbets to receive the top, the vent floor, and the back. As illustration 13–4 shows, the rabbet for the back runs down the inside long edge. The vent-floor rabbet is cut along the lower horizontal edge, and the top rabbet extends along the upper edge of the side unit. Where the vent-floor rabbet and the back rabbet join in the corner you will need to use your wood chisel to clean out the cut and square it up.

Next, the ends of the vertical sides receive a ¾-inch-wide by ⅜-inch-deep rabbet to fit the bottom. Cut these, then check the fit and alignment of all the pieces. Trim the rabbets if necessary.

If everything fits, glue and screw the sides to the bottom, using three 1½-inch #8 flathead wood screws on each side, and countersink the screw heads.

Measure and cut the back from the plywood. The size is measurement A minus 1 inch by 48 inches. Also cut the vent floor, measurement A minus 1 inch by measurement B, from the plywood. Fit these two pieces onto the sides, as shown in illustration 13–5. Use a bead of panel adhesive to seal the plywood to the sides and aluminum tape to seal the inside seams. The back and vent floor are then nailed into the rabbets, using 3d finishing nails. You should now have a fairly rigid assembly to work with.

Cut the ¼-inch plywood top next. This measures A minus 1 inch by measurement B plus 7⅛ inches. Then cut the horizontal divider, measurement A minus 1 inch by measurement B plus 3⁷⁄₁₆ inches, and the vertical divider, measurement A minus 1 inch by 45 inches, from your plywood. The horizontal divider is fitted with a ¾ × ¾-inch baluster stock stiffener cut to measurement A minus 3 inches. The vertical divider is fitted with two ¾ × ¾-inch absorber mounts, one per end, also cut to measurement A minus 3 inches. Fasten the stiffener and the absorber mounts with glue and 2d finishing nails. Center the pieces from side to side, keeping them flush with the ends of the plywood.

Cut and fit the glazing mounts next, keeping them ⅛ inch, or the thickness of the glazing material, from the front edge of the side, top, and bottom pieces. Glue and nail them with 3d finishing nails, and set the nails. See illustration 13–6.

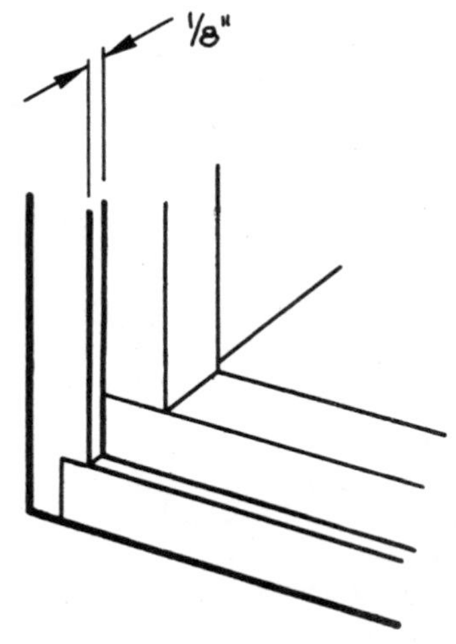

Illustration 13–6—Glazing support detail.

Now cut the rigid foam insulation to fit, using a straightedge and utility knife. The insulation for the top is the same width as the glazing mount but inset ½ inch from the grille edge. Then cut a piece for the back, completely covering the plywood. Cut a piece for the bottom, filling the space between the glazing mount and the back insulation. Glue each piece in place with panel adhesive.

Cut a total of eight pieces of insulation for the vertical and horizontal sides, allowing for the center groove. The vertical sidepieces extend to within ⅝ inch of the top rabbet. The space is needed for the top insulation. The horizontal pieces extend to the edge of the center groove. Their width is also measured to within ⅝ inch of the top rabbet. The lower pieces butt against the vent floor.

Cut the insulation for the middle dividers next. The vertical divider insulation fits within, and is the same width as, the absorber mounts. The horizontal piece butts the stiffener on one edge and extends to within ½ inch of the vent edge.

Now you should have all the insulation pieces glued in place. There is no insulation on the vent floor, since this part of the box will be insulated during installation. A few checks should be made at this point: Check to see that the vertical divider will slide into the vertical groove and that the horizontal divider will slide against it in the horizontal middle groove; also, check the fit of the top. To avoid any problems with sliding the dividers into place, see that the grooves are clear and that the divider edges are sanded smooth.

The absorber panels are now cut from the soffit panel, but are not to be fastened in place until after they are painted. They are equal in width to the length of the absorber mounts, and their length is 45 inches, or equal to the length of the vertical divider. Be careful in cutting these pieces, as the edge left by the tin snips can be very sharp.

Now you can break out the flat black spray paint, and paint the absorber panels on both sides to absorb heat. Every piece of insulation attached to the vertical middle divider should remain shiny to reflect any light that sneaks behind the absorber panels.

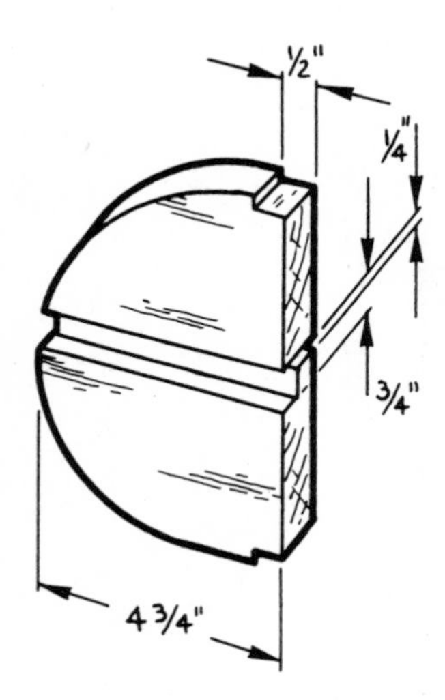

Illustration 13–7—Grille side detail.

Also, the rear, or plywood, side of the vertical middle divider is painted white so that the rear (cold air) side of the divider won't heat up and reverse the airflow. Brush on two coats of a gloss enamel. A wood primer must be applied first to this panel and to all the exterior surfaces.

While the paint is drying, the pieces for the grille can be cut and assembled. Follow chart 13–3 carefully to avoid errors. Basically, the grille is 1¾ inches narrower than measurement A and fits inside the vent opening to a depth of ½ inch.

The grille sides, shown in illustration 13–7, are made first. Cut two lengths of 1 × 8 to 4¾ inches. For safety, use a handsaw for these cuts. Measure 3⁹⁄₁₆ inches in from a short edge, and square a centerline across the piece. From a long edge, measure in 1³⁄₁₆ inches, and square a line across the long axis. With a compass set for a radius of 3⁹⁄₁₆ inches, make an arc from the intersection of the two lines, giving the piece a rounded front. Designate one short side as the top of the piece. *Note:* The ¾-inch-wide dado is not exactly centered. Mark a line ⅛ inch to the top side of the centerline. Then mark another line ⅝ inch below the centerline.

Cut ¼ × ½-inch rabbets, as shown in illustration 13–7, on the top and bottom corners. Check the placement of the center dado by holding the piece against the grille opening of the main assembly. The top line of the dado should match the top of the horizontal divider.

Mark the depth of the dado, ⅜ inch, on the edge of the board, and

Photo 13–2—Cut the dado with a backsaw. The piece is too small to be cut safely with a circular saw.

clamp it to your workbench. To be safe, cut the dado with a handsaw, making a cut to the inside of each line, as shown in photo 13–2, then clean the groove with a chisel.

Cut the middle dado, ¼ inch wide by ½ inch deep, to receive the horizontal grille divider. Then cut the curved edge with a saber saw or coping saw. Repeat the process to make the opposite sidepiece, but before cutting the second grille side, be certain you will end up with a right- and a left-hand piece.

Make the two grille trim pieces next; the rabbet in these boards is cut with a utility knife because the boards are too narrow to be cut safely with a saw. Cut two pieces of 1 × 2 pine to length, measurement A minus 2¾ inches, and mark the rabbet area, ½ inch wide and ¼ inch deep. Clamp one piece to your workbench with a straightedge on the ½-inch mark. Score on the line to the ¼-inch depth, then readjust your straightedge and score the area again along the center of the rabbet. Be sure to keep your free hand away from the utility knife. See photo 13–3.

Turn the board on edge, and cut along the ¼-inch line until the first section of the rabbet is free. Continue cutting until the second section lifts away. Clean the rabbet area with a sharp wood chisel, then check the fit with a piece of plywood. Repeat the procedure for the second trim piece. Finish the pieces by rounding the edge opposite the rabbet with sandpaper or a hand plane.

The grille divider should be made in the same manner as the grille trim. It is ¾ inch longer than the grille trim and 4¾ inches wide, ripped

Photo 13–3—Cut the rabbet for the grille pieces very carefully with a utility knife. Keep your hand well away from the blade.

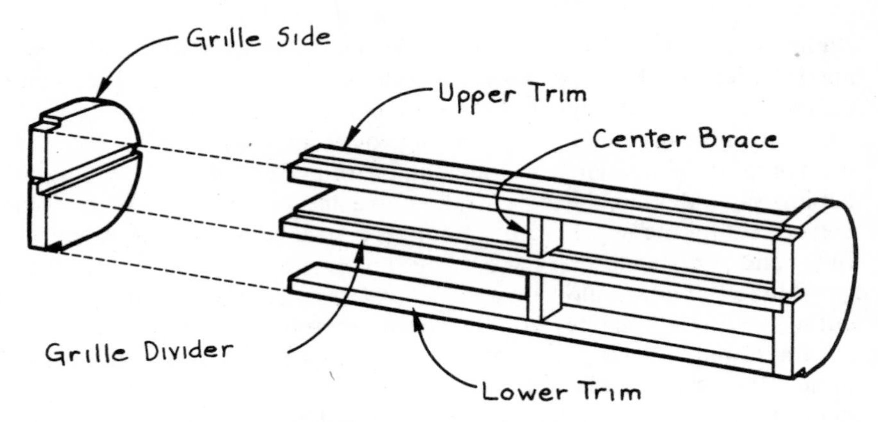

Illustration 13–8—Grille assembly details, shown from the back.

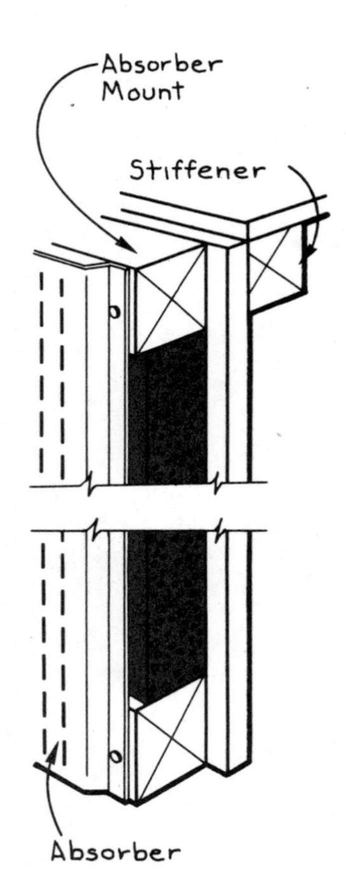

Illustration 13–9—Absorber panel detail.

from a length of 1 × 6. Rabbet the back of the piece as you did the grille trim.

Assemble the pieces of the grille, using waterproof wood glue and 6d finishing nails, as shown in illustration 13–8. The grille trim is butted to the grille side; the grille divider is set into its dado. Align the rabbets on the grille sides to those on the grille trim and divider. Now round off the outer divider edge. When assembled, the unit can be checked by sliding it into the vent opening, although the top piece and horizontal divider have not yet been fitted. The grille should fit snugly and evenly into the sides of the assembly. Then cut the two center braces of 1 × 2 pine to fit, and fasten them with glue and 3d finishing nails. Do not attach the grille at this time.

Assemble the main heater pieces when the paint is dry. Slide the vertical divider into place, then the horizontal divider. Use a bead of panel adhesive to seal between the two parts. Drill two pilot holes to fit 1½-inch #8 flathead wood screws, and fasten the pieces together through the absorber mount into the stiffener. The vent floor, the horizontal divider, and the top must all be flush at the grille edge for the grille to fit properly.

Laying the assembly on your workbench, fasten the absorber panels in place, as shown in illustration 13–9. Use 1-inch aluminum nails, making 1/16-inch-diameter pilot holes in each of the soffit ribs to get the nails started. Use a nail set to drive the nails down completely.

The top can now be glued and nailed in place. Set the 3d finishing nails so that the nail holes can be filled with wood putty. Fill any other cracks, screw holes, and gaps as well, and apply first a primer coat, then the finish coat of exterior paint.

It is easier to paint the grille while it is off the main assembly. Use a primer coat first, then finish the grille unit with paint to match your interior woodwork. Install the grille in the heater with 1-inch aluminum nails through both the top and the bottom. Add 7⅛-inch-long pieces of quarter-round trim to the vent edge on each side of the grille.

From trellis stock, cut two glazing trim pieces the same length as the heater sides and two the same length as the width of the heater. Then cut the miters to join the corners after laying the pieces on the heater and marking where the cuts need to be made. When painting the trim, prime all surfaces, but paint only the outside surfaces. The side resting against the glazing will be sealed with silicone caulk.

Mark the glazing to be cut by resting it inside the heater on the glazing mounts. Because of expansion of the glazing, allow a ⅛-inch gap between the glazing and the heater sides all around. Acrylic sheets are best cut with a saber saw or fine-tooth handsaw used with light pressure.

Clean the glazing before installing it. Apply a bead of clear silicone caulk on the glazing mounts, and lay the glazing on the mounts. Smooth the silicone by pushing down on the glazing until the glazing and edges

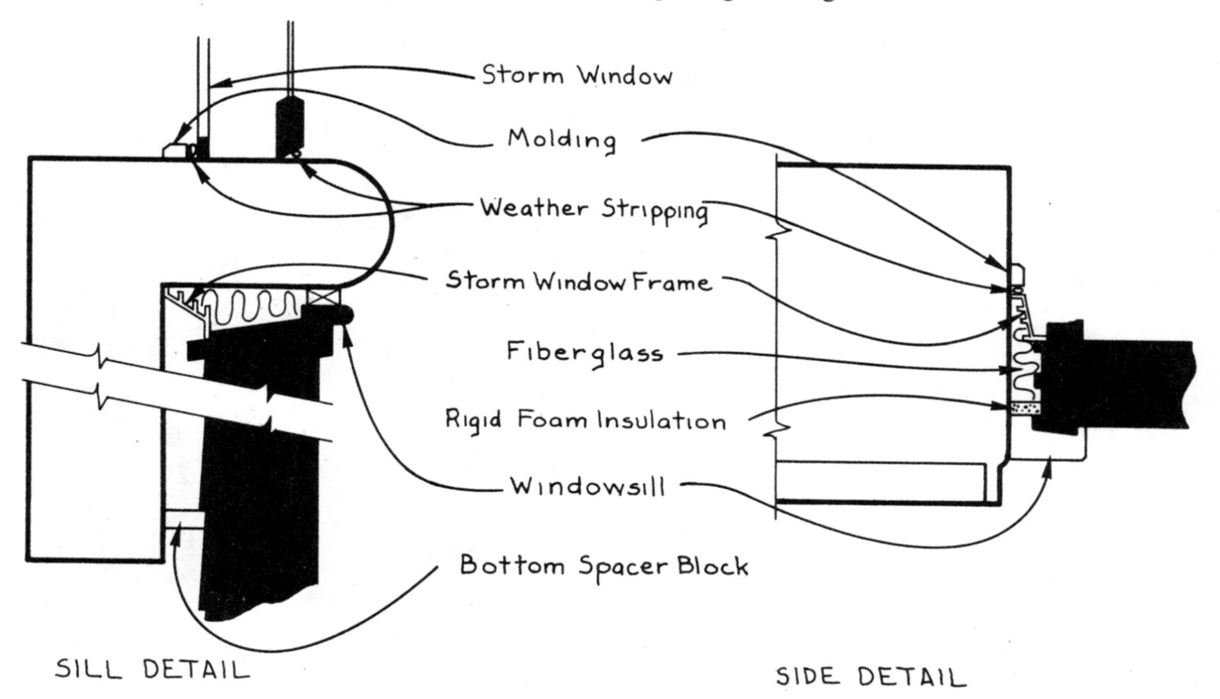

Illustration 13–10—Details of the insulation, weather stripping, and support block after the window heater is installed.

of the heater are flush. Apply more silicone to the front edges of the sides, top, and bottom. Align and nail the trim to the heater, and you are ready to install the heater in your window.

Plan what you are going to do before hoisting the heater to your window. Illustration 13–10 shows how to install the heater. If you have storm windows, chances are that the heater will rest on the aluminum frame. Get the heater in position, and mark a suitable location for the bottom spacer, if needed. An upper spacer block is used to keep the heater from rocking but may be eliminated if the heater rests directly on the inside sill. Remove the heater from the window and pack out the sill with fiberglass insulation. Nail the bottom spacer block to the bottom of the heater, and fasten the sides to the outside sill with metal angle brackets, as shown in photo 13–4. So that the window may be opened in a fire emergency, or simply for ventilation, avoid fastening the heater to the sash.

Apply vinyl tube weather stripping, as shown in illustration 13–10. Pack insulation between the window sides and the heater. Rigid foam can be used to hold the fiberglass in place and to cover the top of the fiberglass insulation on the sides.

Photo 13–4—Metal flashing over the top of the heater will protect it from rain. Note the position of the mounting bracket.

We have found that it is advisable to cover the top of the collector with a piece of metal flashing, unless the top is protected by the eaves of your home. If it is exposed to a lot of rain and snow, the top should be flashed with standard aluminum roof flashing. Be sure to bend the edges down over the sides, so the entire top of the collector has a protective metal covering. The flashing can be painted if you wish.

Cut a molding strip to fit across the top of your heater. Paint it, then nail it to the top of the heater in front of the window sash. In the same way, cut two strips of trim to cover the side joints. Paint them first, then nail them to the sides of the heater.

Anyone with small pets or curious children in the house may want to put screen over the grille front. Use mesh of at least ¼ inch so that the airflow is not restricted. Bend the mesh over the rounded front and secure it with a suitable molding and nails.

The heater is easy to operate—you do nothing except keep the glazing clean. You may want to remove the heater during the summer and reinstall it during the cooler months. If not, cover the glazing with a sheet of ¼-inch plywood painted white to reflect light. When temperatures fluctuate in spring or fall the window sash can still be opened.

14 Solar Air Heater

This solar air heater is low in cost and easy to construct, and it operates without any mechanical devices or input from you. It is designed to be incorporated into a stud frame house with aluminum or wood siding.

The heater is a passive solar device that is known as a thermosiphoning air panel (TAP for short) and that operates on the principle that

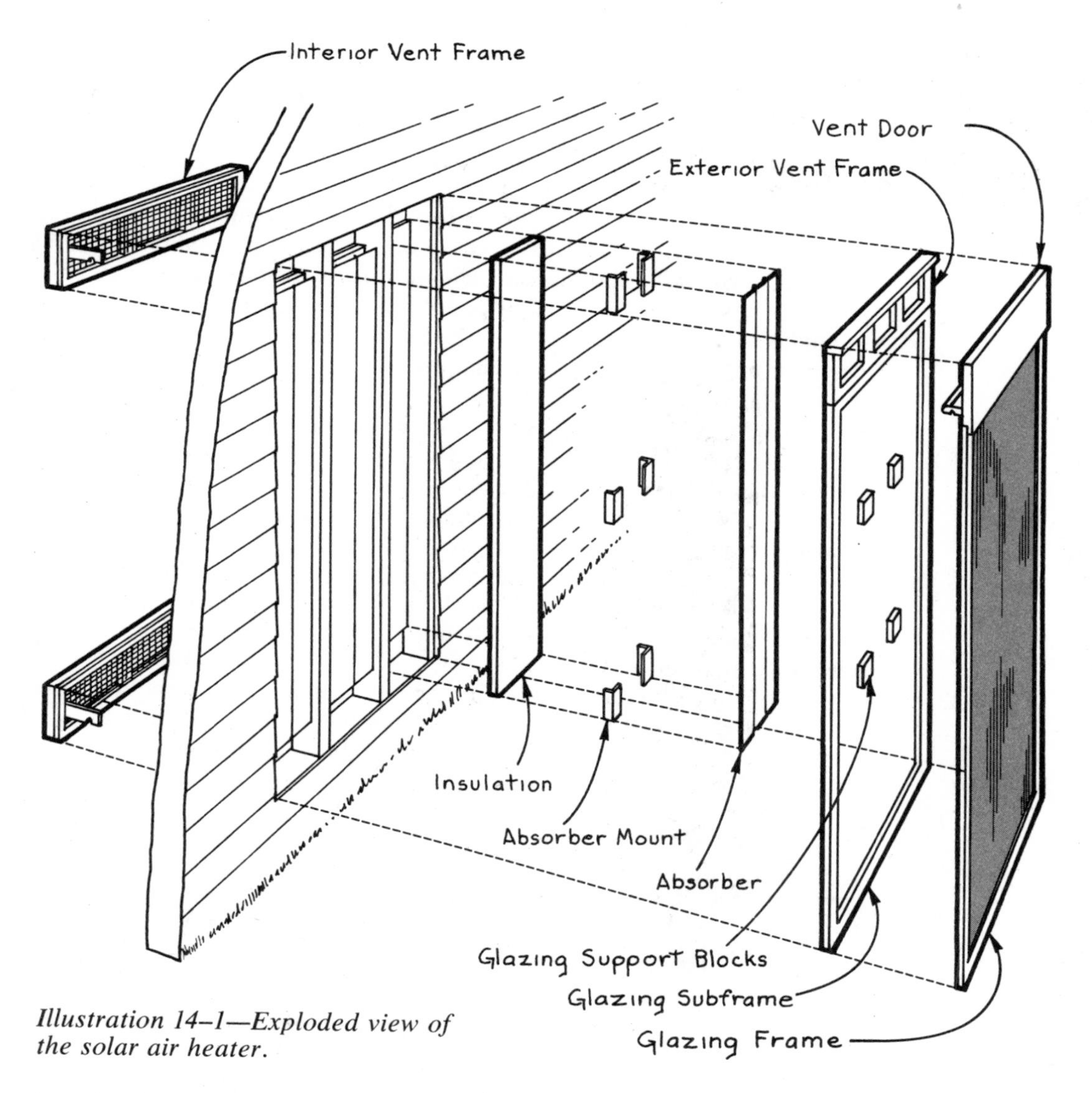

Illustration 14–1—Exploded view of the solar air heater.

hot air rises. It is an insulated box built into the spaces between the studs in a south-facing outside wall, as shown in illustration 14–1. Insulated glazing is fastened to the outside of the box. Three panels of vented aluminum soffit, with air spaces on either side, make up the absorber plate. Sunlight passes through the glazing to heat the aluminum absorber. Temperatures inside the box can reach as high as 120°F, and this hot air rises to vent out into the house. There are six openings in the interior side of the wall, three near the floor and three near the ceiling. When the hot air exits, the panel draws cool house air in, heats it, and releases warm air back to the house. The airflow continues, following the principle of thermosiphoning, as long as the sun shines to heat the absorber.

When the sun is down and the system is not gaining heat, the insulated glazing prevents heat loss to the outside. However, there is a tendency for a back draft to occur as the air in the panel cools off and sinks. By attaching thin damper flaps to the lower room openings, this cooler air is kept within the air panel. In the summer, when you don't want heat in your house, a vent at the top of the air panel is easily opened from inside the house. This vent prevents an excess of heat buildup within the cavity.

The heater may be placed in any wall facing within 30 degrees of south. It does not create glare or privacy problems as extra windows might. It will fit between windows, and, since it is built into the wall, it is a permanent fixture requiring minimal maintenance. Simply keep the glazing clean, and recaulk as required.

When you have chosen a south-facing wall for the thermosiphoning air panel, check to be sure that nearby buildings and trees will not shade the wall in winter. The site evaluator, project 3, is your best way to check an installation site.

In planning the thermosiphon air panel, you may well need to make changes based on different dimensions in the framing of your house. Our plans fit 2 × 4 stud walls with the studs 16 inches on center, but can be adapted to almost any stud wall. The panel covers a wall area

CHART 14–1—
Materials

DESCRIPTION	SIZE	AMOUNT
Lumber		
#2 Pine	1 × 6 × 8′	1
#2 Pine	1 × 2 × 8′	4
#2 Fir	2 × 4 × 4′	1
A-C Exterior Plywood	¾″ × 2′ × 4′	1
Drywall	½″ × 4′ × 8′	1
Hardwood Dowel	¾″ × 26″	1
Screen Molding	¼″ × ½″ × 8′	3

CHART 14–1—*Continued*

DESCRIPTION	SIZE	AMOUNT
Hardware		
Finishing Nails	8d	34
Aluminum Siding Nails	2¼″	36
Aluminum Siding Nails	1¼″	1 lb.
16-Gauge Brads	⅝″	48
Drywall Nails	1¼″	1 lb.
#10 Flathead Wood Screws	3″	4
#10 Roundhead Aluminum Wood Screws	2″	17
#10 Flathead Aluminum Wood Screws	1½″	3
#10 Sheet-Metal Screws	¾″	4
#10 Sheet-Metal Screws	½″	6
#8 Flathead Wood Screws	½″	4
#6 Flathead Wood Screws	½″	8
Lag Bolts	¼″ × 2½″	6
¼–20 Stove Bolts	1″	4
¼–20 Hex Nuts	...	8
Flat Washers	¼″	12
Rubber or Fiber Washers	¼″	4
Staples	⅜″	12
Loose-Pin Butt Hinges with Screws	¾″ × 1½″	3
Aluminum Angle	1/16″ × 1″ × 1″ × 58″	1

[*Continued on next page*]

CHART 14–2—
Tools

- Hacksaw
- Saber Saw
- Circular Saw
- Drill and Bits
- Square
- Level
- Miter Box
- Caulking Gun
- Heavy-Duty Staple Gun
- Tin Shears
- Stud Finder

CHART 14–1—*Continued*

DESCRIPTION	SIZE	AMOUNT
Hardware—*continued*		
Aluminum Flashing	3″ × 50″	1
Vented Aluminum Soffit	13½″ × 80″	3
Aluminum Bar Stock	⅛″ × ¾″ × 40″	1
Aluminum J Channel	¾″ × 12′	2
Window Screen	4½″ × 45″	1
½-Inch-Mesh Hardware Cloth	4″ × 46″	2
Miscellaneous		
Foil-Clad Rigid Foam Insulation	1″ × 4′ × 8′	1
Closed-Cell Foam Weather Stripping	3⁄16″ × ⅜″	10′
Construction Adhesive	...	1 tube
Primer	...	1 pt.
Interior Enamel	...	1 pt.
Exterior Enamel	...	1 qt.
Flat Black Absorber Paint	...	1 qt.
Silicone Caulk	...	1 tube
Qualex Glazing Material*	¼″ × 4′ × 8′	1
2-Mil Polyethylene	12″ × 13″	1
PolyGrowers Aluminum Extrusion*	⅞″ × 1¾″ × 8	3
Aluminum Iron-On Tape	3″	1 roll

**Qualex glazing material and PolyGrowers aluminum extrusion are available from: PolyGrowers, Inc., Box 359, Muncy, PA 17756.*

7½ feet high and 4 feet wide. It is necessary to have an area approximately this large to generate enough heat to achieve effective thermosiphoning. Nearly all the work is done from outside your house, so pick a day or two when you can leave the wall exposed.

The materials are listed in chart 14–1, and the necessary tools, in chart 14–2.

Since the air panel is located within the studs of an exterior wall, the first job is to locate the studs. This can be done by thumping on the interior of the wall with your fist. As you move across the wall, the thump should change from a hollow to a solid sound as you hit a stud. You could also use a magnetic stud finder that locates the lines of nails which hold the drywall, or other interior sheathing, to the studs. Mark an area 48 inches wide from the center of one stud, across two more studs, to the center of the fourth stud. The height of the air panel on the interior of the house is 86½ inches, starting at the top of the bottom plate (the board which runs horizontally along the edge of the floor behind the drywall). The plate is usually 1½ inches thick. Lay out the stud cavities, as shown in illustration 14–2, then lay out two sets of register openings, one set just above the bottom plate and the other set 80 inches higher than the first. The openings are 3¼ × 13½ inches. Drill a ¼-inch-diameter hole at each corner of the openings, and cut out the six openings with a saber saw, as shown in photo 14–1.

At the lower outside corners of both the top and bottom register openings, drill two holes through the exterior sheathing and the siding

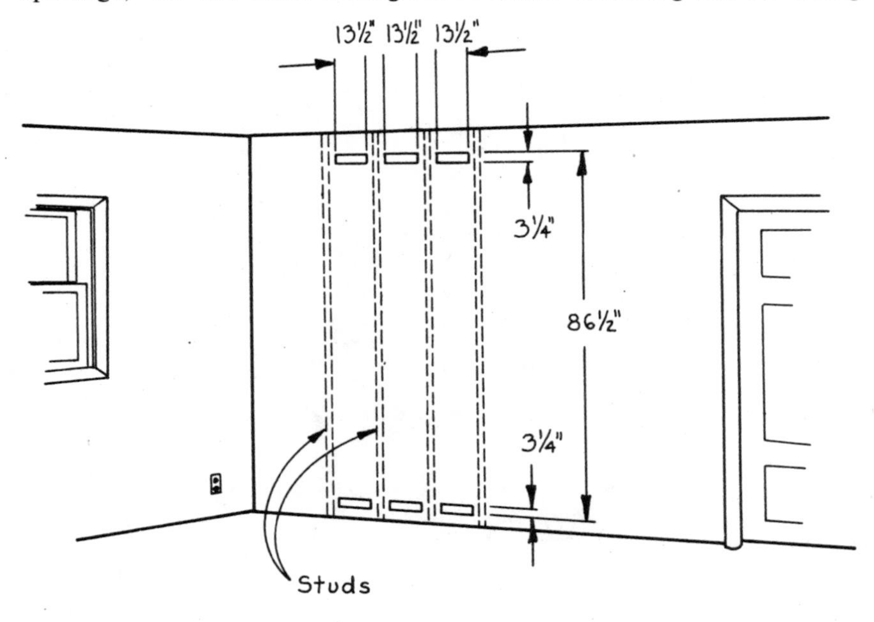

Illustration 14–2—Layout details for the interior vent openings.

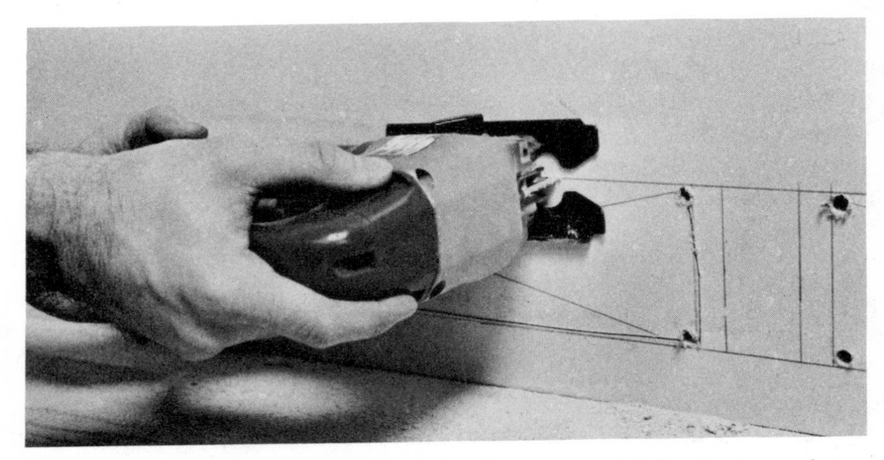

Photo 14–1—After carefully marking the areas, cut the vent openings. Either a saber saw or a keyhole saw can be used.

to help locate the exterior dimensions of the air panel. Now start work on the outside of your house. Lay out an area 48 × 90 inches, measuring up from the lower drilled holes and to a width that reaches the center of the 2 × 4 studs on either side. Use a level to be sure that this area is square and plumb, then mark the lines to be cut.

To cut aluminum siding with a circular saw, the saw should be fitted with a fiber metal-cutting blade. Set the depth to cut through the siding only, not through the sheathing. As shown in photo 14–2, lower the saw

Photo 14–2—A circular saw equipped with a metal-cutting blade is the fastest way to cut through aluminum siding.

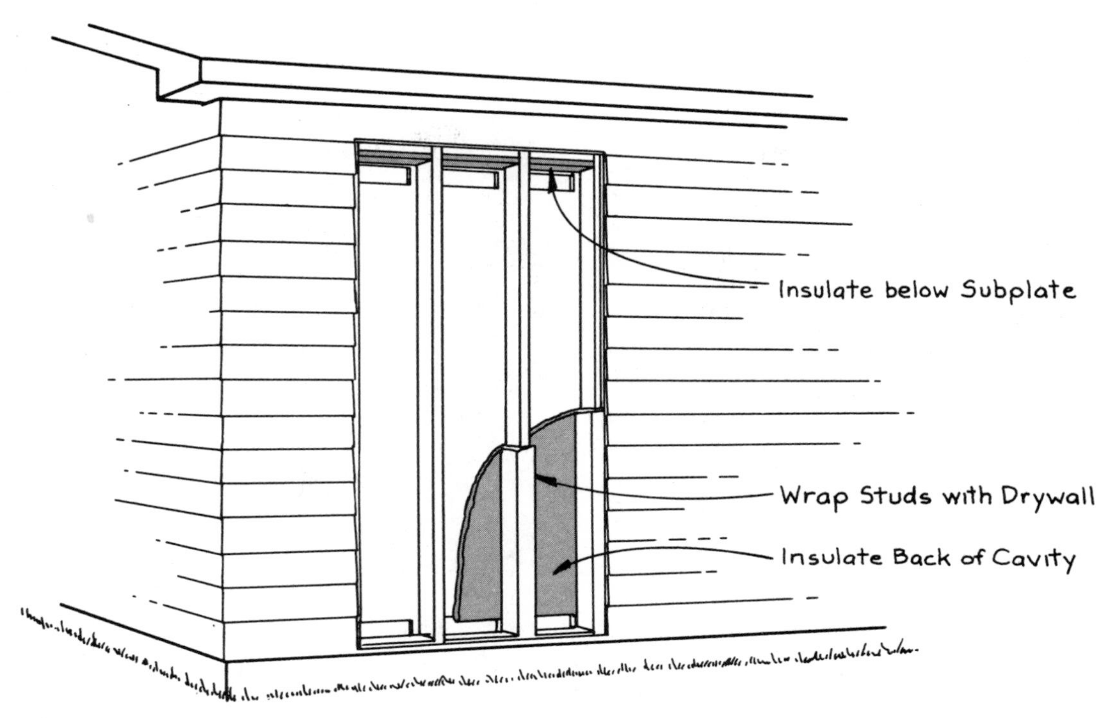

Illustration 14–3—Drywall and insulation are installed in the stud cavities.

into the siding material with the guard held back. Cut along the marked line. The metal-cutting blade will cut through any nails that are found on these lines. Remove the siding from the area. Change to a wood-cutting blade, and cut along the edge of the siding to remove the sheathing. Pull away the sheathing, and remove the insulation from the stud cavities.

Illustration 14–3 shows the air panel exterior. Close each cavity off at the top with a 2 × 4 subplate. Cut three 14½-inch pieces to span the distance between the studs. Toenail the subplates at the top of each opening 1 inch above the interior register opening to allow space for insulation.

The air cavities are lined with ½-inch drywall to keep moisture and heat away from the studs. Cut six 3½ × 87½-inch pieces of ½-inch drywall. Nail the pieces to each side of the exposed studs with drywall nails. Cover the exposed edges of the two center studs with two pieces of drywall, each 2½ × 87½ inches. Cover the edges of the exposed side studs with 1¼ × 87½-inch pieces of drywall. To finish the edges of the drywall, apply 3-inch-wide aluminum iron-on tape with a clothes iron, as shown in photo 14–3. Use the tape according to the manufacturer's suggestions.

Photo 14–3—Aluminum iron-on heat tape should be applied to the drywall-covered studs with a clothes iron.

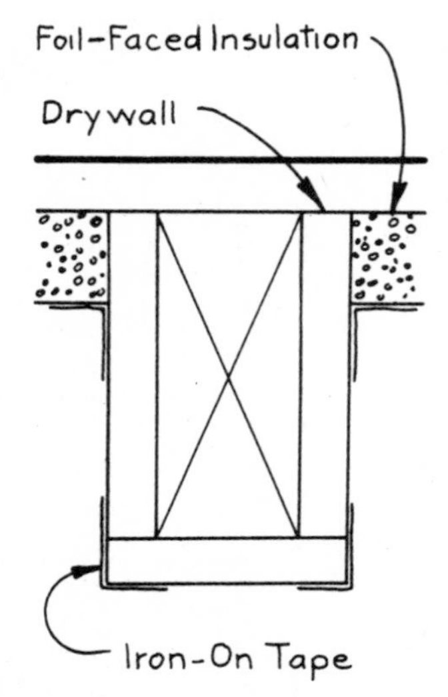

Illustration 14–4—Stud details.

Insulate the cavities with 1-inch foil-clad rigid foam insulation. First cut three pieces, approximately 3½ × 13½ inches, to fill the upper spaces. The stud spaces will probably vary, so cut each one to fit. Apply iron-on tape to all the cut edges of the insulation before installing it, otherwise it will give off gases if the heater gets too hot. Using construction adhesive, fasten the insulation to the bottom of the subplates at the top of the stud cavities. Next cut three pieces of insulation, each approximately 13½ × 79¾ inches, for the back of the stud cavities. Again tape the edges and glue these pieces to the back of the stud cavities and between the two sets of register openings. The edges of the insulation should be wrapped with tape to protect them. Illustration 14–4 shows what the opening should look like.

Now you are ready to build the absorber plate assembly, which absorbs solar heat and warms the air. Start by cutting 18 pieces of 1/16-inch-thick 1 × 1-inch aluminum angle, each 3 inches long, for the absorber mounts. Since the absorber plates will fit vertically between the register openings, plan the location of the absorber mounts accordingly. The top mounts are level with the lower edge of the openings; the lower mounts are just above the bottom openings; and the middle mounts are centered. Position the brackets 1½ inches from the front edge of the studs so they will be separated from the front and back walls of the cavity by an air space, as shown in illustration 14–5. Drill ⅟16-inch-diameter pilot holes in one side of each angle, and nail the mounts to the sides of each stud with 2¼-inch aluminum siding nails.

The absorber plates are cut from vented aluminum soffit material with tin shears. Cut three 13½ × 79½-inch pieces. Measure each cavity before cutting the absorber panels.

Cut ¾-inch plywood to 2¼ × 2¼ inches to make four glazing support blocks. Fasten two blocks to each of the two center studs with 8d finishing nails. Position the lower two 42 inches above the bottom edge of the opening, and the remaining two 17 inches higher. The support blocks serve to hold the glazing steady in windy conditions.

Illustration 14–5—Absorber bracket placement.

Paint the absorber section of the air panel with flat black absorber paint. Prepare the aluminum absorber panels by scrubbing them with cleanser and rinsing them throughly. Then wash them with a diluted solution of vinegar to give some "tooth" to the metal, which will increase the adherence of the paint. Mix 1 cup of vinegar to 1 gallon of water and brush it on the plates. When they are dry, paint the absorber plates on both sides. Also paint the sides and fronts of the studs and the small support blocks. Do not paint the insulation, as it serves to reflect heat back to the absorber plates.

When the paint is dry, hold the absorber plates in position between the vent openings, and drill through both the plate edge and the absorber mount edge with a 5⁄32-inch-diameter bit. Fasten the absorbers to the mounts with ½-inch #10 sheet-metal screws, one at each mount.

To finish the edges of the exterior opening, fit aluminum J channel at the cut edges of the aluminum siding. Cut two pieces of ¾-inch aluminum J channel, each 90¼ inches long, for the sides, and one piece, 49 inches long, for the bottom of the siding. Referring to illustration 14–6, fit the J channel around the edges of the siding and fasten it in place on both sides and at the bottom with 1¼-inch aluminum siding nails.

Build an exterior vent to release hot air in summer at the upper end of the air panel. Start by cutting a vent frame of ¾-inch plywood, 8 × 47½ inches. As shown in illustration 14–7, lay out three vent holes in this frame, each 3½ × 13½ inches, to correspond to the cavities within the studs. Hold the frame against the studs and mark the openings. Cut out the three openings with a saber saw. Cut a 4½ × 45-inch piece of window screen and fit it over the vent openings. Staple the screen to the vent frame with ⅜-inch staples.

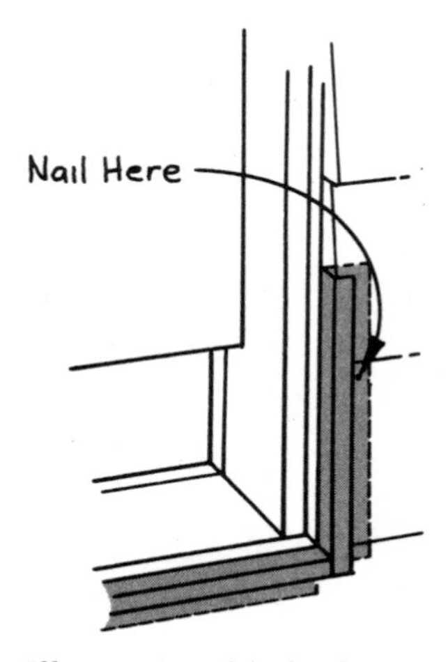

Illustration 14–6—J channel installation.

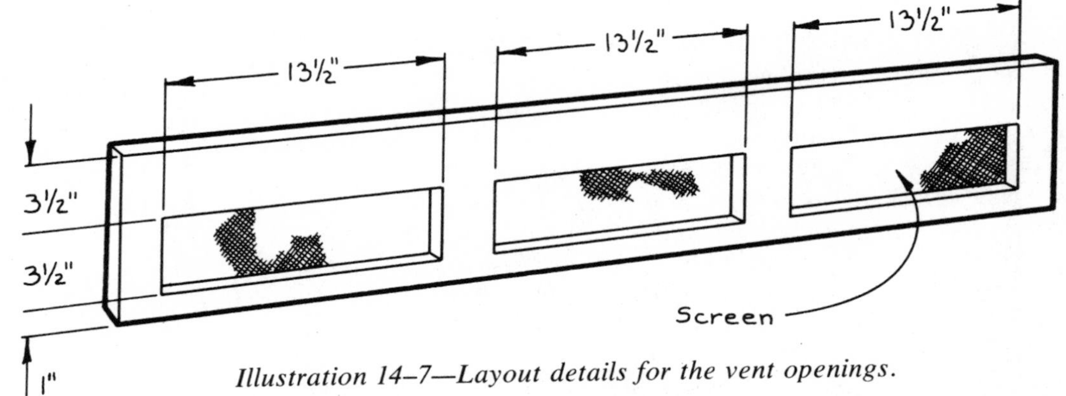

Illustration 14–7—Layout details for the vent openings.

Cut a 47½-inch piece of 1 × 2 pine for a vent door support. Illustration 14–8 shows the vent door assembly. Cut three mortises in the support board to accept ¾ × 1½-inch butt hinges. Cut the mortises with either a router or a wood chisel. Drill four 3⁄16-inch-diameter holes through the board from edge to edge for the mounting screws. These holes should align with the studs. Cut a 8⅝ × 47½-inch piece of ¾-inch plywood for the vent door. Along one edge, designated as the bottom, cut a ¾-inch-wide rabbet 9⁄16 inch deep. The rabbet will help to fit the door over the glazing and prevent air leaks. To cut the rabbet, measure ¾ inch from the bottom edge and square a line across. On the edge, mark the 9⁄16-inch depth. Set your circular saw to cut to a 9⁄16-inch depth. Cut with the saw blade just to the waste side of the line, and make several more passes with the saw to remove excess wood. Finish the rabbet with a wood chisel and sandpaper. Attach the butt hinges to both the support and the vent door.

Before installing the vent door, cut a 3 × 50-inch piece of aluminum flashing. Along one long edge, bend a 90-degree angle ½ inch from the edge. Bend the aluminum between two clamped boards, as shown in

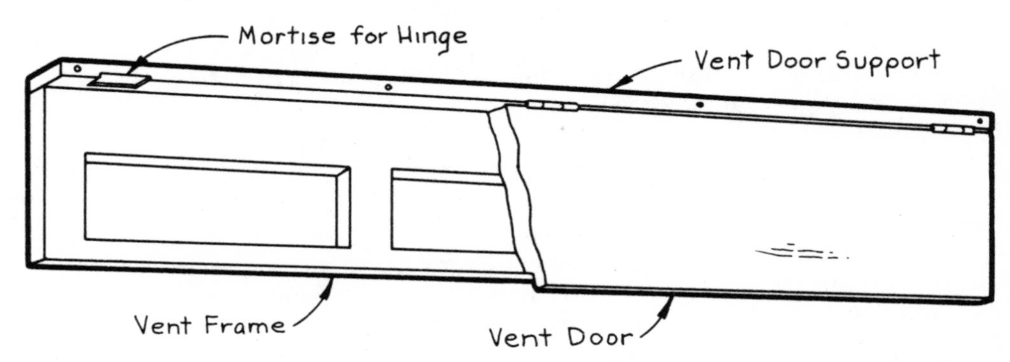

Illustration 14–8—Vent door and vent assembly attachment details.

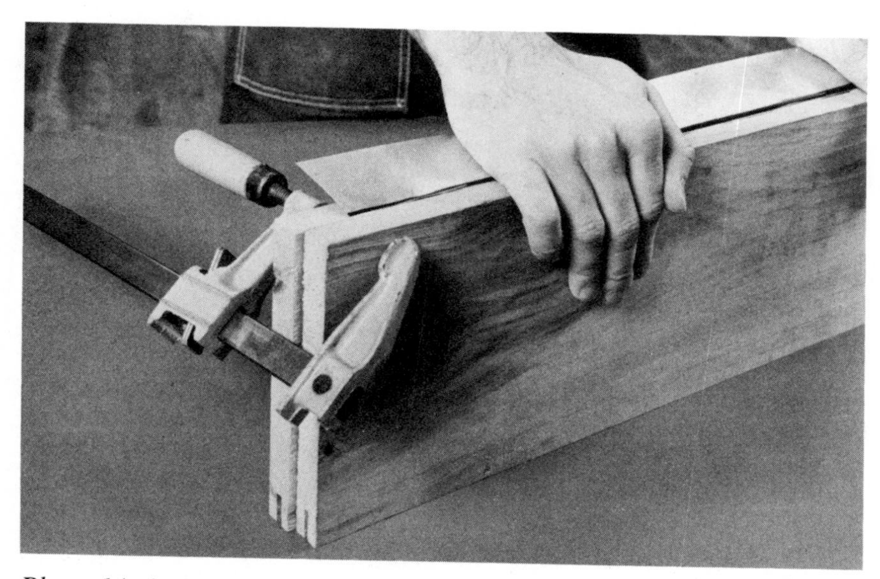

Photo 14–4—Bend the aluminum flashing for the top of the air panel, using two clamped boards.

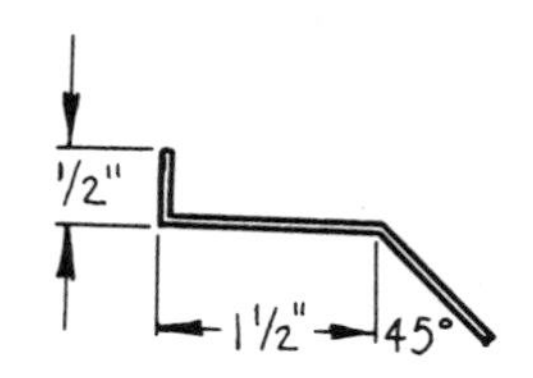

Illustration 14–9—Flashing details.

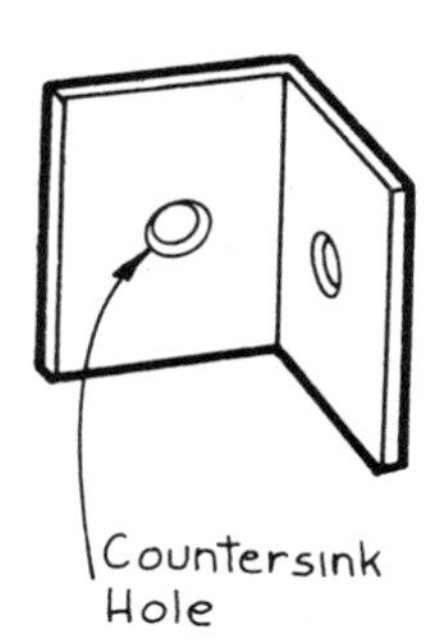

Illustration 14–10—Vent bracket.

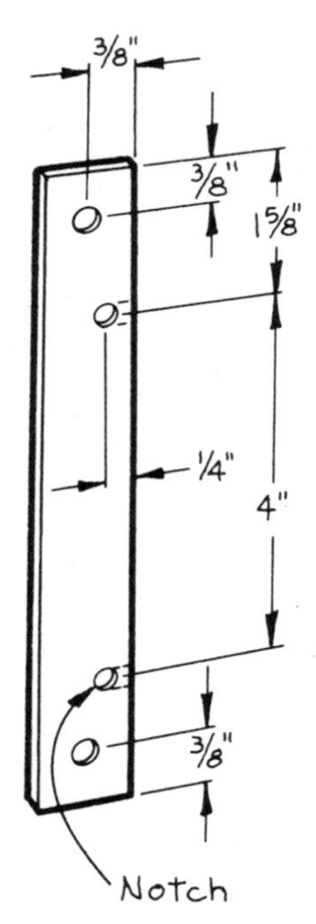

Illustration 14–11—Vent door arm.

photo 14–4. Use a third board to press it down. On the opposite long edge, bend a 45-degree angle, as shown in illustration 14–9. Fit the flashing under the top edge of the house siding by cutting away a little flashing on each side. When installed, the flashing should extend 1 inch beyond the opening on each side. Place the vent door and support against the flashing, making sure the support board is level. Drill ⅛-inch-diameter holes in the studs behind the support board for the mounting screws. Fasten the support to the studs with four 3-inch #10 flathead wood screws. Then, holding the door open, nail the vent frame, with the screen on the back, to the studs, holding it snug against the bottom of the support board.

The vent door is now fitted with a device that holds it open during hot weather to prevent unwanted heat buildup. The device may easily be operated from the inside of the house and cannot be opened from the outside. Start by cutting four pieces of 1 × 1-inch aluminum angle, each 1 inch long, to serve as brackets. Drill a ¼-inch-diameter hole centered in each leg of each piece. Countersink one hole on the inside, as shown in illustration 14–10. Next, cut four 5-inch-long linkage arms of ⅛ × ¾-inch aluminum bar stock. Drill four ¼-inch-diameter holes in the arms, as shown in illustration 14–11. With a hacksaw or file, cut from the edge to the middle holes to form two notches. Next cut two ¾-inch-diameter dowels, each 13 inches long, for handles. Into each end

Illustration 14–12—Vent door assembly with control mechanism.

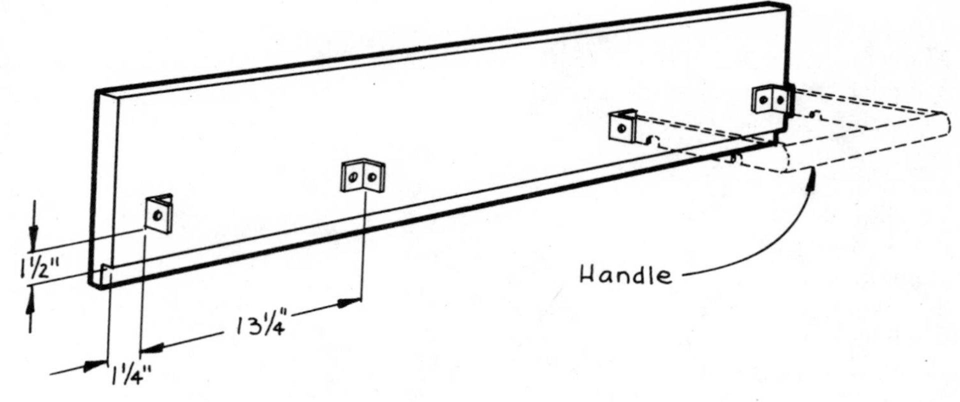

of the dowels, drill a 5/32-inch-diameter hole, ¾ inch deep. Attach the handles to the linkage arms with four ¾-inch #10 sheet-metal screws. Be sure the notches face the same way when the handles are attached.

Referring to illustration 14–12, lay out the brackets on the vent door. Place the brackets 1½ inches above the top edge of the rabbet. The two outer brackets are 1¼ inches from the ends, and the two inner brackets are spaced to fit the handle length, 13¼ inches from the outer brackets. Drill ⅛-inch-diameter pilot holes, and fasten the brackets to the vent door with four ½-inch #8 flathead wood screws.

Attach the linkage arms to the brackets, using ¼–20 × 1-inch stove bolts with two nuts and two washers to fit each one. Cut slots in the window screen and push the handle arms through. As illustration 14–13 shows, to make a freely moving joint, use a washer on both sides of the joint and put two nuts on the bolts.

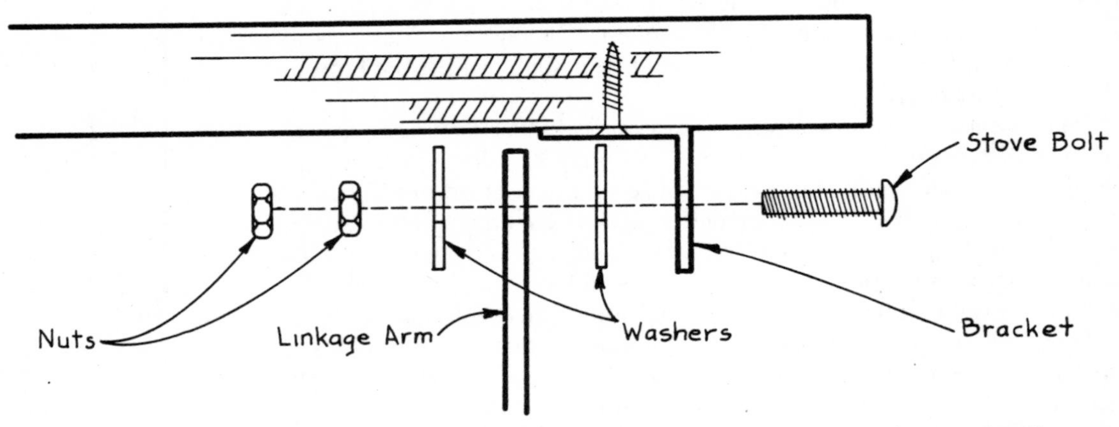

Illustration 14–13—The vent door mechanism attachment details.

Now mark the location of the vent latch bolts that will fit the notches in the linkage arms. With the vent doors closed and the arms level, mark the position of the notch that is nearer to the door. At these points drill 3⁄16-inch-diameter holes for the latch bolts. Turn 1⁄4 × 2½-inch lag bolts into the holes so that 1 inch of each bolt protrudes. The bolt will catch the notch and hold the vent door closed, or, using the second set of notches, hold the door open. On the inside edge of the vent door, attach 3⁄16 × 3⁄8-inch closed-cell foam weather stripping to make a snug seal all around.

Next construct the two vent registers that cover the openings into the house. The frames are made from two 48-inch lengths of 1 × 6 pine. As shown in illustration 14–14, lay out three openings in each frame, 3¼ × 13½ inches each. Allow 2⅝ inches between the openings and 1⅛ inches at the ends. Center the openings top to bottom. With a router, cut a ¼-inch radius around the outside of the frames. If you don't have a router, round the edges with sandpaper or a plane. Cut out the openings with a saber saw. Sand the registers and finish them to match the woodwork in your home.

The screening in the registers is ½-inch mesh hardware cloth. With tin shears, cut two pieces, each 4 × 46 inches. Paint the screens on both sides to match your house wall. Staple the screens to the back of the frames with ⅜-inch staples.

Make brackets to hold the register in place. From ⅛ × ¾-inch aluminum bar stock, cut four pieces, each 5 inches long. In one end of each piece, drill two 3⁄16-inch-diameter holes—one at ¼ and one at ¾ inch. Countersink the holes, but be careful not to make them too large for #6 flathead wood screws. Make a 90-degree bend in the bracket 1½ inches from the end in which you drilled the hole. To position the notch in the bracket, first measure the distance from the surface of the interior

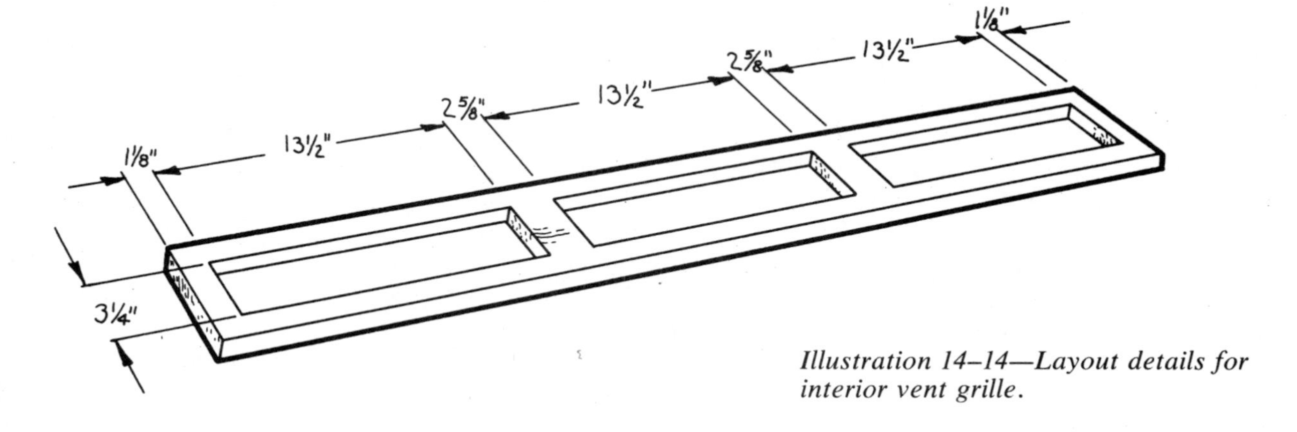

Illustration 14–14—Layout details for interior vent grille.

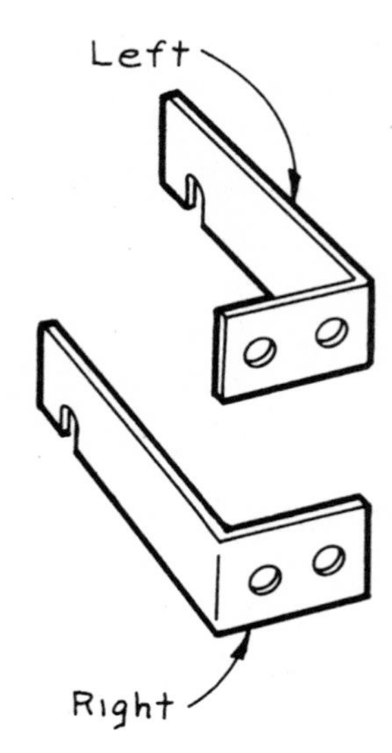

Illustration 14–15—Grille brackets.

wall to the lag bolts behind the vent door. Lay out the notch, to the same dimension plus ¼ inch, measured from the 90-degree bend. Note that the brackets are bent to fit the left and right sides of the register, as shown in illustration 14–15. Cut one ⅜-inch notch in the lower edge of each of the four brackets. On the back of the registers, attach the brackets at the correct height and spacing to match the lag bolts, using ½-inch #6 flathead wood screws.

To finish the registers, cut four pieces, each 48 inches long, and four pieces, each 5½ inches long, from ¼ × ½-inch screen molding. Miter-cut all the ends. Fasten the screen molding over the hardware cloth on the back of the registers, using ⅝-inch 16-gauge brads. You need to cut notches in the moldings to fit the brackets.

Position the lower register in its opening, and have a helper mark the location of the notches in the aluminum latches from outside the house. At the marked points, drill 3⁄16-inch-diameter pilot holes. Turn ¼ × 2½-inch-long lag bolts in these holes to serve as latches.

For the lower register only, cut three pieces of 2-mil polyethylene, each 4 × 13 inches. The polyethylene will serve as back draft dampers to prevent the siphoning system from reversing itself at night. When the air is cool in the air panel cavity, it will fall toward the lower register. If you do not have back draft dampers, the cool air will circulate into the house. Make sure the damper pieces cover the openings entirely. Staple the polyethylene only along the top edge, leaving the sides and lower edge free to float up when the air panel is working. See illustration 14–16, which shows the back draft dampers. To hold the registers in position over the wall openings, hook the brackets over the lag bolt latches attached to the studs.

Now, construct a subframe to hold the glazing for the air panel. Rip three 8-foot 1 × 2s to a width of 1 inch. Cut two of the boards 81½ inches long for the sides and use the third for two pieces, each 45⅞ inches long, for the top and bottom. Place the glazing subframe side-pieces in position against the side studs and use 8d finishing nails to hold them. Nail the top and bottom pieces to the center studs.

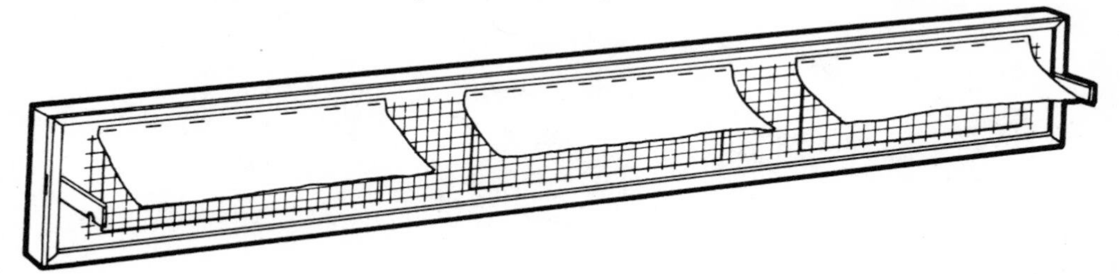

Illustration 14–16—Polyethylene back draft dampers on the back of the grille frame.

The frame which holds the glazing is constructed from extruded aluminum, which provides a tight fit around the double-thick glazing material. Refer to illustration 14–17 to see how the glazing frame fits the glazing and the subframe. With a hacksaw, cut two pieces of 7/8 × 1¾-inch aluminum, each 47⅞ inches long, and two pieces, each 81¼ inches long, and miter-cut all the ends. Drill five 7/32-inch-diameter holes in each sidepiece and three 7/32-inch-diameter holes in both the top and bottom pieces.

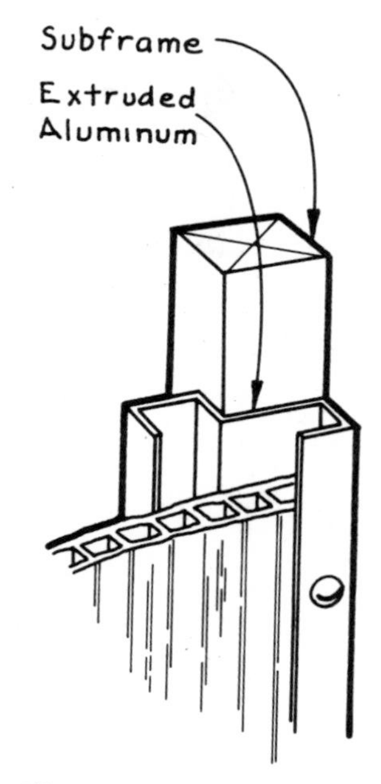

Illustration 14–17—Glazing detail.

At this time you are ready to paint again, using a primer and exterior house enamel to match the house siding. Paint the J channels at the edge of the siding, the glazing subframe, the glazing frame, the flashing, and the vent door assembly.

We have specified an insulated, double-walled acrylic glazing material. It is easy to handle and effective in controlling heat loss. With a circular saw or a saber saw, and using a fine-tooth blade, cut the glazing material 47¼ × 80¾ inches. Assemble the aluminum frame around the glazing. Using the predrilled holes in the frames as guides, drill 7/32-inch-diameter holes through the glazing. Set the glazing frame with the glazing into the exterior wall space, and mark the location of the screw holes. Also mark where the glazing is over the glazing support blocks, and pull the glazing away. Drill ⅛-inch-diameter pilot holes in the subframe, and 7/32-inch-diameter holes in the glazing where marked. Replace the glazing and frame, and fasten the assembly to the subframe on the bottom and sides with thirteen 2-inch #10 roundhead aluminum wood screws, and at the top with three 1½-inch #10 flathead aluminum wood screws. Drill ⅛-inch-diameter pilot holes in the glazing support blocks, and fasten the glazing at these four points with rubber or fiber washers, metal washers, and 1½-inch #10 roundhead aluminum wood screws. Paint the heads of the screws to match the frame.

The final construction detail is applying caulk to the perimeter of the glazing subframe where it meets the wall of your house. Use a good-quality silicone caulk to seal completely around the frame. Do not use caulk to seal the glazing to the glazing frame. The glazing frame will seal the glazing sufficiently and also allows the glazing to expand and contract with the change in temperatures. Do seal the mitered corners of the frame and any other seams that air or water will pass through.

Servicing the thermosiphoning air panel consists only of keeping the glazing clean and opening the exterior vent door in hot weather. Also be sure to keep the vent openings clear of obstacles. If one thermosiphoning air panel is useful, why not put in another?

15 Solar Basement Door

The solar basement door is a remarkably easy project with a lot of potential benefits. Within the space of your basement entrance, you can create a compact growing area. Like many good ideas, this concept is very simple. By replacing your present basement door with one that is glazed with two layers of clear acrylic, you achieve a small greenhouse. The existing basement steps make good shelves for

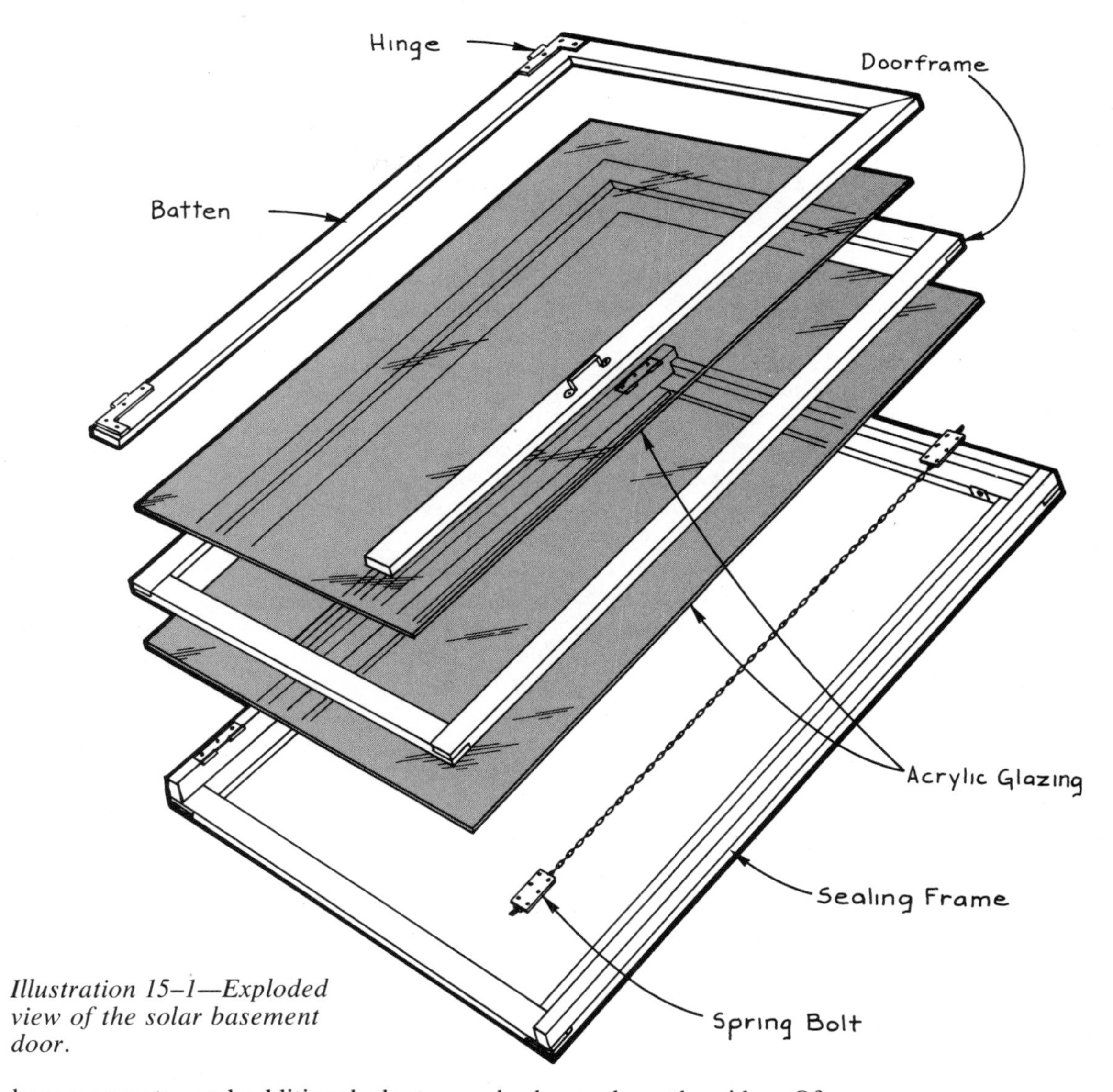

Illustration 15–1—Exploded view of the solar basement door.

boxes or pots, and additional plants can be hung along the sides. Of course, if you don't have a sunny basement entrance this project is not for you.

In warm weather the glazed door is left open during the day to prevent overheating. Sunshine heats the small greenhouse in cold weather. In addition, on cloudy days you can open the inner basement door so that the basement area can moderate the temperature in the greenhouse. You may find that you can grow "spring salads" all winter. Also, the basement-steps greenhouse could be the ideal spot to serve as a cold frame for starting seeds in early spring. On the coldest days, move the plants inside and close both cellar doors to conserve the basement heat.

When you are considering whether you have a suitable location for a solar basement door, keep the following points in mind. First, your door should face within 30 degrees of solar south. We discuss finding solar south in project 1, but in this instance a magnetic compass will give you a fairly accurate reading. Second, the solar basement door must not be shaded by nearby buildings or trees.

Your third question is whether to fit a new door into the present frame. The main consideration is a tight fit so that the warmed air in the greenhouse cannot leak out. The doorframe must form a tight seal with the sealing frame, and the sealing frame must be tightly sealed against the existing top of the entrance. You might need to change the way the door is hinged to fit an existing frame.

If you have a frame, but it is in poor condition, the best course is to replace it. Make any needed repairs to your entranceway structure at the same time. You need to have a firm surface on which to attach a new sealing frame. If the existing frame is in good shape and tightly fastened to the sides, leave it in place and make the solar basement door to fit within it. For security, the door is designed with self-locking catches that can only be opened from the inside.

An exploded view of the solar basement door is shown in illustration 15–1. Chart 15–1 is the materials list. A tools list, chart 15–2, follows.

CHART 15–1—
Materials

DESCRIPTION	SIZE	AMOUNT
	Lumber	
#2 Pine	5/4 × 2 × A – 3″	1
#2 Pine	5/4 × 2 × B	2
#2 Pine	1 × 4 × A	2
#2 Pine	1 × 4 × B	2
#2 Pine	1 × 4 × A – 3¼″	3
#2 Pine	1 × 4 × B – 1⅝″	4
#2 Pine	2 × 3 × B	1
Baluster Stock	¾″ × ¾″ × 8′	1

CHART 15-1—*Continued*

DESCRIPTION	SIZE	AMOUNT
Hardware		
Finishing Nails	4d	4
Aluminum Trim Nails	1″	20
#12 Flathead Wood Screws	1¾″	8
#10 Flathead Brass Wood Screws	1¼″	11
#8 Flathead Wood Screws	⅝″	8
#8 Flathead Wood Screws	1½″	12
#8 Panhead Aluminum Sheet-Metal Screws with Rubber Washers	¾″	42
Spring Bolts	4″	2
HL Hinges with Screws	7″	2
Utility Handle	4″	1
Masonry Anchors	¼″ × 1″	8
#18 Jack Chain	B − 14″	1
Miscellaneous		
Vinyl Tube Weather Stripping	¼″ × ½″	18′
Waterproof Wood Glue	...	4 oz.
Primer	...	1 pt.
Exterior Enamel	...	1 pt.
Latex Caulk	...	1 tube
Wood Putty	...	4 oz.
Acrylic Sheet	⅛″ × A − 3¼″ × B − 1⅝″	1
Acrylic Sheet	⅛″ × A − 8¼″ × B − 4⅛″	1

CHART 15-2—
Tools

Saber Saw
Circular Saw
Drill and Bits
Acrylic-Cutting Knife
Wood Chisel

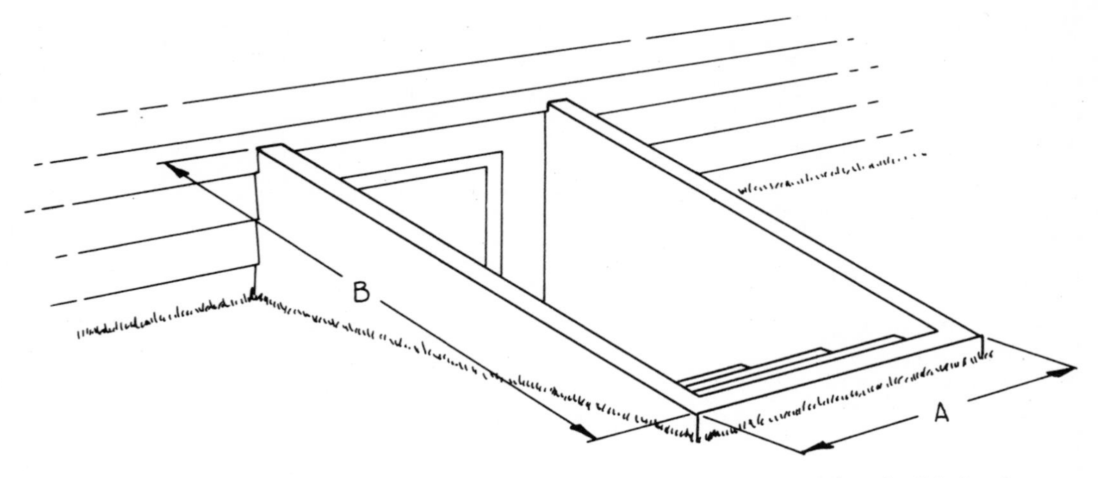

Illustration 15–2—The two measurements required to build the door are A, the width of the entrance, and B, the length of the sides.

Before you purchase materials for your new basement door, measure your basement entrance area. As shown in illustration 15–2, measure the surface where the sealing frame will be attached. Measurement A is the distance from one outside edge of the framework to the other outside edge; measurement B is the distance from the top to the bottom of the framework. The doorframe is smaller than the sealing frame by 3 ¼ inches from side to side and by 1 ⅝ inches from top to bottom. The bottom edges of the sealing frame and doorframe are flush. You can easily put the A and B measurements into the materials list formulas to calculate the size of the pieces you will need.

If you plan to keep your existing frame, measure within it to find the door dimensions. Allow ⅛ inch clearance on the top and ¼ inch from side to side. The bottom edge should be open.

The instructions for constructing the solar basement door begin with building a completely new sealing frame. If you do not need to do this, begin with the directions for the door itself.

The sealing frame is made of pine 1 × 4s, which you will fasten to your existing basement entranceway. The first step is to cut two pieces of pine to measurement A for the top and bottom. Next, cut two pieces equal to measurement B for the sides. They are joined at the corners with half-lap joints, as shown in illustration 15–3.

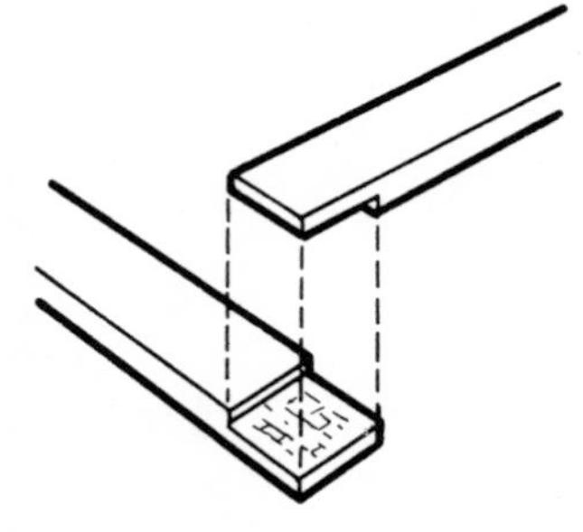

Illustration 15–3—Half-lap joint.

Make the half-lap joint by placing the four frame boards in position on a flat surface, with the ends overlapped at the corners. Square each corner and carefully mark where each board crosses the other, as shown in photo 15–1.

Photo 15–1—Lay out the frame pieces in a large, flat area, and mark the positions of the half-lap joints.

Disconnect your circular saw and set the depth guide to cut ⅜ inch, or half the thickness of the frame boards. Make the first pass with the saw just to the cut side of the line, and then make several more cuts across the half-lap areas. Remove the excess wood from the joint area with a wood chisel. Check the fit as you finish both pieces of a joint, trimming them if necessary.

On the sides and top of the sealing frame, attach a board as a stop, to make a tight fit around the new door and to mount the hinges. The stop on the hinge side is a piece of 2 × 3 ripped to 1¾ × 1½ inches. The hinges can be set on either side of the door, depending on your site. The other two stops are pieces of 5/4 × 2-inch pine, which has actual dimensions of 1⅛ × 1½ inches. Cut one piece of 5/4 × 2-inch pine and the piece of ripped 2 × 3 to measurement B for the side stops, and one piece of 5/4 × 2-inch pine to measurement A minus 3 inches for the top stop.

Fasten the sealing frame together with waterproof wood glue and 1½-inch #8 flathead wood screws. Assemble the frame on top of the stops. Scrap ¾-inch boards are needed to support the top and handle side. Line up the pieces with the corners square and the stops flush along the outside edges. The 2 × 3 is on the hinge side. Clamp the pieces in position. Drill and countersink two ⅛-inch-diameter pilot holes in each corner. Then drill three more holes evenly spaced on each side and two more along the top stop. Be careful not to drill completely through the unit. Glue the joints and the length of the stops, then fasten the pieces together with the 1½-inch #8 flathead wood screws through the lapped joints of the frame and into the stops.

Sand and paint the sealing frame unit. Use an oil-based primer and two coats of high-quality exterior enamel.

Drill four ¼-inch-diameter holes evenly spaced in each side of the sealing frame. Set the frame in position over the entry wall, and mark the location of the frame holes on the sill. Remove the frame. If your entry is wood covered, drill ⅛-inch-diameter pilot holes. If the entry is all masonry, drill ¼-inch-diameter holes with a masonry bit, and tap an expansion shield into each hole. Before fastening down the frame, apply a bead of latex caulk on the wood or concrete surface to seal the joint from water and air leaks. Set the frame in the caulk, and secure it with eight 1¾-inch #12 flathead wood screws.

At the top of the frame, where it joins the house siding, it is necessary to caulk or use flashing to prevent water from leaking into the stairway. As shown in photo 15–2, shape the flashing with a vise or clamps and two boards. Although different houses will call for a variety of solutions, the flashing must go up under the house siding whenever possible, and it should cover the siding-frame joint. Nail the flashing to the siding and the sealing frame with 1-inch aluminum nails.

Construction of the basement doorframe is very similar to the sealing frame. The doorframe is also 1 × 4 stock, which holds a double layer of acrylic glazing. Start by cutting two pieces, each measurement A minus 3¼ inches, for the top and bottom pieces. Then cut two pieces

Photo 15–2—Carefully shape the flashing for use at the top of the basement door, using a large vise or two boards clamped together.

of 1 × 4, each measurement B minus 1⅝ inches, for the door sides. Cut the single middle glazing support board from ¾ × ¾-inch baluster stock, measurement B minus 8⅝ inches.

Lay out the doorframe pieces as you did the sealing frame, and mark the ends of the boards for half-laps. It is important that the door fits into the frame, so measure carefully. Cut the half-lap joints. Assemble the doorframe, using waterproof wood glue and ⅝-inch #8 flathead wood screws. Place the doorframe within the sealing frame while you fasten the pieces together.

The middle glazing support, ¾ × ¾ × B minus 8⅝ inches, is toenailed into the center of the frame with four 4d finishing nails. Drive the nails in, then set them and fill the holes. Sand the door and paint it with a coat of primer and two coats of exterior paint.

The double glazing of the basement door is made of two pieces of ⅛-inch clear acrylic. The acrylic sheet material is light and can be cut with a saber saw fitted with a fine-tooth blade. If the acrylic has a protective paper cover, leave the cover on until you install the glazing, as it does scratch easily, but then remove the paper at once, because it will adhere to the plastic sheet when it is exposed to heat and sun. You can also use any translucent plastic glazing, but avoid glass, as it would be too vulnerable to breakage.

First cut the bottom piece of glazing to size, measurement A minus 3¼ inches by measurement B minus 1⅝ inches. As shown in photo 15–3, support the length of the glazing with 2 × 4s or 2 × 3s, adjusting

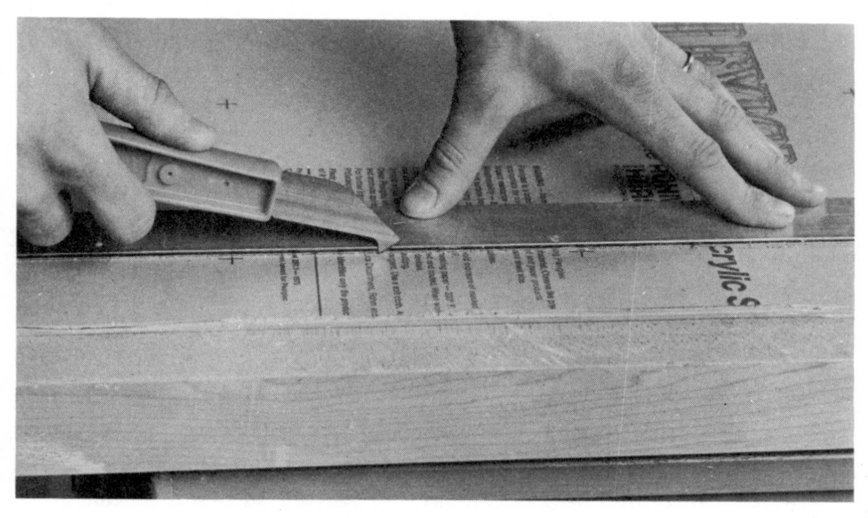

Photo 15–3—The best way to cut the glazing accurately and safely is with a straightedge guide and an acrylic-cutting knife, as shown.

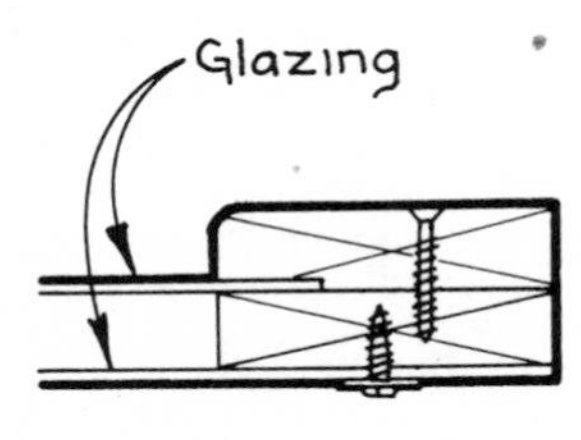

Illustration 15–4—Glazing detail.

them for the saw blade. Keep the saw close to a firm edge to minimize vibration, which could crack the acrylic sheet. Next, drill twenty 5⁄16-inch-diameter holes approximately 1 inch from the outside edge of the sheet and approximately 10 inches apart. These holes need to be oversize to allow for expansion and contraction of the acrylic. Set the glazing onto the frame temporarily and mark the holes. Remove the glazing, and drill 1⁄16-inch-diameter pilot holes in the frame. Apply a bead of latex caulk on the doorframe and set the glazing onto it, checking to see that you have a tight seal all around. Do not use silicone caulk on the glazing, since it will not allow the sheet to expand and contract. Hold the piece in place with ¾-inch #8 panhead aluminum sheet-metal screws fitted with rubber washers. Use faucet washers if screws with prefitted washers are unavailable. Tighten the screws carefully in the acrylic since the glazing should be snug but not so tight that it can't expand slightly.

Battens hold the top piece of acrylic to the doorframe and are also the finish trim of the door. The battens cover the edge of the glazing on the top and the two sides, as shown in illustration 15–4. The bottom of the glazing is flush with the bottom of the doorframe and has no batten. This allows water to run off the glazing rather than be retained by a batten. Two screws through the glazing and into the bottom edge of the doorframe through oversized holes in the glazing hold it from slipping out of the battens.

Cut two pieces of 1 × 4 pine equal to the length of the sides of the doorframe, and one piece equal to the width of the doorframe. Clamp the pieces one at a time to a workbench and prepare to cut a rabbet into one edge of each piece. The rabbet is 1¼ inches wide and ⅛ inch deep. Lay out the rabbet, and set the circular saw to cut ⅛ inch deep. Clamp or nail the board securely to the bench, and cut along the 1¼-inch line, then make several more passes in the rabbet area. The clamps will probably have to be moved in order to allow the saw to pass the complete length of the board. Use a sharp wood chisel to clear the rabbet. The rabbet does not have to be exact or finished in appearance, since it will be on the underside of the board and is necessary only to hold the acrylic glazing.

Miter the corners of the boards to 45-degree angles. Then bevel the inside top corners, if desired, to add a decorative touch to the door. The bevels are not necessary, however, and can be omitted. Set the battens on the doorframe to check the fit, and trim where needed. Finish-sand the battens and apply a primer coat. The finish coats of paint can be applied after the battens are fastened to the doorframe.

When the primer has dried, lay the battens in place so that pilot holes can be drilled to fasten the battens to the doorframe. Clamp the battens before drilling so that the holes remain aligned, or drill the first hole in each piece, insert a screw, then drill the remaining holes. Use a countersink bit for 1¼-inch #10 brass flathead wood screws. Drill at least four holes in each sidepiece and three in the top piece. Keep the holes in the outside 2 inches of the battens. Do not drill within the rabbeted areas.

With the battens in place, mark the area for the acrylic sheet. Measure the rabbeted area of the battens, and transfer the marks to the doorframe. Mark the rabbeted area completely around the sides and the top of the doorframe.

The acrylic sheet must have room under the battens to expand and contract, and to move if the door should twist, otherwise the sheet will break under pressure. Measure the distance across the doorframe in the rabbeted area, then subtract at least ½ inch. The ¼ inch each side of the sheet will be sufficient for the expansion and contraction. Allow ¼ inch at the top of the sheet, measured from the bottom edge of the doorframe to the rabbet line. The estimated coefficient of expansion for ⅛-inch acrylic sheet material is 1/32 inch per linear foot of material.

Cut the acrylic sheet to the determined size with a fine-tooth blade, either with a handsaw or a circular saw.

File the edge of the sheet to remove burrs and to put a slight bevel on the edges. Lay the sheet in position, and set the battens in place. Drive in the batten screws until the battens are snug to the sheet. Fill the countersink holes in the battens with wood putty and sand the battens smooth. Install two #8 panhead aluminum sheet-metal screws with rubber washers along the bottom edge of the glazing to keep the sheet from slipping out. Apply two coats of exterior enamel to the battens, and you are ready to install the hardware.

Now you are ready to mount the door in its frame. Use two 7-inch HL hinges (so called because of their shape). With the door in position, set the hinges in place, as shown in illustration 15–5. Be sure the hinges are on the side of the sealing frame with the ripped 2 × 3. Center them in line with the top and bottom of the doorframe pieces. Mark the screw holes and drill pilot holes. Fasten the hinges with the screws provided. You might want to damage the screw slots, after the screws are seated, for security reasons. Cut the screws off if they come through the frame. Then mount a 4-inch utility handle on the door.

For security reasons, use two 4-inch spring bolts that automatically lock the door closed so that it can only be unlocked from the inside of the stairwell. Mount the spring bolts at the top and bottom of the door on the underside, as shown in illustration 15–6. The bolts usually come packaged with screws. The bolts are just within the door side at a point where they will extend into the top and bottom frame pieces. Within the sealing frame, drill two holes to receive the bolts, and use a wood chisel

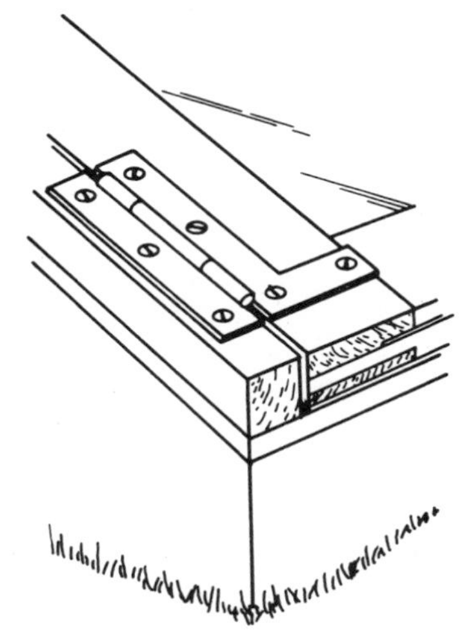

Illustration 15–5—Hinge position.

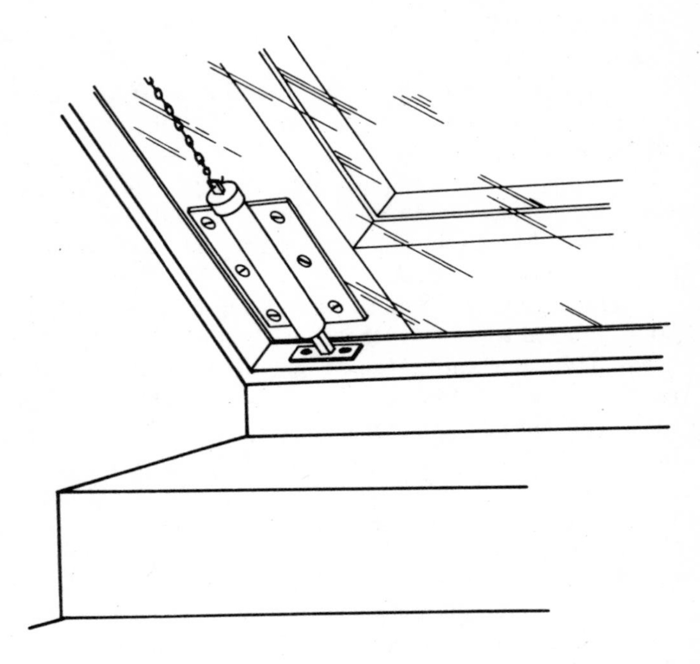

Illustration 15–6—Spring bolt lock on bottom interior of door. Another lock should go at the top of the door.

to mortise a space deep enough to hold the striker plates. Fasten each plate so that it lines up with the spring bolt. Test the catch and the fit from outside to be sure the bolts are securely in place with the door shut. Then by opening, then reclosing, the links with pliers, attach a release chain between the bolts to operate them together.

Install vinyl tube weather stripping so the door will close against it around the perimeter of the sealing frame. Place the weather stripping so that it won't interfere with the panhead screws on the door.

Now you have a place to start seeds, raise your winter greens, or simply enjoy flowering plants. You will need to experiment with temperatures, adjusting the outer and inner basement doors in various weather situations. We recommend painting the inner door white or another light color. With the inner door closed, there will be a good deal of reflected light on your plants, and heat will build up in your greenhouse, which masonry steps will absorb and store to some degree. This heat will help to keep the greenhouse warm at night, but you may also wish to cover the glazing with insulation in months of most severe cold.

16 Solar Cooling Chimney

It is possible to cool your house with heat generated by the sun, although that might sound like a contradiction. The solar cooling chimney is a device that passively cools a house without fans or other mechanical devices. It works like this: The sun's heat warms air in a rooftop collector; a large vent above the collector releases the heated air; room air is pulled into the collector; and cool air is pulled into the house from a low, north-side window or doorway. Your house, then, becomes a thermosiphon to expel hot air.

The collector is a large, rectangular box installed within the rafters of your home. Illustration 16–1 is an exploded view of the cooling chim-

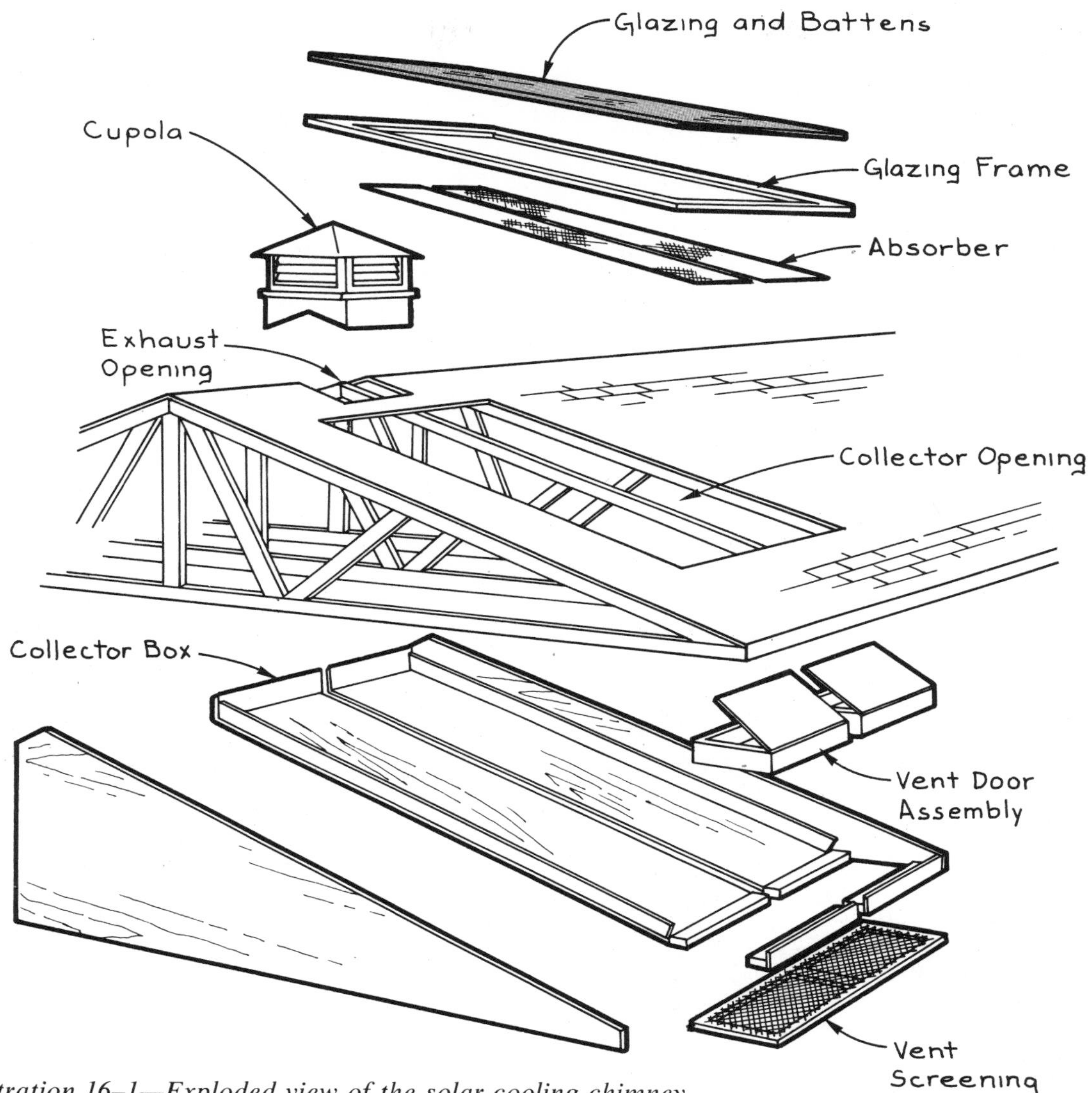

Illustration 16–1—Exploded view of the solar cooling chimney.

ney. Metal lath absorbers raise the air temperature inside the box.

The collector box is the "motor" of the cooling system. As air in the collector box warms and rises, it pulls house air through the ceiling vents at the bottom of the collector box.

Two factors produce the cooling effect. First, air movement is started, which increases evaporation and cools by the wind chill factor. Second, warm air is exhausted through the cupola and is replaced with cooler

CHART 16–1—
Materials

DESCRIPTION	SIZE	AMOUNT
	Lumber	
A-C Interior Plywood	¾″ × 2′ × 4′	1
A-C Exterior Plywood	½″ × 4′ × 8′	3
#2 Pine	2 × 4 × 8′	2
#2 Pine	2 × 3 × 8′	2
#2 Pine	5/4 × 4 × 8′	1
#2 Pine	5/4 × 2 × 8′	1
#2 Pine	1 × 4 × 10′	1
#2 Pine	1 × 3 × 10′	7
#2 Pine	1 × 2 × 2′	1
	Hardware	
Galvanized Common Nails	16d	25
Galvanized Common Nails	10d	4
Galvanized Common Nails	6d	26
Aluminum Flat Stock	⅛″ × 2½″ × 17′	1
Aluminum Flashing	7″ × 26′	1
Aluminum Mesh	18″ × 2′	2
Finishing Nails	6d	26
Aluminum Nails	1¼″	½ lb.
Aluminum Nails	1″	40
Cement-Coated Box Nails	6d	16
#12 Roundhead Wood Screws	1½″	2

CHART 16–1—*Continued*

DESCRIPTION	SIZE	AMOUNT
Hardware—*continued*		
#8 Flathead Wood Screws	½″	2
#8 Panhead Screws	1″	20
#6 Flathead Wood Screws	1¼″	73
Wire Staples	⅞″	35
Staples	⅜″	32
Continuous Hinge with Screws	1½″ × 21″	2
Screw Eyes	⅜″	6
Pulleys	½″ × 1¼″	2
Miscellaneous		
Cupola	24″ × 24″ (height optional)	1
Tempered Patio Door Replacement Glass	¼″ × 46″ × 76″	1
Waterproof Wood Glue	...	4 oz.
Primer	...	1 pt.
Exterior Enamel	...	1 pt.
Flat Black Absorber Paint	...	½ gal.
Silicone Caulk	...	3 tubes
Roof Cement	...	1 tube
Nylon Cord	¼″ × 8′	1
Butyl Glazing Tape	¼″ × ½″ × 34′	1
Expanded Wire Plaster Lath	22″ × 8′	2

CHART 16–2—
Tools

Tools
Backsaw
Circular Saw
Drill and Bits
Router
Caulking Gun
Tin Shears
Wood Chisel
Wrecking Bar

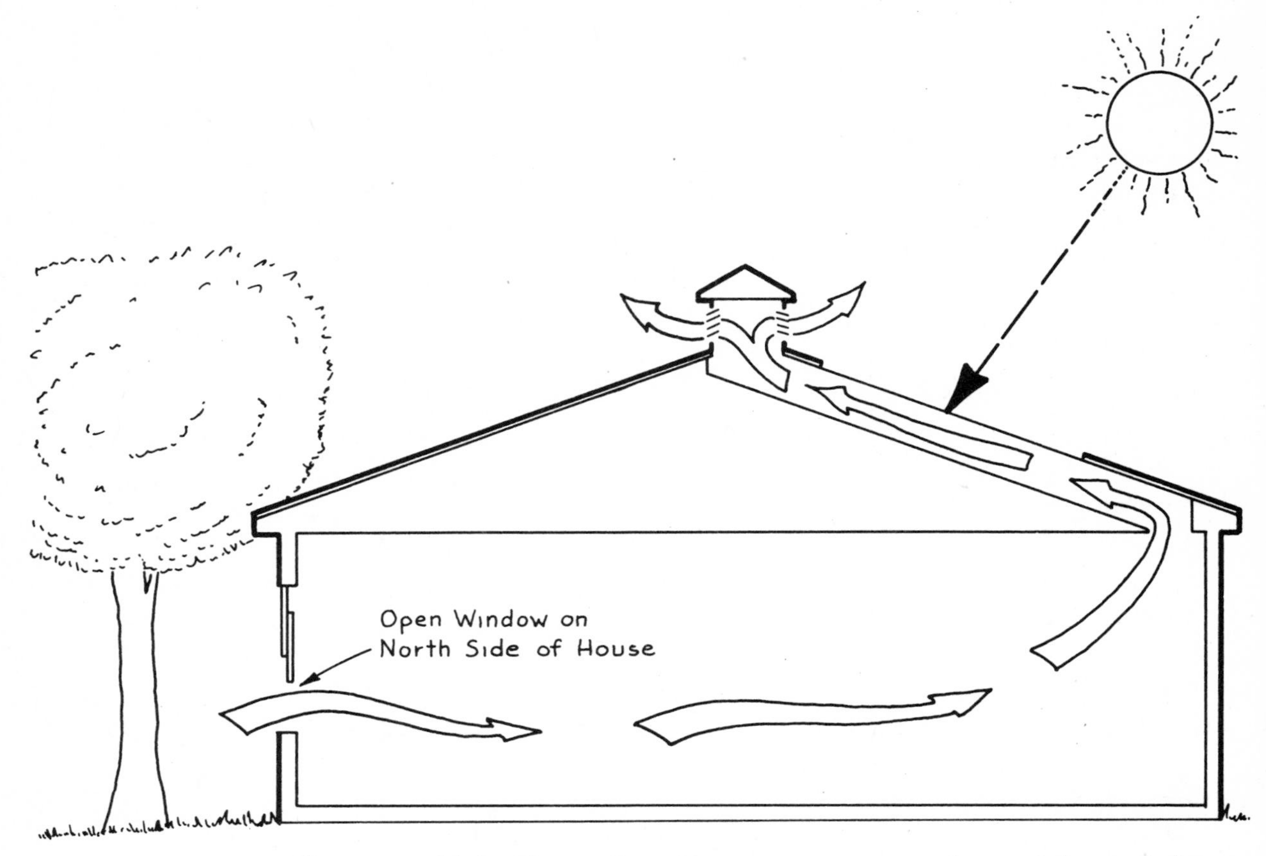

Illustration 16–2—The cooling chimney operates on a thermosiphon principle.

air through a north-side window or door. Illustration 16–2 shows the air movement through the house and cooling chimney.

Most solar chimneys are attached to the side or roof of a house in an upright position. Our chimney is mounted flush in a section of south-facing roof so that it does not appreciably alter the appearance of the house. The exhaust opening is topped with a purchased cupola, which can be found in a style to complement your home. Within the house, two small ceiling vents close off the cooling chimney when it is not needed.

The solar chimney is particularly easy to install in new construction, or it can be fitted into an existing roof. The glazing is standard sliding patio door replacement glass and can be a "second," or flawed piece, as it is not easily seen after installation.

The instructions given here are for a house with truss rafters spaced 24 inches on center. You will need to alter some of the dimensions if your house construction differs. The design and performance of the cooling chimney will not be appreciably altered, and the solar chimney can be adapted to fit almost any roof construction. Note the warnings in the text when dimensions must be changed. It is necessary to use standard patio door replacement glass, however, and you may need to "pack out," or add lumber, to rafters so that you have a continuous surface on which to mount the glass and a good support for the plywood collector sides.

A purchased, prebuilt cupola is the topmost part of the solar cooling chimney. Of course, a home-built cupola will work equally well. Installed on the ridge of your house in the center of the area above the collector box, it will allow the warm air to flow out of the top of the collector box and protect the opening from rain and snow. Install the cupola, following the manufacturer's recommendations, at the same time the collector box opening is made.

The materials needed are listed in chart 16–1 and the tools in chart 16–2.

Begin building the chimney with a frame to fit around the glazing. The frame is set into an opening cut into the roof and sealed to be weathertight. The glazing, a single layer of ¼-inch tempered glass, measures 46 × 76 inches.

Cut the top piece of the glazing from 49½ inches long from a standard 2 × 4. Cut two 79½-inch sidepieces from two 2 × 3s. The bottom piece and the middle glazing support are both made from 5/4 pine, which actually measures about 1⅛ inches thick. From 5/4 × 4 stock, cut the bottom piece 49½ inches long, and from 5/4 × 2 stock, cut the middle glazing support 72½ inches long. *Note:* Rafters set 16 inches on center call for a second 72½-inch glazing support.

Using a router or a circular saw, cut a rabbet ⅜ inch deep by 1 inch wide along one edge of the two sidepieces and along the top piece, as shown in illustration 16–3. A straightedge guide clamped to your bench will help make a straight cut. Also be sure to clamp the board securely before cutting.

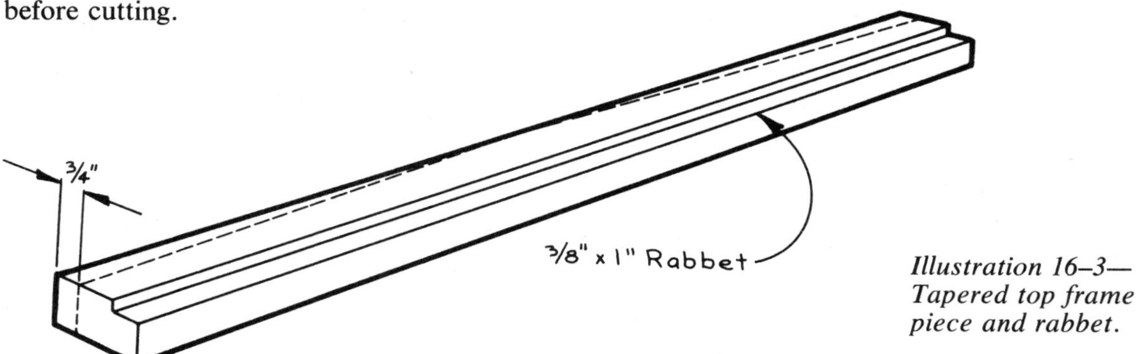

Illustration 16–3—Tapered top frame piece and rabbet.

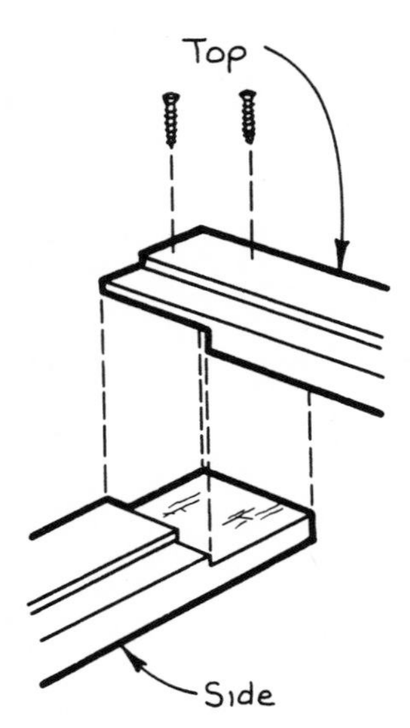

Illustration 16–4—Top corner.

Mark a taper on the top piece of the glazing frame, as shown in illustration 16–3. The taper will help rainwater flow around the frame. Mark ¾ inch from the top edge at each end of the board, draw lines from the top center of the frame to the ¾-inch marks, and cut the taper.

Lay out a half-lap joint at each corner to join the four frame pieces. Set the top piece over the sidepieces with the ends overlapping, the corners square, and the rabbets to the inside. Carefully mark where each board crosses the other. Set a circular saw to cut half the thickness of the framing boards. Clamp a marked board to your bench and cut just to the waste side of the line, then make several more passes across the half-lap area and remove the excess wood with a wood chisel. Be sure to reset the saw to 9⁄16 inch before cutting the half-lap in the bottom piece (see illustration 16–5).

Check the fit of all joints before assembling the grazing frame. Use a square to check that the corners are at right angles. Drill 1⁄16-inch-diameter pilot holes and countersink the screw holes. Fasten the frame pieces together, as shown in illustrations 16–4 and 16–5, with waterproof wood glue and 1¼-inch #6 flathead wood screws.

The middle glazing support fits into the frame so that it will rest on the center rafter and support the glazing. Toenail the middle support (or supports for 16-inch rafter spacing) into the glazing frame with 6d galvanized common nails. The support is ⅜ inch below the top edge of the frame, flush with the surface of the rabbet. Check the fit of the glass in the frame and make adjustments in the rabbets if necessary.

Paint the glazing frame with primer and two coats of exterior enamel.

From inside the attic, mark the position for the openings in the roof for the collector frame and cupola. The collector frame opening is 44½ inches wide and 72½ inches long. Position the top of the frame approximately 8 inches below the cupola opening. The cupola opening is dependent on the size of the cupola purchased. Drive a large nail through the roof at each corner of the intended openings for both the collector and cupola; the nails should be 1 inch to the side of the rafters so that the frame and cupola sides rest directly over the rafters, as shown in illustration 16–6. A drill can be used for slate roofs instead of the nails to avoid breaking the slate. Note the 1 inch of sheathing extending over the rafters on the sides of the opening.

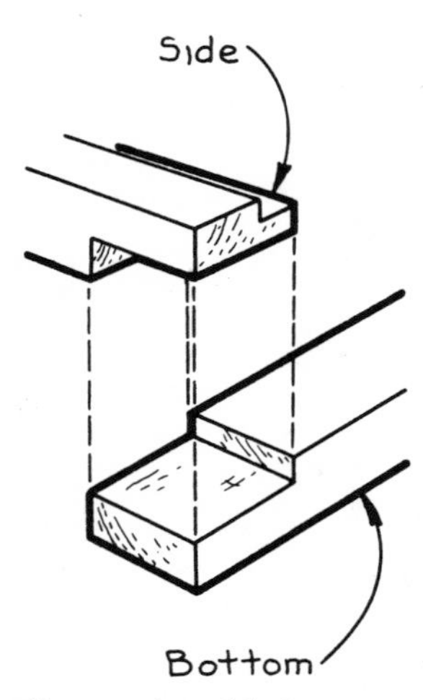

Illustration 16–5—Bottom corner.

On the outside of the roof, center the collector frame over the four nails or marking holes. Use chalk or a chalk line to draw a line on the roofing material ¼ inch from the outside of the frame. Remove the frame and cut away all the top roof-coating material, but not the wood sheathing, within the chalked line, using a cutter suitable for your roof.

Draw a line from nail to nail or from hole to hole with a chalk line. Within this second line is the area of roof sheathing that is to be cut out and removed. Set the circular saw to cut just the depth of the sheathing. Cut along the chalk line with enough overlap at the corners to free the sheathing. From the discarded sheathing, or with wood strips, cut a filler strip to place over the center rafter (or two rafters if they are 16 inches on center).

Cut the opening for the cupola in the same way, or place the cupola on the roof peak and cut the opening according to the manufacturer's instructions. Clean the areas by pulling nails, staples, and other obstacles from the rafters.

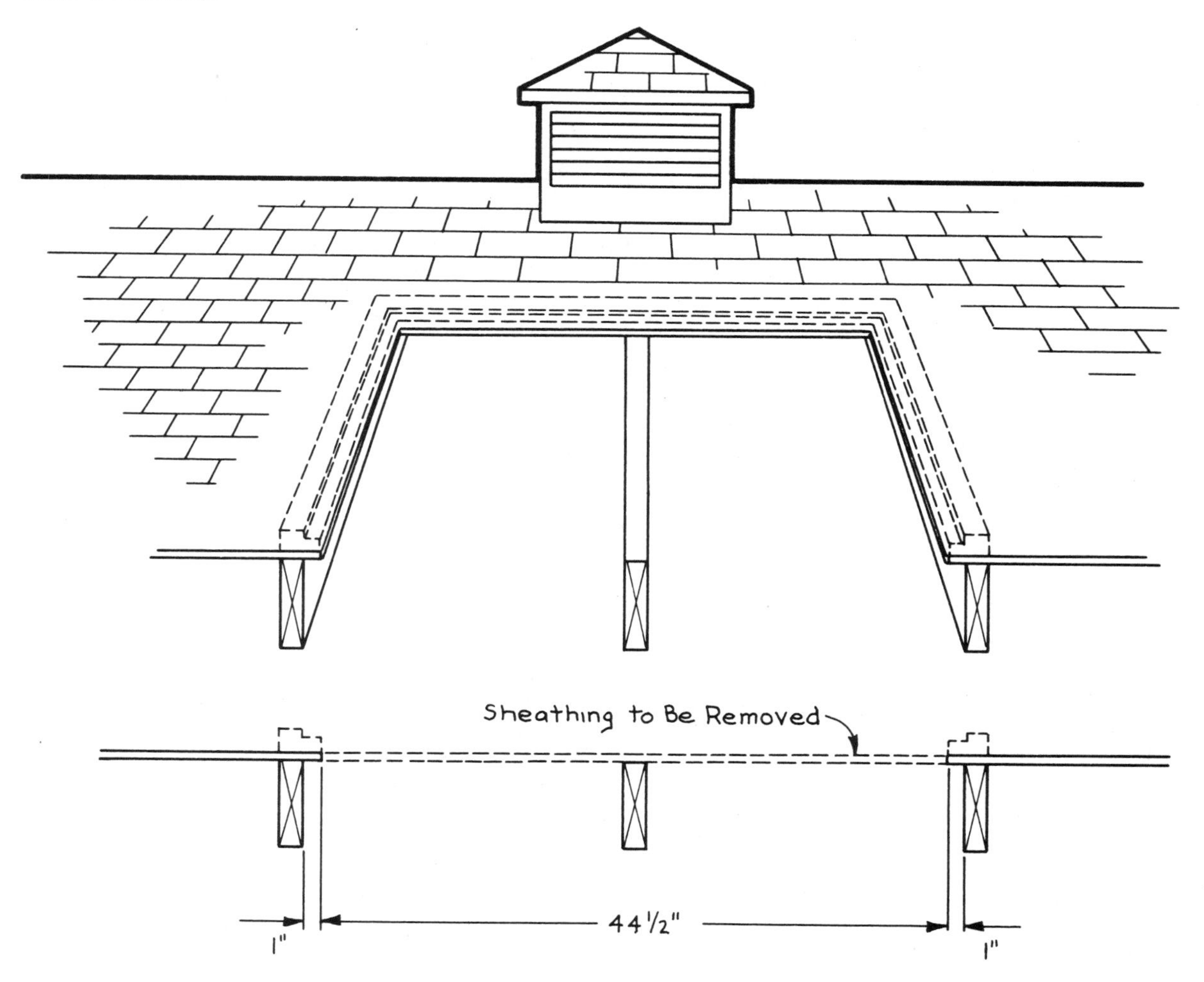

Illustration 16–6—Glazing opening to be cut out of roof sheathing and position of the glazing frame.

Place the frame gently over the opening. Block the frame away from the roof, temporarily, just high enough to apply a bead of silicone caulk under the frame. Remove the blocks and fasten the frame to the roof sheathing and rafters with 16d galvanized nails.

Roof flashing covers the joint between the glazing frame and the roof. Proper installation at this time will mean years of leak-free use, so it is important to follow these steps. The size of the flashing used is dependent upon the type of roof surface. Roll roofing materials require longer pieces of flashing than do standard roof shingles. Both types require step flashing. Step flashing means every shingle or layer of roll roofing has an individual piece of flashing to repel water effectively, as shown in illustration 16–7.

Cut enough pieces of flashing for the sides to match the number of rows of shingles or roll roofing in the glazed area. Cut the pieces 10 × 11½ inches for shingles, or 10 × 36 inches for roll roofing. A few short pieces will have to be measured and cut to finish rows that are above or below the top or bottom of the glazed area. Also, cut two pieces 10 × 51½ inches each for the top and bottom of the glazing frame.

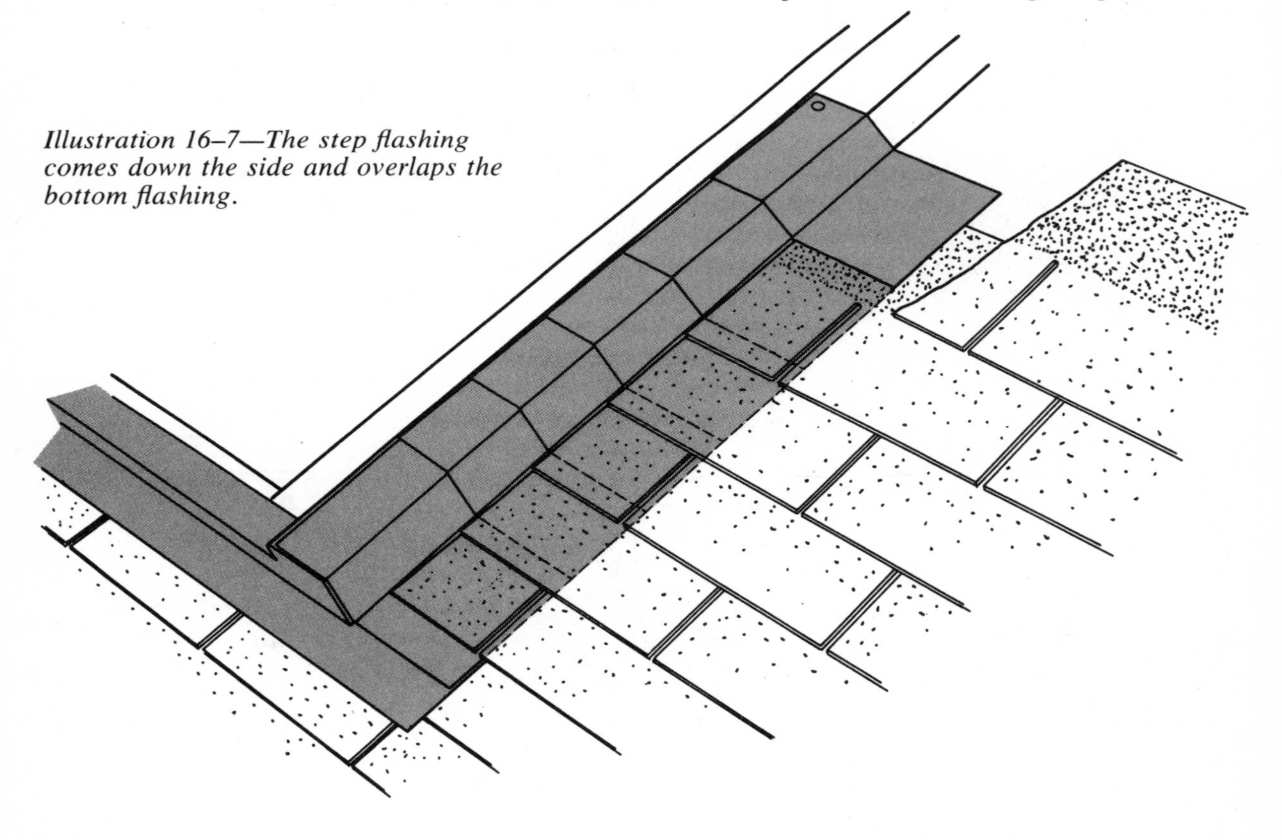

Illustration 16–7—The step flashing comes down the side and overlaps the bottom flashing.

Bend the flashing accurately by clamping the pieces between two boards, as shown in photo 16–1. Use a third board to press the flashing over the edge of the clamped boards. The pieces cover the top of the frame (up to the rabbets on the sidepieces), bend to cover the edge of the frame, and bend again, but in the opposite direction to extend over the roof.

At the lower edge, set the piece of bottom flashing over the frame and over the roofing materials as shown in illustration 16–8. Cut the flashing to fit at the corners, and seal the joints with roof cement. Nail the flashing to the frame with 1-inch aluminum nails.

Continue the flashing up along the side of the frame with the pieces of side flashing, pulling nails from the roofing material as needed. Overlap the bottom flashing with the side flashing, as shown in illustration 16–7. Nail all the pieces of flashing to the frame with one 1¼-inch aluminum nail in the upper corner of each flashing piece.

The absorber box is built next, since access is easier without the glass in place. The construction may vary somewhat depending on your rafter slope and construction. Our instructions are for trussed rafters, which are assembled in one plane. Truss roofs are the easiest to deal with, and additional considerations may be required with other roofs. Overlapped rafters and ceiling joints will need to be filled out so that you can fasten the absorber sides to a flat surface.

Photo 16–1—Shape the step-flashing pieces for the sides between two clamped boards, or have the flashing bent at an aluminum siding supplier.

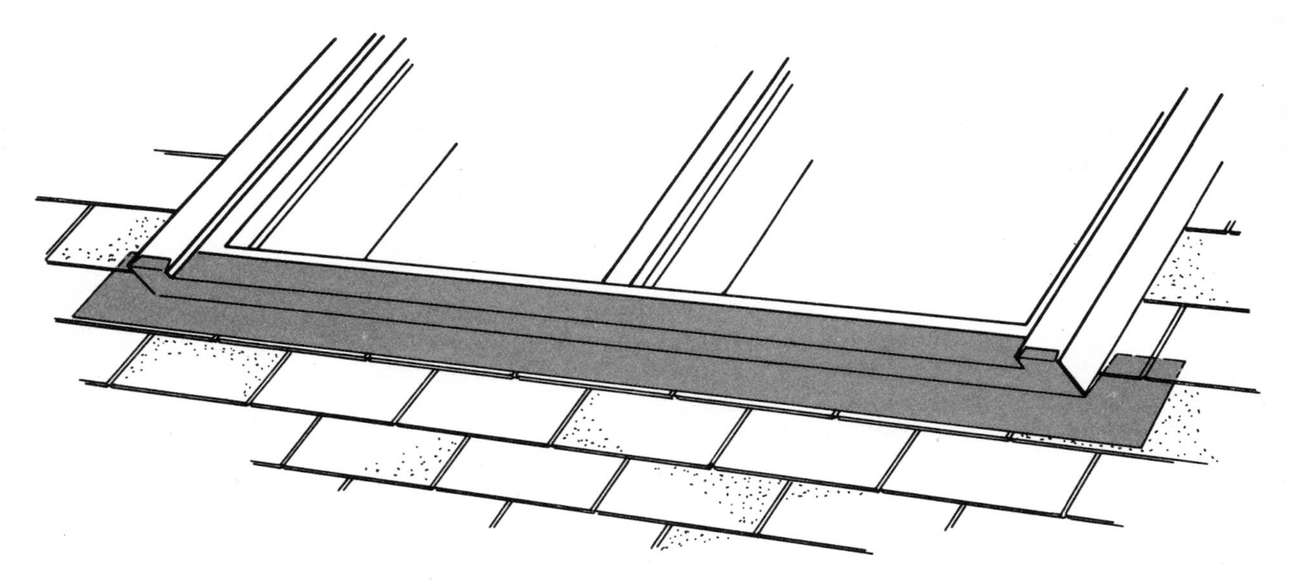

Illustration 16–8—The bottom flashing goes on first and overlaps the shingles.

Measure and cut two tapered pieces of ½-inch plywood for the sides of the collector box. These are labeled *A* in illustration 16–9. The collector box is 12 inches deep measured from the roof sheathing to the collector bottom. The sidepieces extend below the collector box to the bottom of the trusses to give better support. Measure and cut the sidepieces to fit your roof. Then cut two pieces, labeled *B*, tapering from the top of the sidepieces, to fit the remainder of the space under the cupola. Two more sidepieces, labeled *C*, are necessary to fill in the area near the vent doors. Cut these to fit, also. *Note:* The collector box must stop 12 to 18 inches short of the end of the rafters or it will be impossible to nail in the endpieces. Cut plywood endpieces, labeled *D*, to fit from the top of the rafters to the bottom of the collector box bottom. Then, cut two pieces for the box bottom, shown in illustration 16–1, page 207, to fit between the rafters the length of the collector box, and two additional pieces for the upper end of the collector box. The bottom and top pieces are also installed later. The size of the pieces is dependent upon the length and width of your collector box and the 12-inch depth of the collector box.

The collector box bottom is supported by 1 × 3s nailed to the plywood sides. Cut the 1 × 3 collector box bottom supports to fit. Rafters on 24-inch centers require four pieces; 16-inch-center rafters require six pieces. In order to increase heat absorption in the collector box, paint all the surfaces that will be inside the box with two coats of flat black absorber paint.

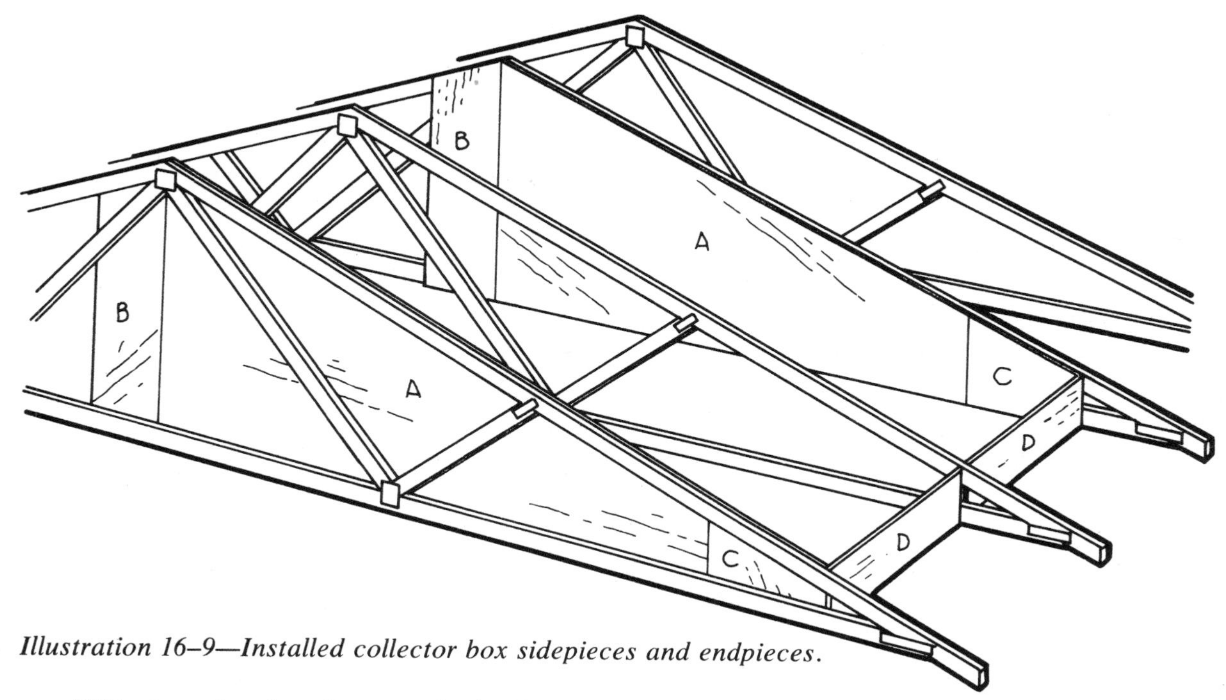

Illustration 16–9—Installed collector box sidepieces and endpieces.

With the painted surfaces to the inside, nail the plywood sides to the rafters and ceiling joists, as shown in illustration 16–9. This will provide a solid surface on which to nail the other pieces of the collector box.

Next, for the lower end of the collector box, build vent door assemblies to close and open the chimney. Build the frames first, then the doors, and then install the assemblies. The doors close against jambs made from 1 × 3s. For rafters set on 24-inch centers, cut four 22½-inch pieces of 1 × 3 for the long sides and four 16-inch pieces for the short sides to make two frames. For rafters set on 16-inch centers, measure and cut six 14½-inch pieces and six 16-inch pieces of 1 × 3 to make three frames. As shown in illustration 16–10, nail the jambs together with the short pieces butted into the longer ones to fit between the rafters. Use 6d cement-coated box nails.

Next, cut pieces of ¾-inch plywood for the vent doors. The width is 17½ inches, and the length should be cut to fit the vent frame. Use primer and two coats of interior paint to finish the jambs and the doors. Again referring to illustration 16–10, attach the doors to the jambs with a continuous hinge cut to the width of the doors. Centered in each door, and approximately 12 inches from the hinge, screw a ⅜-inch screw eye into the upper surface. This will hold the cord to raise the vent doors.

Cut 2 × 4 headers to fit between the joists to support the vent doors. Install the headers with their faces parallel to the floor, between the

To Fit
16"

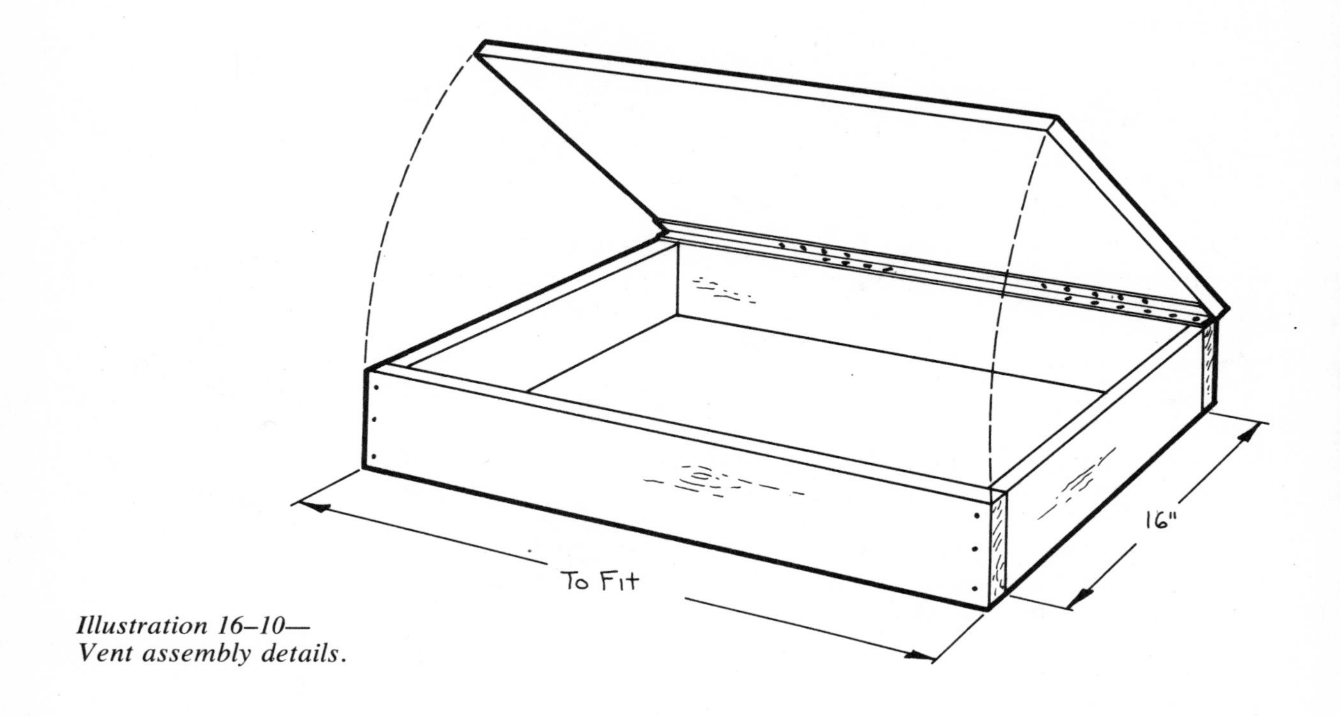

Illustration 16–10—
Vent assembly details.

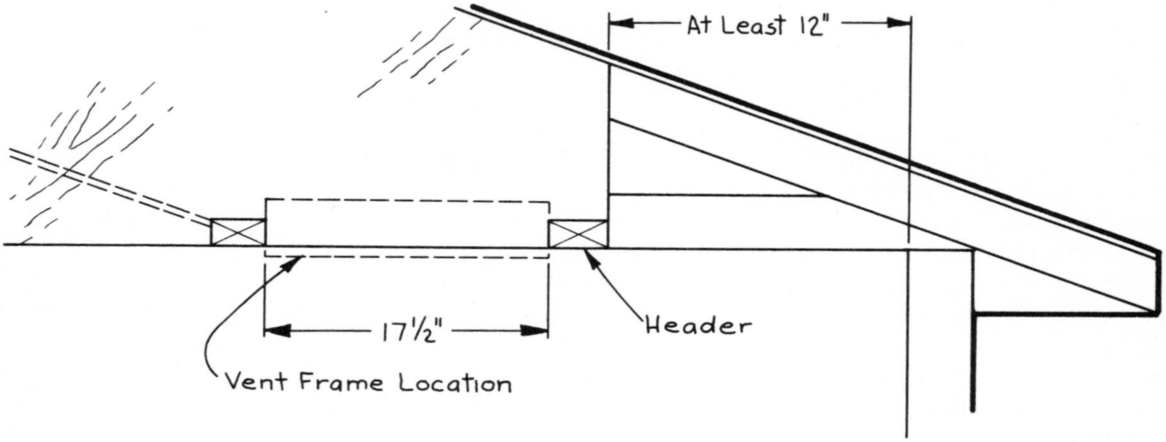

At Least 12"
17½"
Header
Vent Frame Location

Illustration 16–11—Vent and header positioning within the rafters.

collector sides and the center ceiling joist, at the lower end of the collector box. The lower headers are flush with the end of the collector sides. Allow a 17½-inch space between the headers to fit the vent assemblies. Nail through the outer rafters into the headers with 16d common nails, and toenail the headers to the center rafter or rafters with 6d common nails, as shown in illustration 16–11.

Install the endpieces at the lower and upper ends of the collector box. Nail them to the collector sides and to the headers with 6d common nails. Nail the complete vent assemblies to the headers and to the ceiling joists with 6d finishing nails. Be sure to fit the frames ½ inch below the joist to match the ½-inch drywall, or whatever material you wish to use, to finish the ceiling. This will give a finished edge to the vent openings. Use a scrap of ½-inch plywood held against the joist to help you keep the spacing. Also be sure that the door hinges are on the south side of the opening, as shown in illustration 16–12, so that when the doors are raised they will not block the airflow into the collector.

The system for raising each vent door uses a small pulley, a cord, and simple latch blocks, as shown in illustration 16–12. Drill a ¼-inch hole in each header farthest from the exhaust vent to hold the cords. Above each vent door, near the hinges, install a 1-inch pulley in the roof sheathing with ½-inch #8 flathead wood screws. Cut two pieces of ¼-inch nylon cord, each 48 inches long. Tie one end of each cord to the screw eye in each door and thread it through a pulley and down through the header. Latch blocks are constructed later to hold the door in the open position.

The absorber plates are made of expanded wire lath. Using tin shears, cut the lath the length of the collector box and a width that will

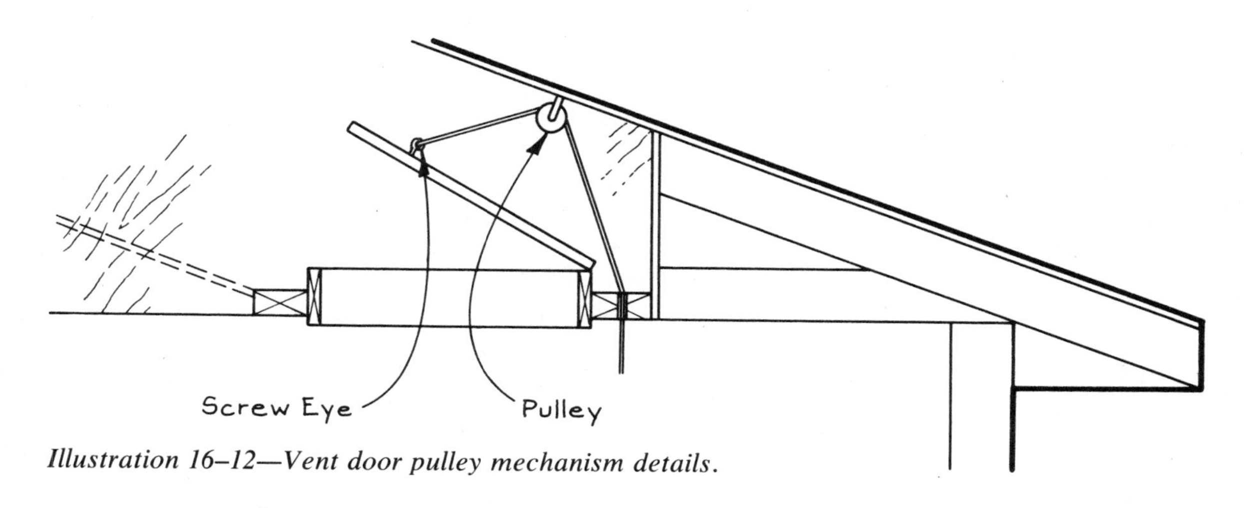

Illustration 16–12—Vent door pulley mechanism details.

fit between the rafters. Position the plates at the lower ends of the collector box bottom supports so that the absorbers will be directly under the glazing. As shown in photo 16–2, staple the absorber plates to the top edge of the supports with ⅜-inch wire staples.

Cut bottom pieces of ½-inch plywood to fit over the rafters and from end to end of the collector box. The sheet edges meet on the center of the rafters. Paint the inside surfaces of the sheets with flat black absorber paint and nail them to the rafters. At this time, make sure all other pieces inside the collector box are also painted flat black, since the collector opening will be accessible only from the glazing opening.

Move to the roof once again, and carefully install the glass in the glazing frame. First apply ¼ × ½-inch butyl glazing tape in the rabbet on all sides and on the center support. Lifting the glass to the roof and setting it in place usually requires several helpers. Lay the top edge of the glass on the glazing tape in the rabbeted top frame first. Then, with one worker on each side of the glass, gently lay it in position.

Aluminum battens, shown in illustration 16–13, finish the glazing details. The ⅛ × 2½-inch aluminum flatstock can be replaced by 1 × 3 wood battens if you prefer. Use 1¾-inch screws with the wood battens, and prime and paint the battens before installing them on the glazing frame.

Cut a 49½-inch batten for the top of the frame, and two lengths, each 76 inches, for the sidepieces. Drill 3/16-inch-diameter mounting holes precisely 1 inch from the edge of the battens to avoid having the screws

Photo 16–2—Fasten the absorber pieces in place with wire staples.

touch the glass. Space the holes every 10 inches along the sidepieces. Drill a hole 1 inch from each end. Center a hole 24¾ inches on the top piece, and then drill holes 8 and 16 inches from the center hole in both directions. Also add holes 1 inch from each end on the top piece.

Align the battens to the glazing frame, and drill 1/16-inch-diameter pilot holes in the frame. Fasten the battens in place with 1-inch #8 panhead screws. Apply a small amount of silicone caulk to each screwhead just before tightening them to the battens. Apply a bead of silicone caulk to seal along the battens on both the glazing side and the outside of the battens. The roof details are now completed, and you can move inside to finish the job.

Fit, cut, and install 1 × 4 filler pieces over the middle joint in the collector bottom so it will be airtight. Caulk all cracks and joints that may be open or leak air into the collector box with silicone caulk. Use fiberglass batts to insulate the ceiling around the vents and collector box. It is important for the warm air in the collector to move out of the house and not radiate back into your living space.

The vent opening is covered with a 1 × 3-framed grille. Make this to fit the vent jambs. Cut two pieces, each 49 inches long, for the sides,

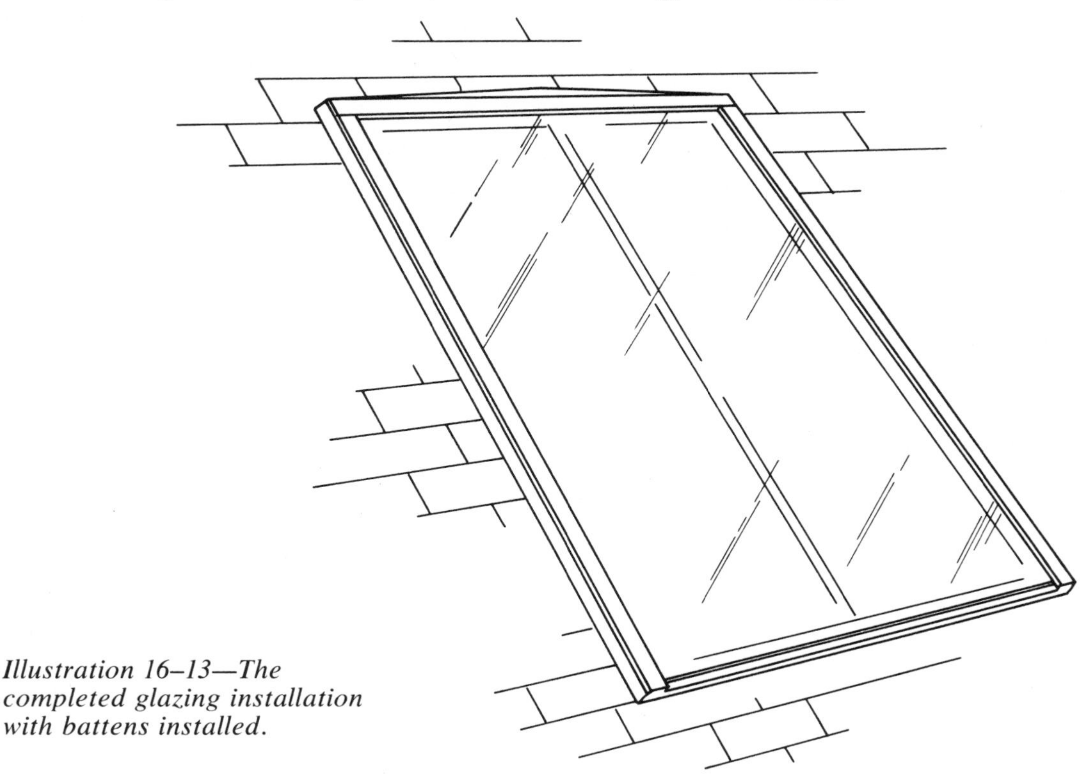

Illustration 16–13—The completed glazing installation with battens installed.

Photo 16–3—The vents should be finished with mitered molding, cut with a backsaw.

and two pieces, each 21 inches long, for the ends. At each corner, cut a miter joint, as shown in photo 16–3. Using wood glue and 6d finishing nails, fasten the frame pieces together. Cut a 1 × 4 board 16 inches long (two for 16-inch centered rafters) for the divider of the vent grille. Glue and nail this board on its face in the center of the grille face, as shown in illustration 16–14. Using a router, cut a ¼-inch radius on all the edges on the side of the frame that will be facing your room.

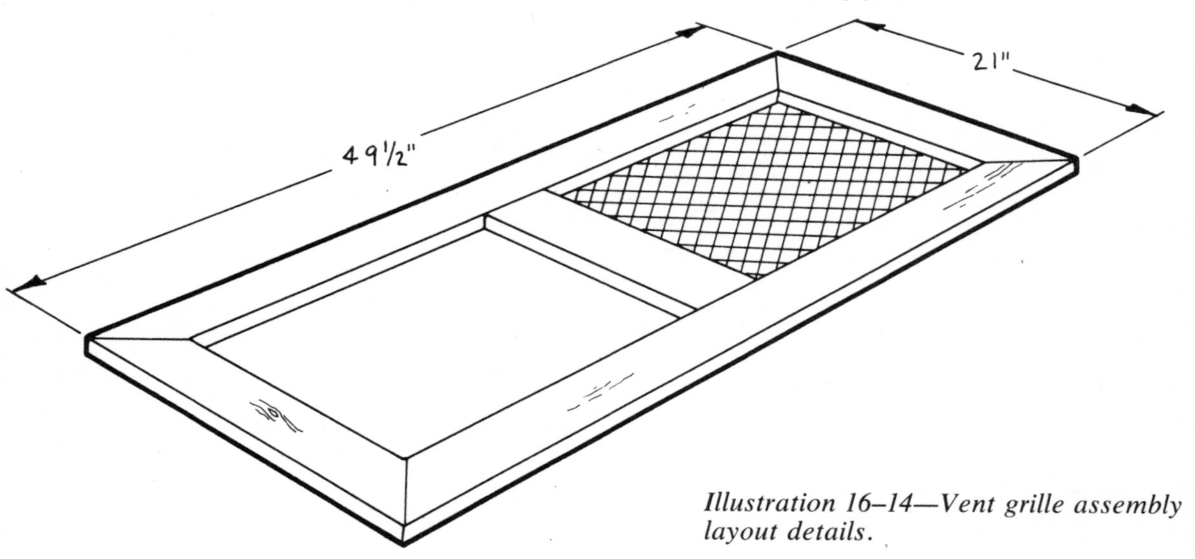

Illustration 16–14—Vent grille assembly layout details.

Cut two 18 × 24-inch pieces of aluminum mesh for the grille, and, using ⅜-inch staples, fasten the mesh to the back of the frame. Nail the grille frame over the vent opening with 6d finishing nails. Paint the frame with primer and two coats of interior finish paint.

The final step in your cooling chimney is to make two block latches, shown in illustration 16–15, to hold the vent door cords. Cut four pieces of 1 × 2 scrap, each 4 inches long. Mark and cut a 4-inch radius on two pieces to make the movable blocks. Using wood glue and 6d finishing nails, fasten the stationary blocks onto the grille frame. Center them within each frame section at the point where the nylon cord emerges. In each movable block, drill a 3/16-inch-diameter hole at the end opposite of the curve, and screw the movable blocks to the grille frame with 1½-inch #12 roundhead wood screws. Place the two blocks close enough so that the nylon cord is held fast between them to keep the vent doors open.

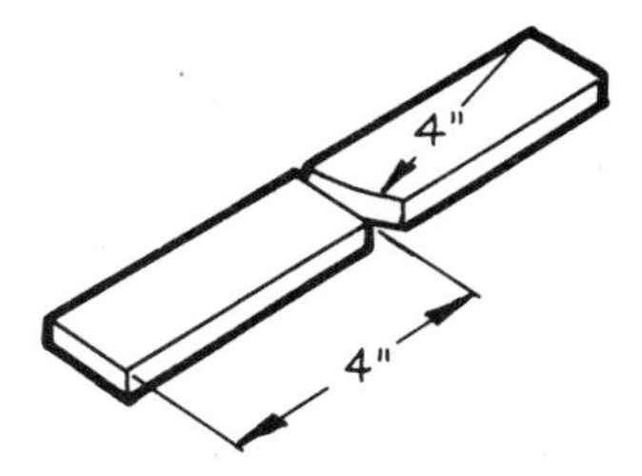

Illustration 16–15—Block latches.

The solar cooling chimney will operate best on the days of hot, bright sun, but even a hazy sun will heat the absorbers so that an air current develops. For best results, close all the house windows on the upper levels and only open a lower-level window or vent that draws air from the coolest side of the house. Remember to shield sunny windows and to open the house to cooler night air.

17 Solar Electric Toy Boat

This photovoltaic (PV) toy boat will make a very special introduction to the solar age for your youngster. What could be more fun than messing about with boats and water, unless it is tapping sunshine to help?

The PV toy boat carries six photovoltaic cells, which power a small motor. PV cells, which are made of silicon, generate electricity when placed in sunlight. The motor is attached to a propeller that moves enough air to push the boat. The cells are mounted above the boat's

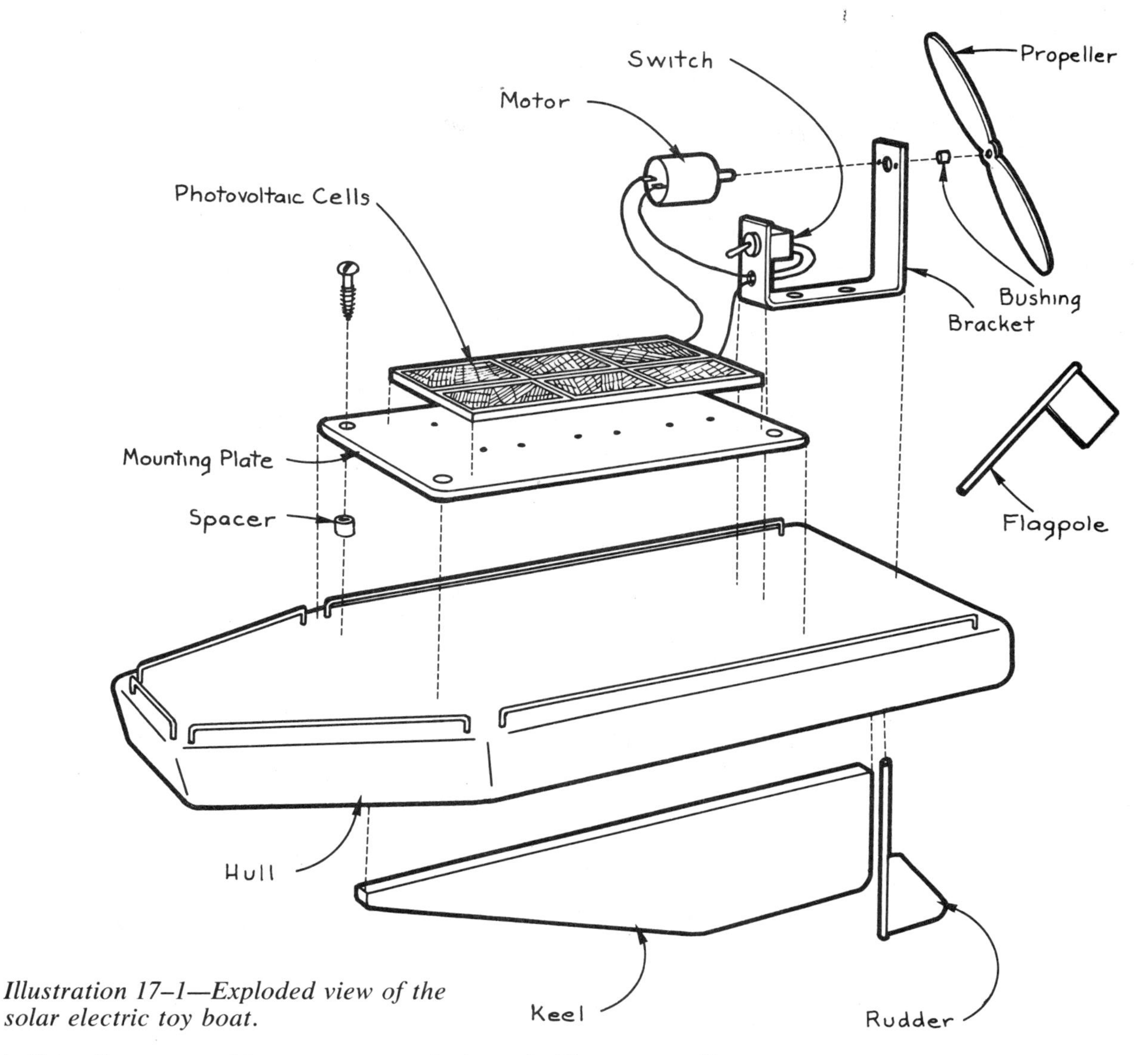

Illustration 17–1—Exploded view of the solar electric toy boat.

hull to allow any sudden waves to wash through. The motor will run as long as the cells are in sunlight, or it can be turned off and on by a switch. The propeller and motor are mounted high enough to stay dry, barring a major shipwreck. A brass rudder sets the boat's direction. The addition of brass side rails and a flag on a brass pole gives the boat a sleek and nautical look.

An exploded view of the PV toy boat is shown in illustration 17–1. The materials needed for this project are listed in chart 17–1. The tools list is chart 17–2.

Building the PV boat starts with shaping the boat hull from a piece of balsa or pine. Balsa is the best choice; if you can't find a piece in the required size, glue several pieces together with waterproof wood glue.

CHART 17–1—
Materials

DESCRIPTION	SIZE	AMOUNT
	Lumber	
Balsa or Pine	2″ × 6″ × 16″	1
Trellis Stock	¼″ × 2″ × 11″	1
	Hardware	
#10 Brass Flathead Wood Screws	¾″	2
#6 Brass Roundhead Wood Screws	¾″	4
6–32 Roundhead Machine Screws	¼″	2
Aluminum Tube	⅛″ dia. × ¼″	1
	Electrical Supplies	
Photovoltaic Cells*	0.5 volts #2–200 Solar Cell	6
Permanent Magnet Motor	Radio Shack #273–208	1
22-Gauge Wire	...	10″
SPST Toggle Switch	Radio Shack #275–612	1
#6 Stud Ring Connectors	...	2
Electrical Tape or Heat-Shrink Tubing	...	2″

**The PV cells may be obtained from Free Energy Systems, Rockdale Industrial Park, P.O. Box 3030, Lenni, PA 19052.*

Cut the boat hull 5¼ × 16 inches from a piece of 2 × 6. Plane or cut the piece to a thickness of 1⅛ inches. Cut a dado, or groove, centered on the underside of the hull, to accept the keel. Use a router to cut the dado ¼ inch wide and ⅜ inch deep, or cut the groove with a wood chisel. The groove is 11 inches long and positioned ¾ inch from the stern (rear) of the boat.

As shown in illustration 17–2, lay out the bow shape on the hull, and cut the hull to shape. Clamp the hull to your workbench and use a plane to cut a 15-degree chamfer on the sides and the bow of the boat, as shown in photo 17–1. With a plane or rasp, and then sandpaper, round all the edges and corners of the hull to approximately ¼-inch radii to give the boat a sleek look.

The keel is a 2 × 11-inch piece of pine trellis stock or other ¼-inch-thick wood. Referring to illustration 17–3, lay out the keel's shape, then cut it to shape with a saber saw. Sand any sharp edges, and fasten the keel into the groove in the hull with waterproof wood glue. Use a square to be sure it is set at a 90-degree angle to the underside surface of the hull.

Use fine sandpaper to give the entire hull a finished surface. Paint the hull and keel with primer and two coats of enamel.

CHART 17–1—*Continued*

DESCRIPTION	SIZE	AMOUNT
Miscellaneous		
14-Gauge Brass	2″ × 8″	1
Brass Rod	1/16″	4′
Maple Propeller	5¼″	1
Rolled Pin	1/16″ I.D. × ¼″	1
Waterproof Wood Glue	…	trace
Epoxy	…	trace
Primer	…	1 pt.
Exterior Enamel	…	1 pt.
60/40 Tin-Lead Flux-Core Solder	…	trace
Acrylic Sheet	⅛″ × 3½″ × 8¼″	1
Paper or Cloth Flag	…	1

CHART 17–2—

Tools

Hacksaw
Saber Saw
Drill and Bits
Router or Wood Chisel: ¼-inch
Plane
Soldering Pencil
Tap: #6–32
Rasp

Illustration 17–2—Layout details for shaping the hull.

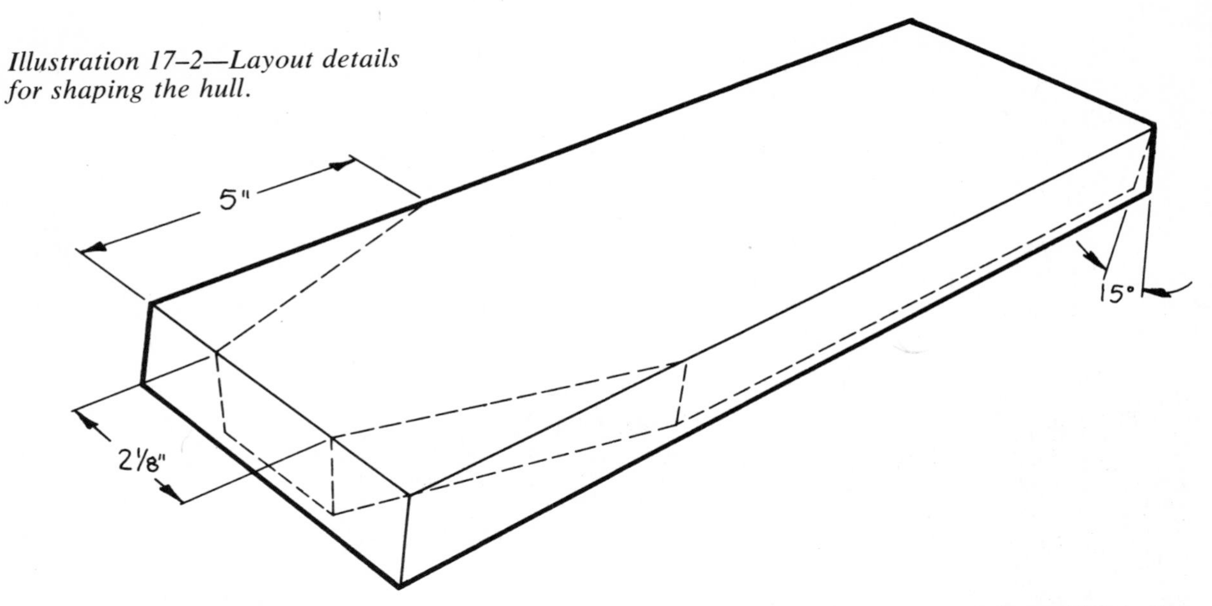

The PV cells are fastened to a sheet of clear acrylic, which is then supported ¼ inch above the boat's hull so that water will not be trapped around the cells. When they are exposed to sunlight, the cells generate electricity to run the motor attached to a propeller. The cells are mounted horizontally so that they will receive sunlight regardless of the direction the boat travels.

For the mounting plate, cut, with a fine-tooth handsaw, a 3½ × 8¼-inch piece of ⅛-inch acrylic. Lay out twelve 5⁄32-inch-diameter holes to fit the terminals on the back of the cells. The most accurate method is to place the six cells on their faces on your workbench. As

Illustration 17–3—Layout details for shaping the keel.

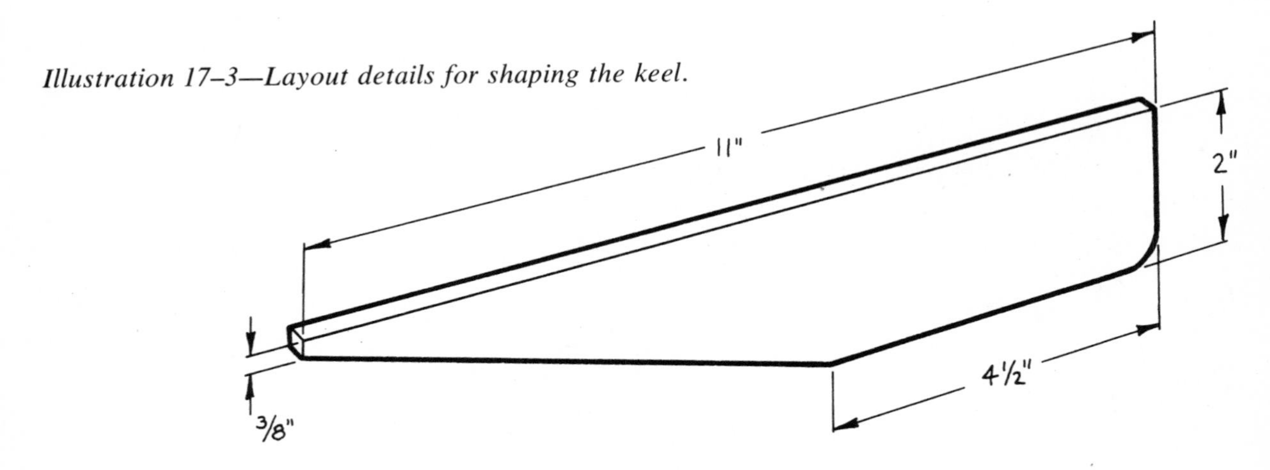

shown in photo 17–2, place the mounting plate over the cells, and mark the terminal locations with a scribe. Also, where needed, lay out four mounting holes ½ inch from the front and back, ¾ inch in from each side of the plate. Drill the holes ⅛ inch in diameter.

Attach the six PV cells to the mounting plate by inserting the terminal bolts through the drilled holes. Following illustration 17–4, connect the cells in series, with each negative terminal fastened to the positive terminal of the adjacent cell. Use the washers, brass plates, and nuts provided with the cells. Leave open one positive and one negative terminal at one end of the series, as these will be fastened to the motor.

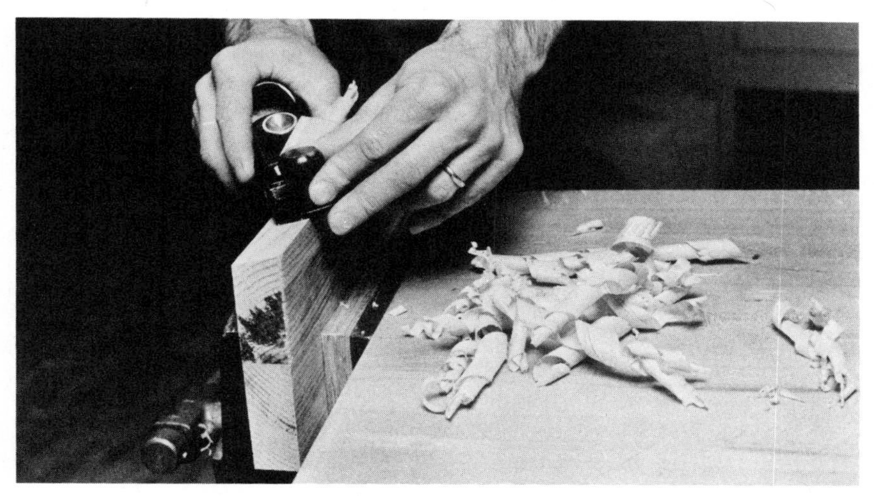

Photo 17–1—Use a plane to form the 15-degree chamfer on the sides of the boat hull.

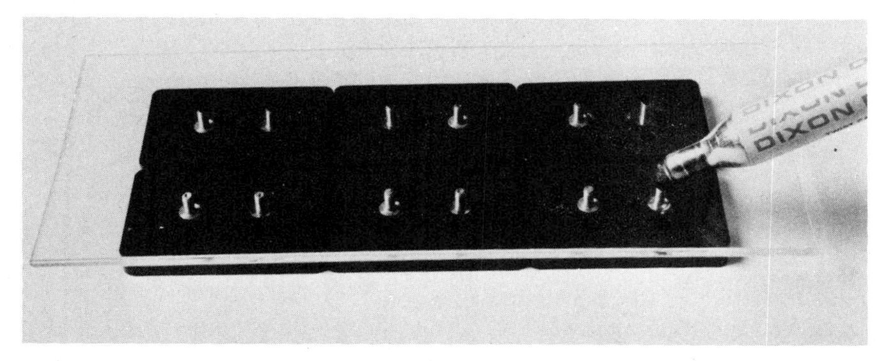

Photo 17–2—Mark the terminal ends by placing the cells upside down and the acrylic sheet on top of them, marking the posts' positions with a pen.

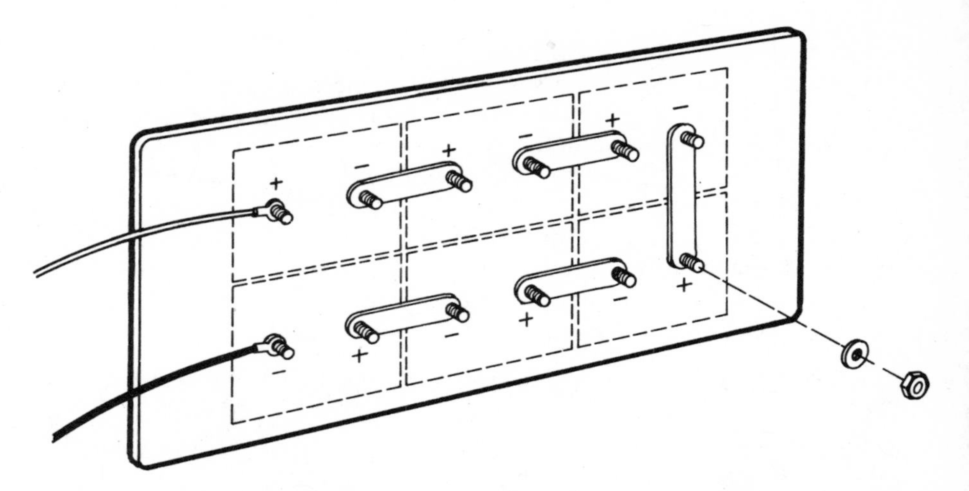

Illustration 17–4—The PV cells must be wired in this fashion to develop enough power to move the boat.

Cut two pieces of 22-gauge wire, each 5 inches long, for the connectors to the motor. Strip the insulation back ⅜ inch on one end of each wire, insert the ends into a #6 stud ring connector, and crimp the connector closed on the wire. Fasten the two connectors to the open terminals, one positive and one negative, and tighten the terminal nuts over them.

Cut four ¼-inch pieces of ⅛-inch-inside-diameter aluminum tubing for spacers. The spacers support the mounting plate above the hull's surface. Set the mounting plate onto the hull 3¼ inches from the stern. Mark the mounting holes, and drill four ¹⁄₁₆-inch-diameter pilot holes in the hull. Fasten the mounting plate to the hull with four ¾-inch #6 brass roundhead wood screws, which go through the spacers before being screwed into the hull, as shown in illustration 17–5.

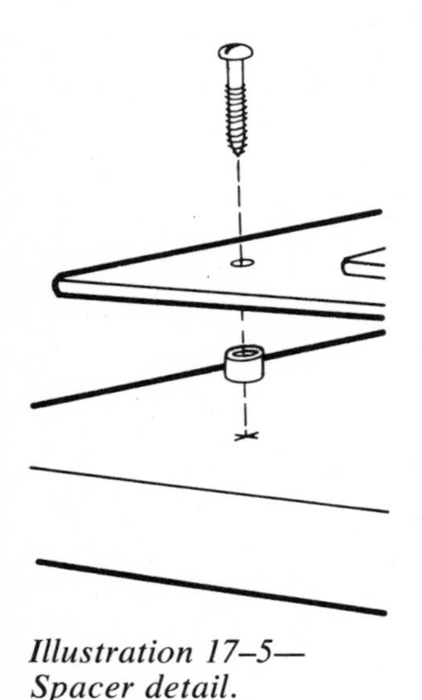

Illustration 17–5—Spacer detail.

Cut the ⅞ × 7¾-inch bracket that holds the switch, motor, and propeller from a piece of 14-gauge brass. Smooth the edges with a file and slightly round the corners. As shown in illustration 17–6, position seven holes in the bracket. First mark a centerline lengthwise down the strip. On this line, drill three ¼-inch-diameter holes. The first, for the motor propeller connection, is ⅜ inch from one end. For the switch and wires, drill one hole ½ inch and the other 1¼ inches from the *opposite* end. Drill two ⁵⁄₃₂-inch-diameter holes in line with the first hole and ⁹⁄₁₆ inch (on center) apart from each other for the motor mounting screws. The final two holes, ⁷⁄₃₂ inch in diameter, are on the centerline, to be used to fasten the motor mount bracket to the boat. Drill the first mounting hole 2½ inches from the switch end, and the second 3½ inches from

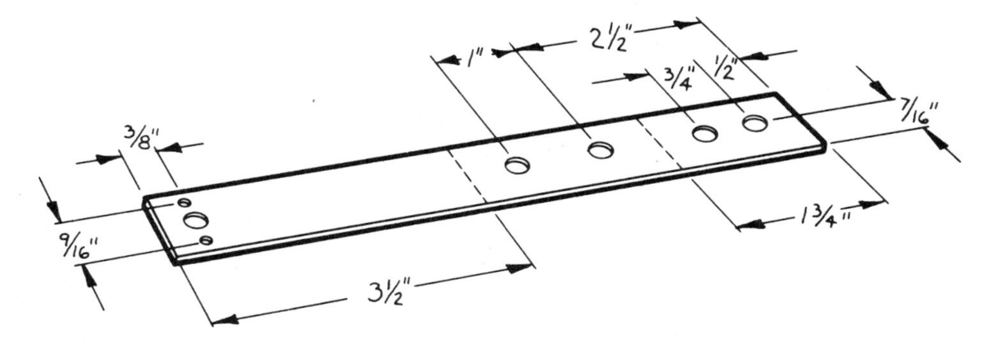

Illustration 17–6—Layout details for the motor mount bracket.

the switch end. Countersink these two holes to accept #10 flathead wood screws.

Referring again to illustration 17–6, bend the bracket at two places to hold the switch and the motor. As shown in photo 17–3, place the brass strip in a vise, and bend it with pliers. Make the first 90-degree bend 1¾ inches from the switch end and the second bend 3½ inches from the motor end. Be sure to make the second bend in the right direction.

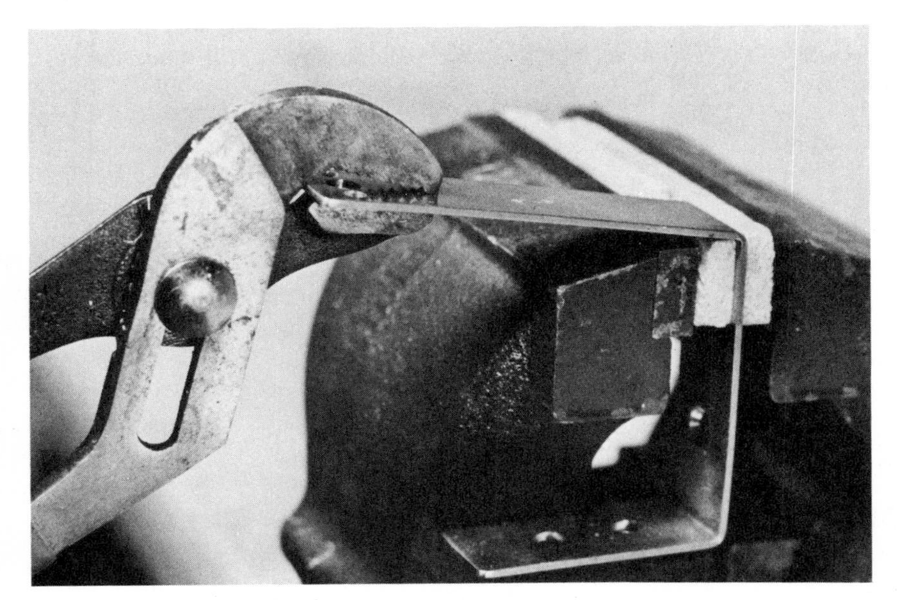

Photo 17–3—Bend the brass strip to hold the motor and switch in a vise with large pliers.

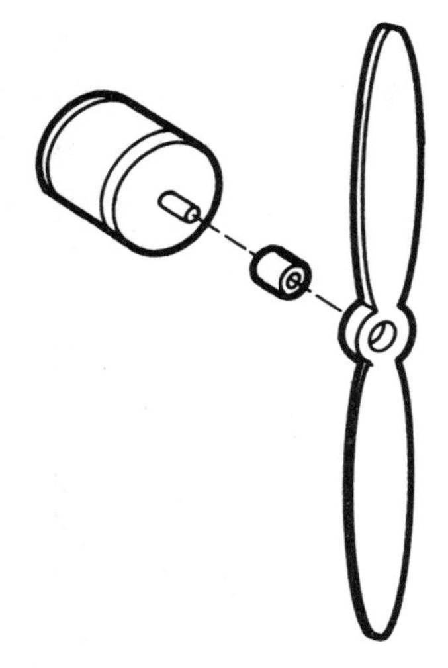

Illustration 17–7—Motor and bushing assembly.

Place the brass bracket ¼ inch behind the acrylic mounting plate, and mark the mounting holes on the hull. Drill ⅛-inch-diameter pilot holes and fasten the bracket to the hull with two ¾-inch #10 brass flathead wood screws.

If there are no threaded mounting holes in the motor housing, you will need to cut threads so the motor can be attached to the bracket, or you may use self-tapping mounting screws. If you are tapping the motor holes, remove the housing from the motor by bending the clips and slipping the housing off. Cut the threads with a No.6–32 tap. (See page 31 for a description of how to cut threads.) Replace the housing and test the motor with a battery to be sure the interior connections are working. Mount the motor to the bracket with two ¼-inch 6–32 roundhead machine screws.

Install the propeller on the motor shaft. A bushing to make the correct fit might be needed. Make the motor shaft bushing from a ¼-inch piece of 1/16-inch-inside-diameter rolled pin. As shown in illustration 17–7, fasten the bushing to the motor shaft with epoxy, and then epoxy the propeller to the bushing.

Complete the electrical hook-up by fastening the switch into the upper hole of the bracket. Run the wires from the PV cells through the lower hole in the bracket. Temporarily connect one of these to the switch, the other to the motor. Complete the circuit with a wire from switch to motor, as shown in illustration 17–8. Place the boat in sunshine to test the direction in which the propeller rotates. If it does not push

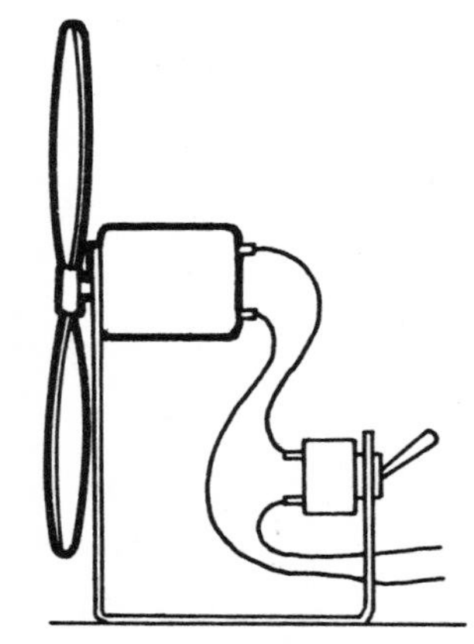

Illustration 17–8—Motor wiring.

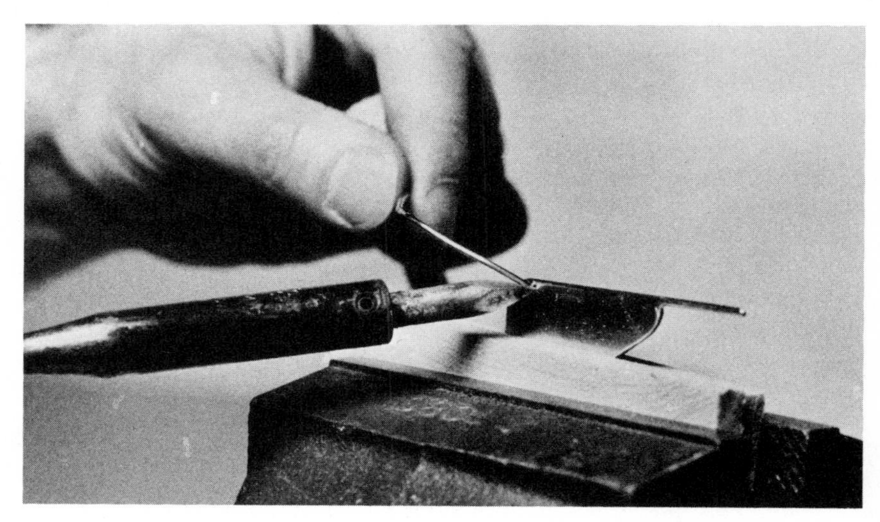

Photo 17–4—Solder the rudder to the shaft with 60/40 solder and a soldering pencil.

air to the back, reverse the wires and test it again. Finally, solder the connections at the switch and in the line with a soldering pencil and 60/40 tin-lead flux-core solder. Cover the connection between the motor and the PV cells with electrical tape or heat-shrink tubing.

Construct and install the rudder next. On a piece of 14-gauge brass, lay out the 1 × 1½-inch shape, as shown in illustration 17–9. Cut the brass and round the corners with a file, also filing the edges to remove burrs. The 3-inch-long shaft is cut from a 1⁄16-inch-diameter brass rod. Solder the two together with a fine-tipped soldering pencil and 60/40 tin-lead solder, as shown in photo 17–4. Use a file to trim the lower end. Drill a 1⁄16-inch-diameter hole centered on the keel line for the rudder shaft. If the shaft is loose, flatten it slightly with a hammer to make a press fit.

When the boat has been constructed and the electrical system is complete, decorate it with side rails and a flag. For the rails, cut a 1⁄16-inch-diameter brass rod to the following lengths: two pieces 12 inches long, two pieces 5 inches long, and one piece 2½ inches long. At each end of each rail, make a ½-inch bend. Lay out the rails along the sides and the bow of the boat ¼ inch from the outside edge. Mark the position of the holes, then drill ten 1⁄16-inch-diameter holes. Epoxy the side rails in place.

Cut a 6-inch-long piece of brass rod for the flagpole, and fasten a paper or cloth flag to one end. Making sure that it will not interfere with the propeller, drill a hole in the stern of the boat at a 45-degree angle and install the flagpole.

The PV boat is now ready for its maiden voyage. The best time to sail the boat is when the sun is bright. Choose a smooth pool of water and a calm day for the best indication of what the PV cells and motor can do to power the boat.

Some experimenting with the boat is possible through the use of mirrors to reflect sunlight onto the PV cells. Use a plastic mirror, or one made of polished metal, to be on the safe side. Hold the reflector as high as possible, and direct the beam of light onto the cells. The boat should increase in speed when the intensity of light is increased. This is also a good technique to use when the sun is low on the horizon.

If your boat should capsize, nothing will happen to the PV cells, but the motor and switch may need drying out. The best treatment for a wet motor is to keep it running to dry it out completely.

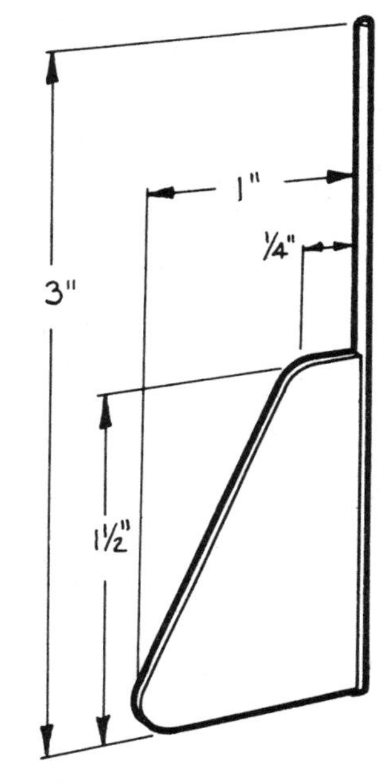

Illustration 17–9—Rudder layout details.

18 Solar Hitchhiker

The solar hitchhiker is made for fun. It is a toy that serves the main purpose of toys—to give pleasure. It is also an instructive example of how solar electricity works.

The base of the hitchhiker is stationary; only the arm moves, and once it starts going, it will continue to go as long as the sun is shining on the photovoltaic (PV) cell. Simply holding your hand to shade the cell will either slow the arm down or stop it completely. The cell and motor are very carefully sized to be able just barely to drive the arm, and this fine balance between success and failure enables the hitchhiker to be most instructive concerning solar energy. Watch the hitchhiker in

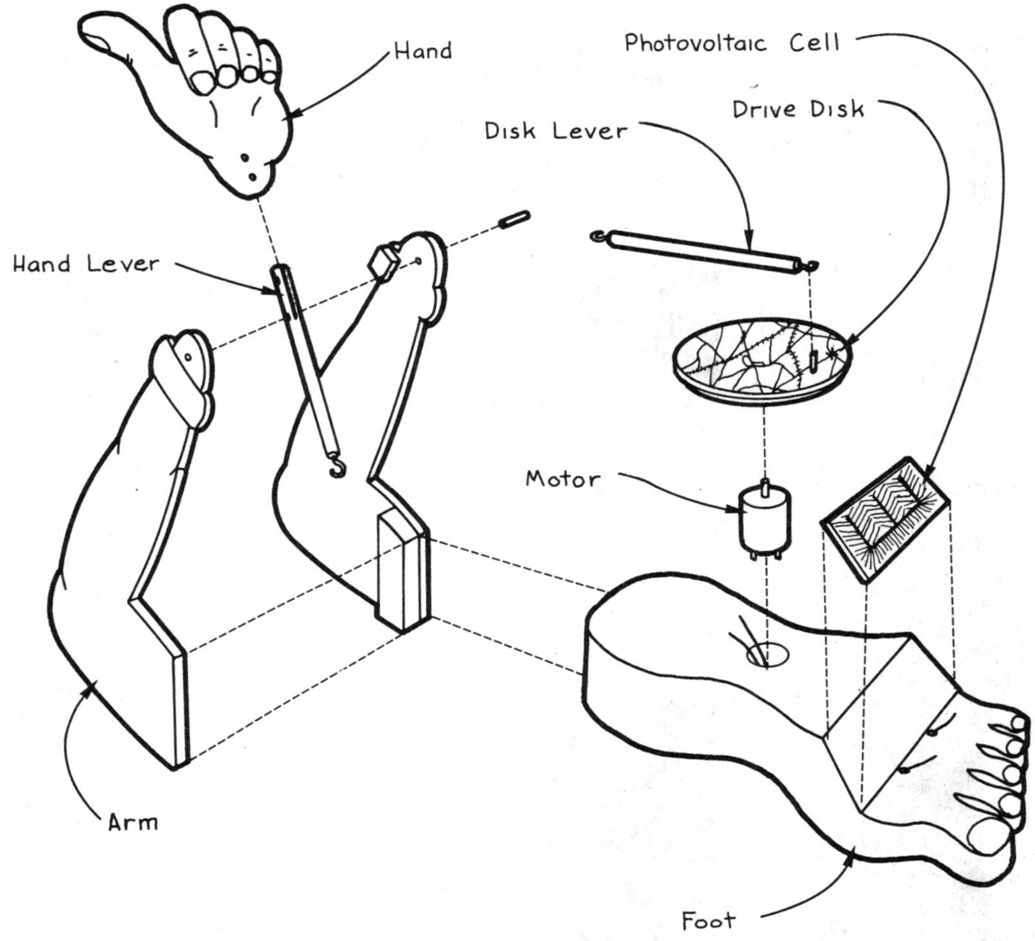

Illustration 18–1—Exploded view of the solar hitchhiker.

January and again in July, and you'll get a good lesson about the strength of the sun during different months of the year.

The hitchhiker has a small motor harnessed to one PV cell. The motor operates a rotating disk equipped with a map of your favorite travel spot. An eccentric, or off-center, pin on the disk gives a back-and-forth motion to the connecting linkage, which is attached to a lever, which, in turn, moves the hitchhiker's fist with his thumb protruding. The hitchhiker's fist is, of course, attached to the hitchhiker's arm, which is directly connected to the hitchhiker's foot. The foot is built with a 45-degree angle to hold the cell so that it can be directed most effectively into the sun.

CHART 18–1—
Materials

DESCRIPTION	SIZE	AMOUNT
Lumber		
#2 Pine	1 × 2 × 4″	1
A-C Exterior Plywood	¾″ × 12″ × 18″	1
Tempered Hardboard	⅛″ × 10″ × 15″	1
Hardwood Dowel	5/16″ × 1′	1
Hardwood Dowel	⅛″ × 3″	1
Hardware		
#6 Flathead Wood Screws	1½″	2
#6 Flathead Wood Screws	½″	2
#4 Screw Hook	¾″	1
#4 Screw Eyes	¾″	2
Electrical Supplies		
Photovoltaic Cell	Radio Shack #276–126	1
Permanent Magnet Motor	Radio Shack #273–208	1
18-Gauge Bell Wire	···	1′
Miscellaneous		
Wood Glue	···	trace
Silicone Caulk	···	trace
Self-Adhesive Felt	6″ × 1′	1
Road Map	···	1
Carbon Paper	···	1 sheet
Black, Pink, Red, and Blue Enamel	···	small amounts

CHART 18–2—
Tools

- Saber Saw
- Backsaw
- Drill and Bits
- Router (optional)
- Soldering Pencil

All this is not quite perfect anatomy, but this hitchhiker's only job is to stop traffic, making it a nifty little item for the person who travels. The only problem? The PV cell has to face south, so it is only good for catching rides north!

An exploded view of the solar hitchhiker is shown in illustration 18–1. The materials you will need are listed in chart 18–1 and the tools in chart 18–2.

Start building the solar hitchhiker by laying out the wooden parts which make up the foot, hand, and arm. These shapes are printed on grids to make it easier for you to transfer them to your wood. To make your designs, you will need three pieces of 8½ × 11-inch graph paper, marked in ½-inch squares. Following illustrations 18–2, 18–3, and 18–4, transfer the shapes of the three parts to the graph paper.

Using carbon paper, transfer the full foot shape and the two partial upper foot pieces to a 12 × 18-inch piece of ¾-inch plywood. Lay out the hand and two arm pieces on a 10 × 15-inch piece of ⅛-inch tempered hardboard. With a saber saw, cut out all the pieces from the plywood and hardboard.

Glue the upper parts of the foot together, keeping the edges flush. When the glue has set, measure 1½ inches from the square end, and draw a line across, parallel to the square end, as shown in illustration 18–5. Set your saber saw to cut a 45-degree angle. After clamping the upper foot to your workbench, cut along this line to give you an angled edge. The PV cell will be glued to this angled area, so that it will receive more sunlight than a flat surface.

Illustration 18–2—Duplicate the foot shape on ½-inch graph paper for tracing on your wood.

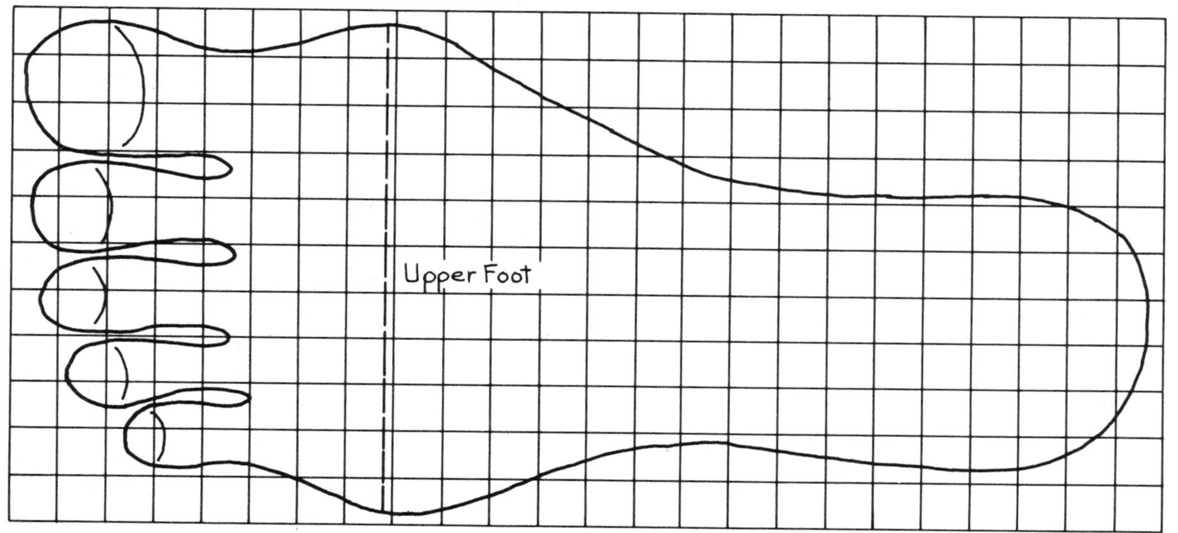

Illustration 18–3—Duplicate the hand shape on ½-inch graph paper for tracing on your wood.

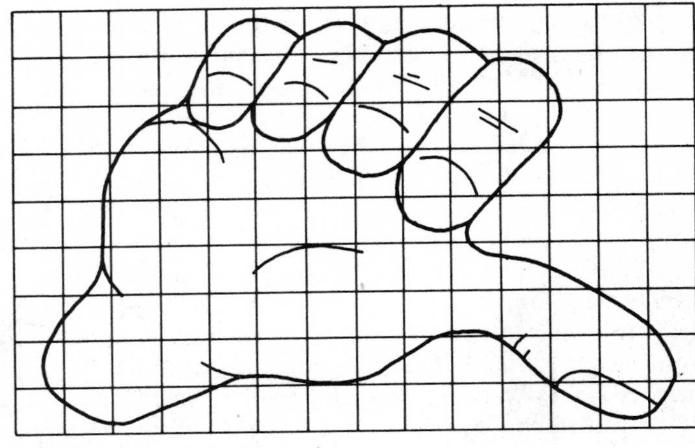

Illustration 18–4—Duplicate the arm shape on ½-inch graph paper for tracing on your wood.

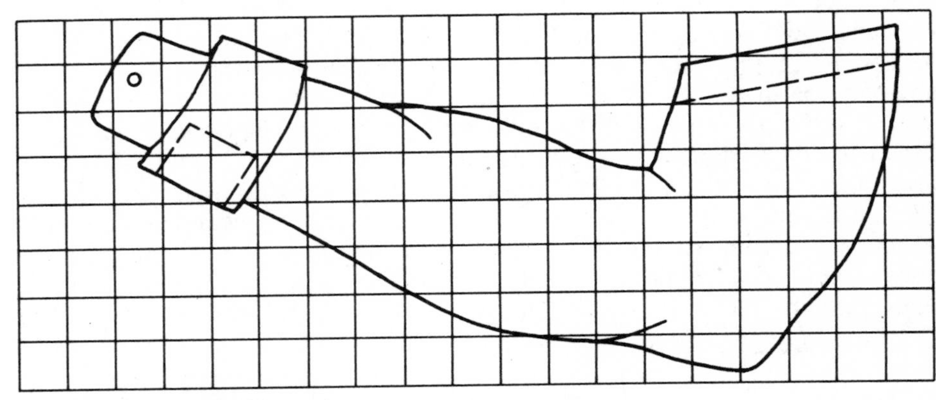

After the glue has dried, smooth out any irregularities between the two layers to give the piece a unified form. Use sandpaper wrapped around a stick or a file, as shown in photo 18–1, to finish the toes, then

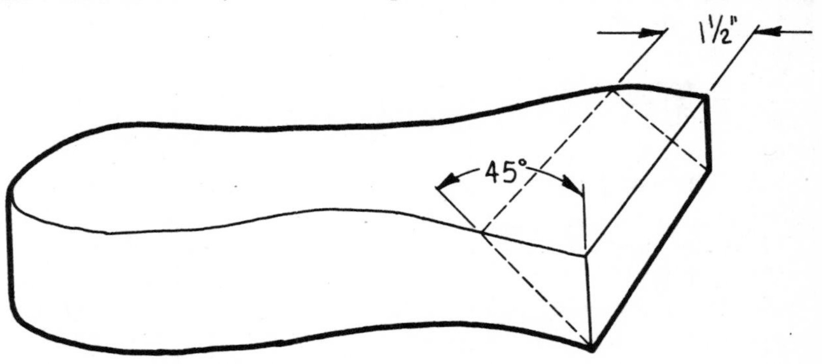

Illustration 18–5—Layout details for the bevel on the upper foot piece.

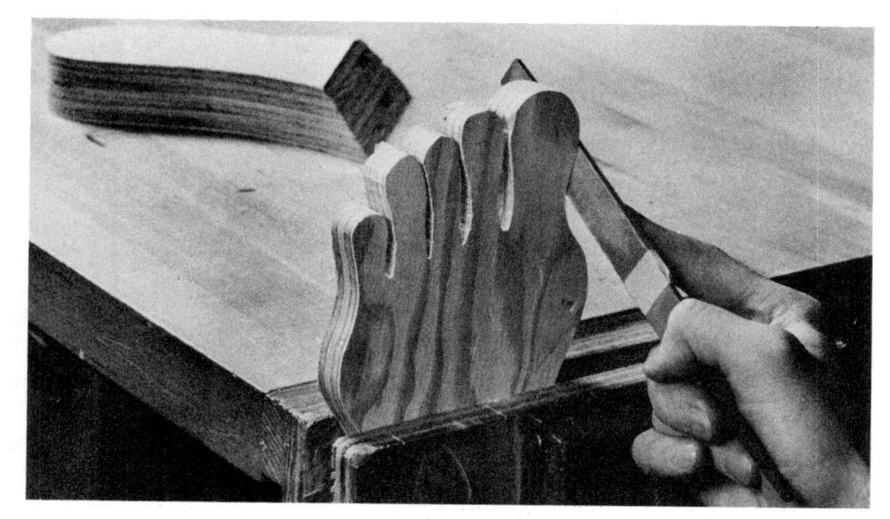

Photo 18–1—Make a shaper from sandpaper and a file to smooth out the toes and the foot pieces.

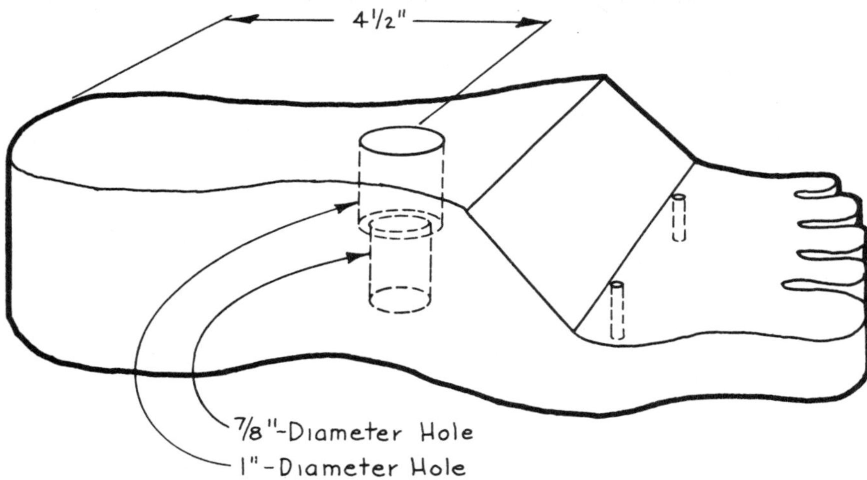

Illustration 18–6—Motor and wire hole details for the upper and lower foot pieces.

glue the upper foot part to the foot base, aligning the sides and the heel. Also sand the angled surface smooth to facilitate mounting the solar cell.

Refer to illustration 18–6 to lay out the motor hole in the foot. On the top side of the foot, measure 4½ inches from the heel. Center the hole side to side. With a 1-inch spade bit, or a size suitable to hold your motor, drill a hole 1⅜ inches deep. (As shown in photo 18–2, wrap a strip of tape around the shank of the drill bit to indicate the depth you need.) Then drill the rest of the way through the foot with a ⅞-inch spade bit.

At the base of the angled section of the foot, drill two ⅛-inch-diameter holes for the wires from the solar cell, as shown in illustration 18–6. The holes are approximately 1¾ inches apart and 1¾ inches from either side of the foot.

Photo 18–2—A temporary depth marker to drill the motor hole can be made by wrapping tape around the spade bit.

Turn the foot upside down. On the underside of the foot, cut two grooves from the wire holes to the motor hole, for the wires from the PV cell to the motor. You can use a router to cut these grooves ⅛ inch wide and ⅜ inch deep, or carve them out with a utility knife and a straightedge.

Now assemble the arm. Put the two arm pieces together, and drill a hole ⅛ inch in diameter centered in the wrist area, as shown in illustration 18–4. Cut two pieces of 1 × 2 to fit between the arm pieces, one near the base and the other at the wrist. These pieces serve to space the two sides of the arm the correct distance and to act as a bracket to attach the arm pieces to the foot. The upper spacer is ⅜ × ¾ × ¾ inch, and the lower spacer is ⅜ × ¾ × 3 inches, as shown in illustration 18–7. Set the lower spacer block flush with the shoulder end of the arm and mark the angled ends, then cut the block so it is flush with the arm pieces. Now lay the lower spacer on edge. Drill and countersink two ⅛-inch-diameter holes for attachment screws. The position of the spacers is shown in illustrations 18–4 and 18–7. Put the arm pieces together by gluing and clamping them with the lower ends of the arms flush with the edge of the lower spacer and the upper spacer flush with the outside edge, nearest the point of the wrist, within the cuff area.

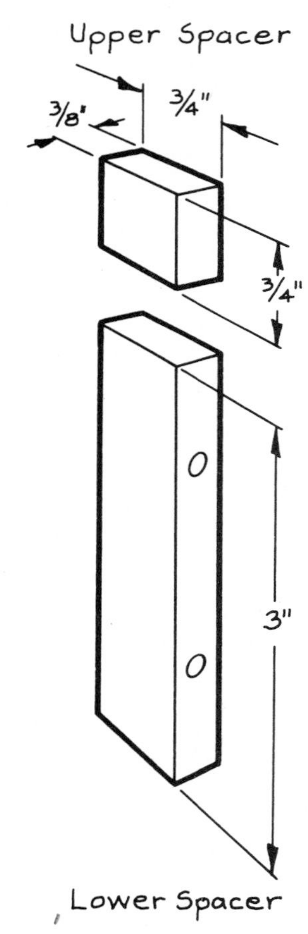

Illustration 18–7—Spacer blocks.

The connecting linkage is made out of two pieces of 5/16-inch-diameter wood dowels. Cut one 5-inch piece for the hand lever. Drill a ⅛-inch-diameter hole ¼ inch from one end, as shown in illustration 18–8. Cut a lengthwise slot in the lever, in the end that has the hole, wide enough to fit the hand piece. The slot is perpendicular to the hole. Clamp the dowel in a vise to cut the slot accurately, as shown in photo 18–3.

Slip the hand piece into the slot, centering the lever within the wrist

Photo 18–3—The slot in the linkage dowel should be cut carefully with a backsaw.

area. Using the ⅛-inch-diameter hole drilled in the linkage as a guide, mark the hand piece. Remove the hand piece from the lever, and drill a ⅛-inch-diameter hole in the wrist area of the hand. Fasten the hand to the lever with wood glue and a length of ⅛-inch-diameter dowel. Cut and sand the dowel flush with the lever.

Next, drill a pivot hole through both the hand and the lever. Measure ⅝ inch from the end of the slotted end of the lever and drill a 5⁄32-inch-diameter hole. At the opposite end of the lever, drill a 1⁄16-inch-diameter hole lengthwise to the end of the dowel, and screw in a ¾-inch #4 screw hook. Position the hook parallel to the plane of the hand.

Next, cut a piece of 5⁄16-inch-diameter dowel 3¾ inches long for the disk lever. Drill 1⁄16-inch pilot holes in each end, and screw in two ¾-inch #4 screw eyes. Position the screw eyes at right angles to each other, as shown in illustration 18–9.

The drive disk rotates to move the hitchhiker. It is made from two disks of hardboard. Cut one disk 2½ inches in diameter for the bottom of the drive disk and another 3⅜ inches in diameter for the top. In the top disk drill two holes, as shown in illustration 18–10. A 3⁄32-inch-diameter hole goes in the center, and a ⅛-inch-diameter hole goes 1¼ inches from the center point. This hole holds the drive pin off center to give the hitchhiker's hand its motion. Drill another 3⁄32-inch-diameter hole in the center of the bottom disk. Glue the two disks together with the center points matching. Cut a ¾-inch piece of ⅛-inch-diameter dowel for the drive pin. Glue the drive pin into the off-center hole in the top disk.

Sand all the parts until they are smooth. Use a primer first, then paint the pieces as whimsically as you wish. Our own solar hitchhiker is an attention-grabber with pink skin, blue sleeves, red toenails and fingernails, and black lines for emphasis around each colored area.

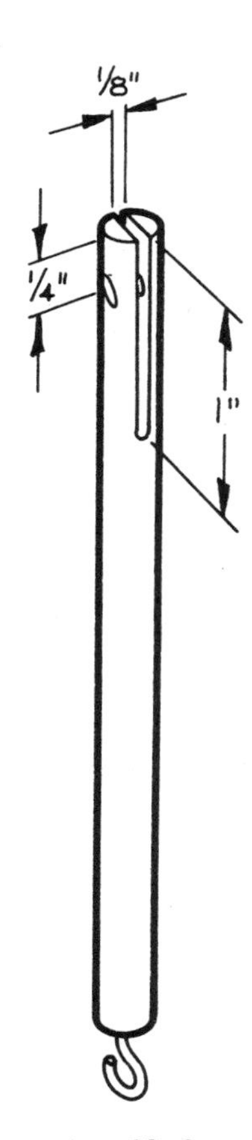

Illustration 18–8—Hand lever.

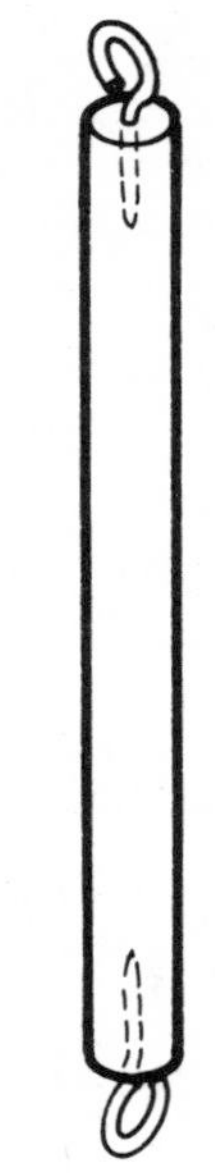

Illustration 18–9—Disk lever.

Choose an area from an old road map and cut out a 3⅜-inch circle. Glue the map piece to the top of the drive disk after punching a hole in the paper for the drive pin.

Now you are ready to wire the solar hitchhiker. Refer to illustration 18–11. Cut two 6-inch pieces of bell wire and strip the insulation ½ inch from each end of each piece. Using a soldering pencil with a fine tip, solder one wire to each connector on the PV cell. Push the wires through the holes in the foot, and glue the cell to the angled portion of the foot with silicone caulk or epoxy.

Insert the motor into the hole in the top of the foot, with the motor shaft facing up. Turn the foot upside down to solder the wires from the motor to the wires from the PV cell, as shown in photo 18–4. Place the wires in the grooves cut in the underside of the foot. To finish the foot underneath, cut a piece of self-adhesive felt to the shape of the foot. Be sure to trace the shape onto the adhesive side of the felt before cutting.

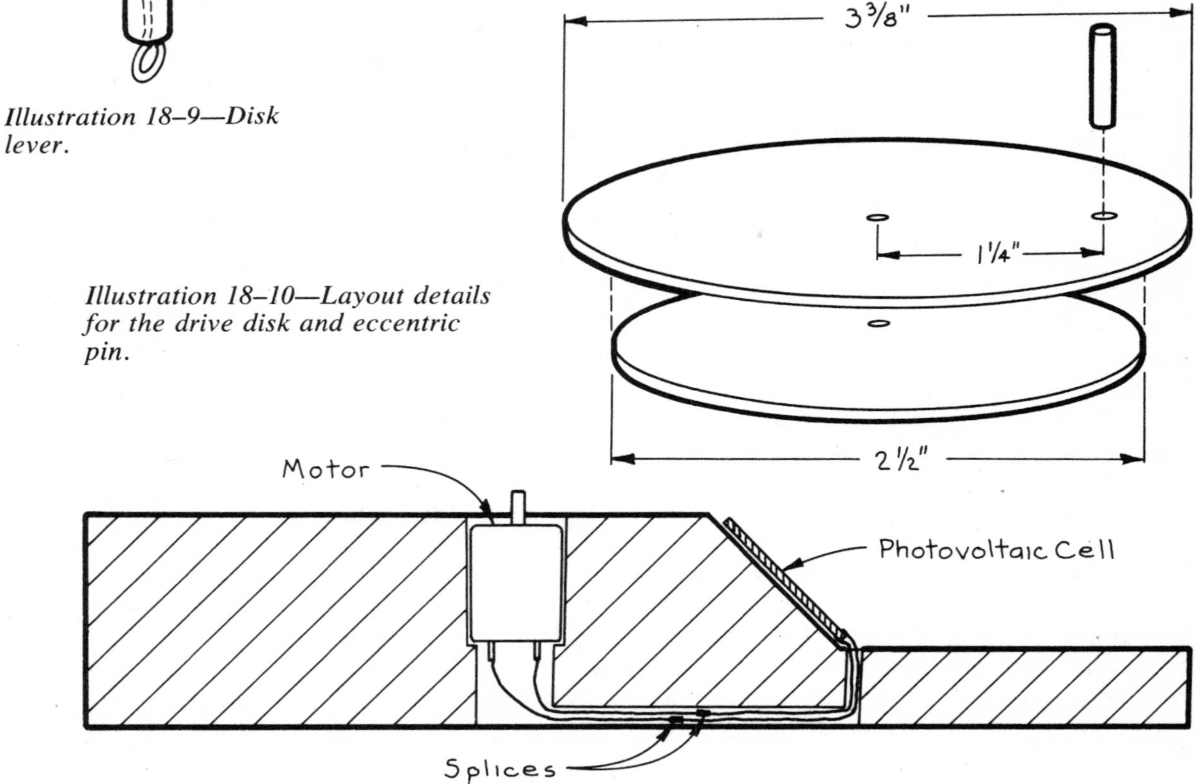

Illustration 18–10—Layout details for the drive disk and eccentric pin.

Illustration 18–11—Motor placement and wire routing details.

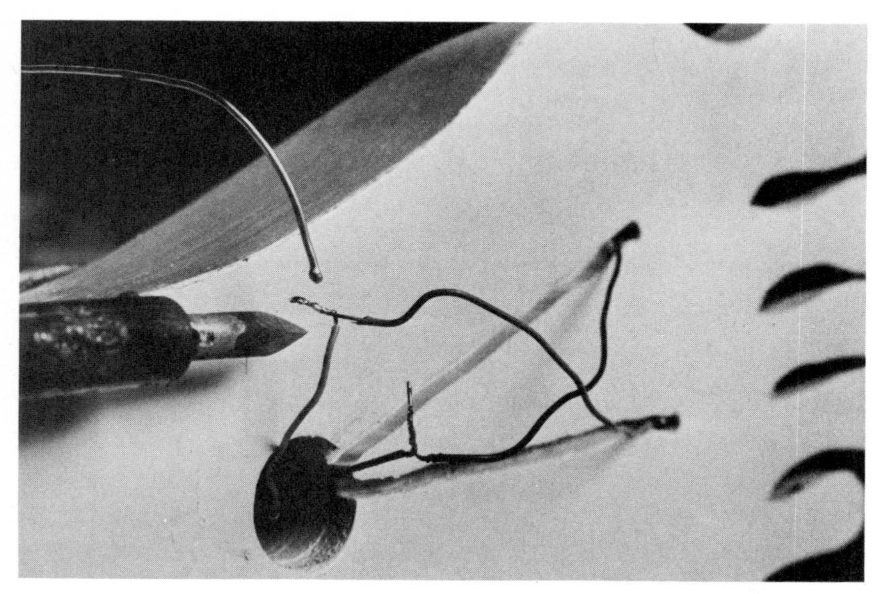

Photo 18–4—Solder the wires from the motor and the PV cell by turning the foot upside down.

To complete the linkage system, cut a ⅝-inch piece of ⅛-inch-diameter dowel for a pivot wrist pin. Place the hand section between the arm sides, with the thumb pointing in the same direction as the elbow, lining up the wrist holes. Put a drop of wood glue in each wrist hole at the top of the arm, and fasten the hand to the arm with the pivot pin. Do not get glue on the dowel where it passes through the hand piece. The hand should rotate fully on the pin.

Place the arm assembly against the heel of the foot. Use 1½-inch #6 flathead wood screws to attach the arm assembly to the foot. Next, press the disk onto the motor shaft. It should be a firm press fit. Finally, connect the hand to the drive disk by slipping a screw eye at one end of the disk lever over the hand lever screw hook and the other screw eye over the drive pin in the disk.

Take the solar hitchhiker outdoors or put it in a window with bright direct sun. Position the PV cell toward the sun, give the hand a starting push, and the hitchhiker will move vigorously. On cloudy days it can stay indoors while you plan the next trip.

Note that the motor has very little torque, and if you do not get the linkage and the hand balanced exactly right, you will have to give the hand a slight helping shove to get it going. Once running, it should continue to do so, as long as it gets enough sun. This is a fair-weather friend indeed.

19 Solar Battery Charger

The photovoltaic (PV) solar battery charger is an excellent introduction to using the sun's energy for a practical use. PV cells convert sunlight into electricity. The use of PV cells is increasing in popularity as an alternative source of energy. This PV-powered battery charger uses 60 cells and can be used for a multitude of small tasks calling for direct current. It will power a 12-volt radio, recharge a 12-volt battery, or maintain a 6-volt battery for an electric fence charger.

The long-range possibilities for photovoltaic conversion of sunlight into electricity are most exciting. The systems are extremely durable and maintenance-free, and they operate without damage to the environment.

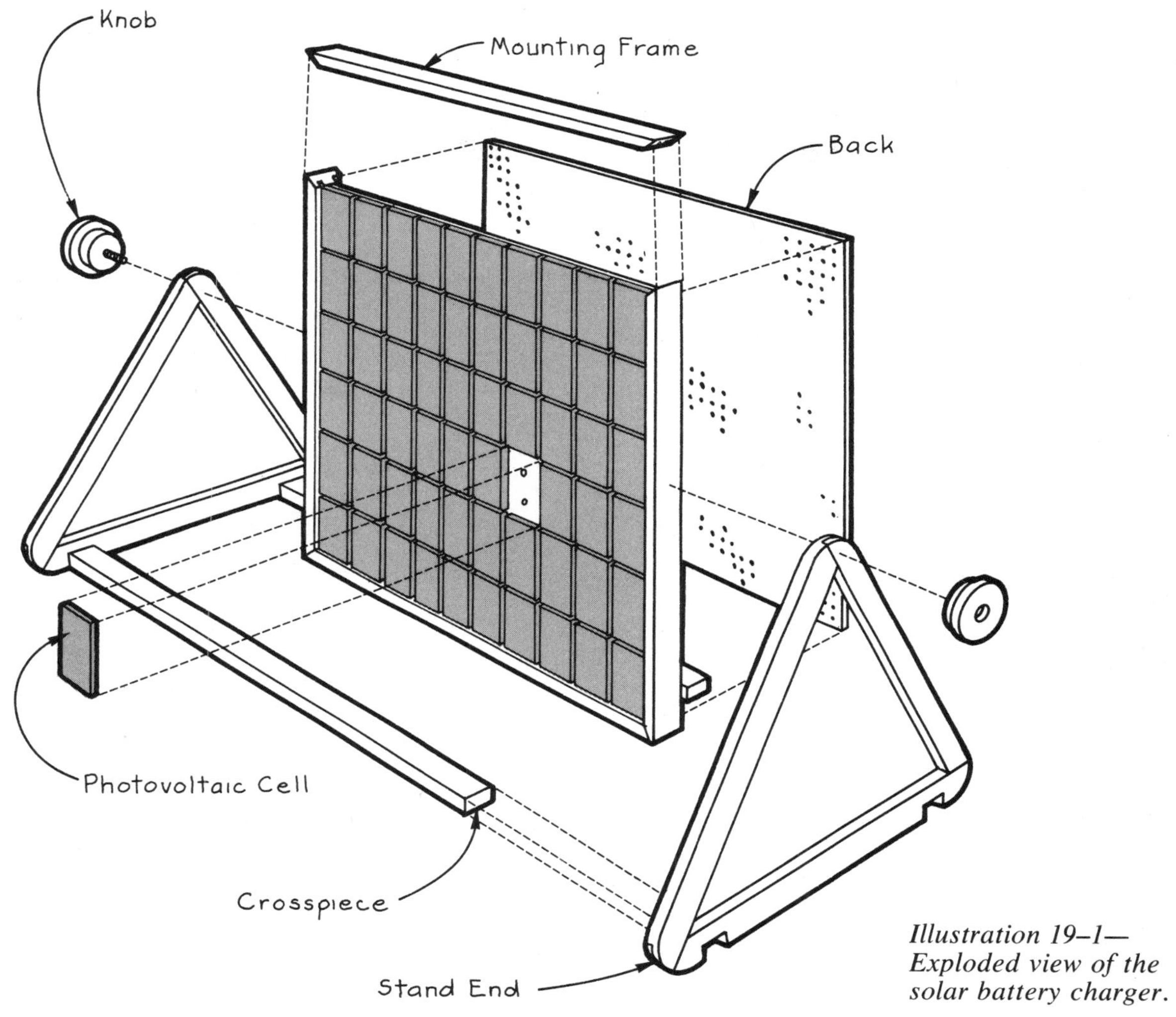

Illustration 19–1—Exploded view of the solar battery charger.

The battery charger includes an array of PV cells mounted within a frame. The array is supported by a triangular stand that allows adjustment of the angle of the frame to maximize usable sunlight. By plugging into different circuits that tap different numbers of cells, you can vary the amount of power you receive from the panel. We include complete instructions for wiring the cells and for building the frame and stand. You do not need any experience with electrical work to complete this project. When complete, the unit will charge 1.5-, 3.0-, 6.0-, 7.0-, 9.0-, and 12.0-volt batteries.

Illustration 19–1 is an exploded view of the solar battery charger. The materials you need are listed in chart 19–1 and the tools needed are in chart 19–2.

CHART 19–1—
Materials

DESCRIPTION	SIZE	AMOUNT
Lumber		
#2 Pine	5/4 × 4 × 6′	1
#2 Pine	1 × 2 × 10′	2
Pegboard	¼″ × 24″ × 27¼″	1
Hardware		
#8 Brass Flathead Wood Screws	1½″	8
#6 Brass Flathead Wood Screws	¾″	12
#6 Brass Flathead Wood Screws	⅝″	6
5/16–18 Stove Bolts	3″	2
Fender Washers	¼ I.D. × 5⁄16″ O.D.	2
5/16–18 Tee Nuts	⅜″	4
Electrical Supplies		
Photovoltaic Cells*	0.45-volt; 700 M/a; ¼″ × 2½″ × 3¾″	60
15-Volt Blocking Diode	Radio Shack #276–1141	1
18-Gauge Speaker Wire	···	6′
Red 18-Gauge Bell Wire	···	11′
Black 18-Gauge Bell Wire	···	32″
Red Alligator Clip	···	1
Black Alligator Clip	···	1
Phone Plug	⅛″	1
Phone Jacks	⅛″	6
#6 Solderless Ring Connectors	···	19

**The PV cells may be obtained from Free Energy Systems, Rockdale Industrial Park, P.O. Box 3030, Lenni, PA 19052.*

CHART 19-1—*Continued*

DESCRIPTION	SIZE	AMOUNT
Miscellaneous		
Wood Glue	...	trace
Polyurethane or Enamel	...	1 pt.
60/40 Tin-Lead Flux-Core Solder	...	trace
Acrylic Sheet	0.10 × 24″ × 27¼″	1

CHART 19-2—
Tools

Handsaw
Drill and Bits
Router
Miter Box (optional)
Wire Cutters
Wire Crimper
Acrylic-Cutting Knife
Soldering Pencil

In the solar battery charger, we have 60 PV cells attached to a rigid board in an array in combination to each other so that electricity is produced when sunlight falls upon the cells. Check your cells to see that the terminals will fit the dimensions given, and, if needed, make the necessary changes in the dimensions. Cells that measure ¼ × 2½ × 3¾ inches and have terminals centered lengthwise 1³⁄₁₆ inches apart will fit this design. Illustration 19-2 shows a PV cell of this design.

If your cells measure other than ¼ × 2½ × 3¾ with terminals 1³⁄₁₆ inches apart, your unit will require slightly different mounting details. Arrange all your cells upside down on a clean workbench in the configuration explained. Lay the clear acrylic sheet on the mounting posts,

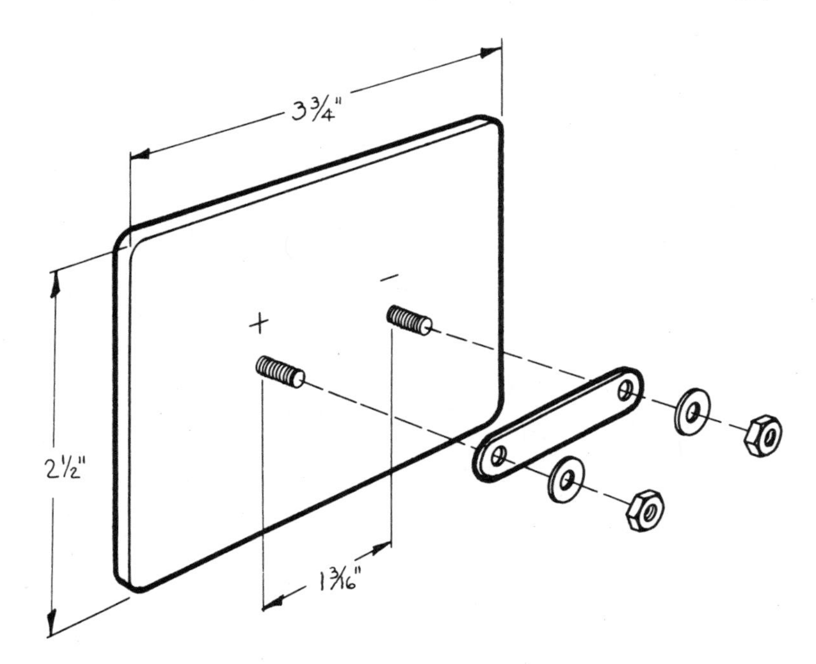

Illustration 19-2—Use PV cells of this size and configuration if possible.

and mark where each posthole needs to be drilled. Or, measure each cell and mounting post location, and lay out the acrylic sheet accordingly.

The mounting board is made from a sheet of acrylic mounted in a pine frame. First cut the acrylic to 24 × 27¼ inches. Use an acrylic-cutting knife and a straightedge, or a saber saw with a fine-tooth blade. Lay out 120 holes to fit the PV cells, as shown in illustration 19–3. Measure 2³⁄₃₂ inches from one 24-inch side of the board, and square a line down to designate the first row of cell holes. Measure in increments of 2⁹⁄₁₆ inches from the first line to the opposite side of the board, where you should again have a space of 2³⁄₃₂ inches. Square lines vertically at these points. Working along the side of the mounting board, measure 2¹⁄₃₂ inches from the top edge, then 1³⁄₁₆ inches (between the terminals of one cell), then 2⁹⁄₁₆ inches (between the terminals of two adjacent cells). Repeat the measurements 1³⁄₁₆ inches and 2⁹⁄₁₆ inches along the remainder of one side. Square lines across at these points. At the intersection of the lines, drill the ⁵⁄₃₂-inch-diameter holes to fit the cell terminals with a slight expansion space. Be careful to hold the acrylic down firmly, as it tends to catch on the drill bit.

The correct assembly of the PV cells in the mounting board is very important. Note that on the back of each cell there is a positive and a negative terminal. Arrange the cells in two modules of 30 cells each. Mount each module in series, then join the two modules together in parallel. Since the cells are provided with flat bars that slip over the

Illustration 19–3—Carefully measure and drill the holes in the acrylic sheet to mount the PV cells.

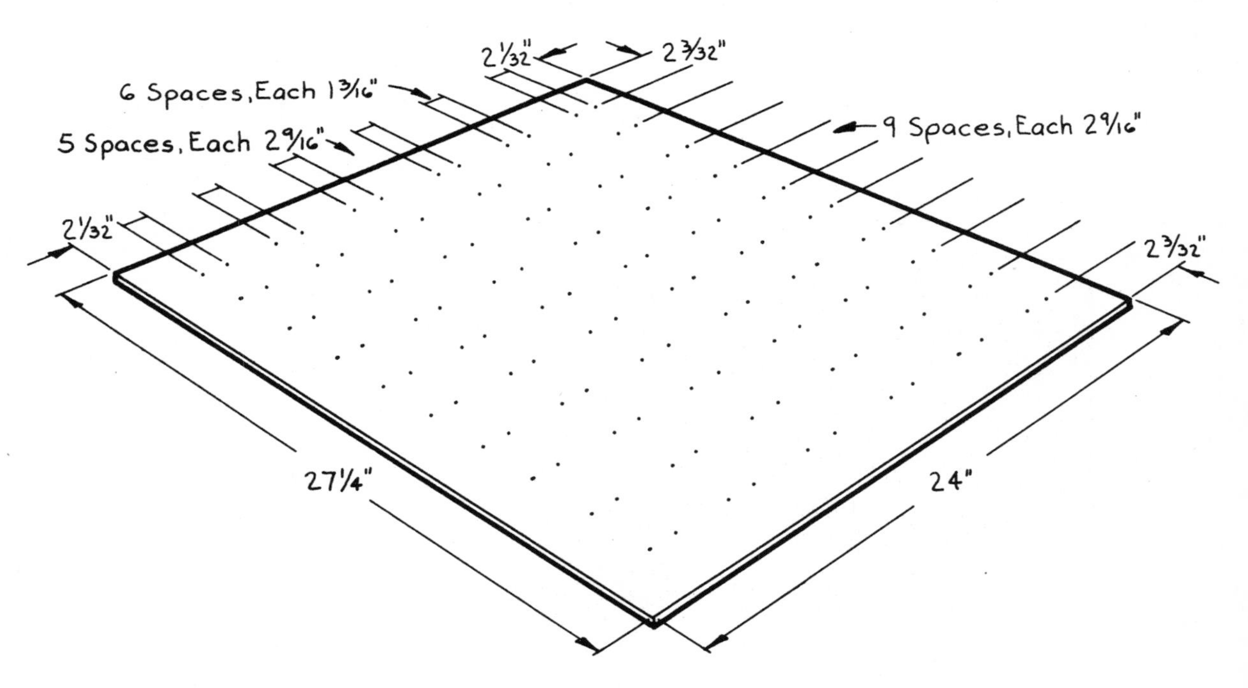

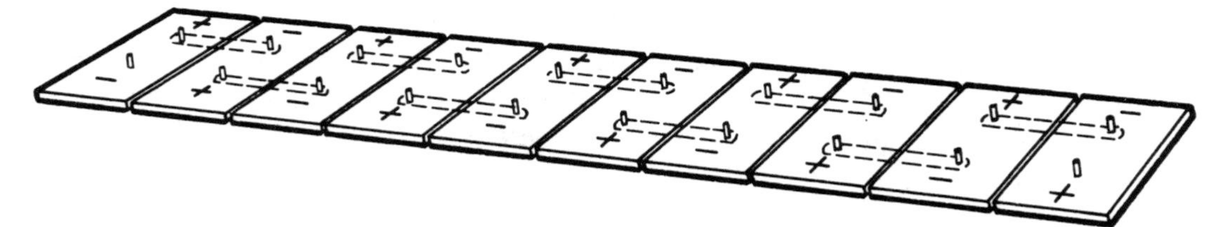

Illustration 19–4—Assemble the PV cells so that positive terminals connect with negative terminals.

terminal bolts, it is easy to fasten them without soldering any wires. Where bars are not used, black wire represents the negative side of the circuit, and red wire represents the positive side of the circuit. To join the PV cells in series, connect the positive terminal of one cell to the negative terminal of the next. Do this for two panels of 30 cells each.

Begin with the top row of cells lying face down on your workbench. Arrange them as shown in illustration 19–4, alternating positive and negative terminals. Then, maintaining the same arrangement, pick them up one at a time and fasten them into the mounting board, as shown in photo 19–1. Remove the terminal nuts, washers, and bar, and slip the bolts through the holes in the mounting board. To the positive terminal add a washer, the connecting bar, and a matching nut. Replace the washer and nut on the negative terminal, which will be added to the circuit later.

When you add the second PV cell, attach the bar from the first cell to the terminal of opposite polarity. When you are wiring electrical components in series, always join one positive terminal to one negative terminal. Continue adding cells along the top row in this manner. As you proceed to the second row, again continue the series, placing the

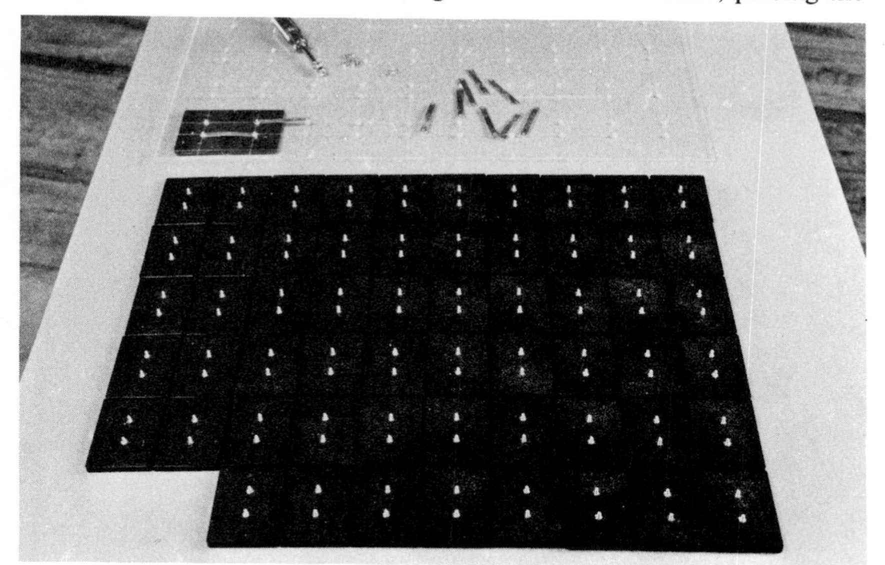

Photo 19–1—First arrange the cells in order, then fasten them to the mounting board with the connector strips in place.

cell so that its bar reaches from positive to negative. Fasten the nuts down firmly so that you will have solid connections. Complete the three top rows of cells in one continuous series with an empty terminal at each end. The top three rows, 30 cells, are referred to as Module 1. Next, begin a second series, Module 2, which is made up of the 30 cells in the three lower rows. As you begin Module 2, notice the positions of the free terminals in Module 1. When you attach the bars to the first cell in Module 2, leave the negative terminal open. Continue adding the cells in series, ending again with an open positive terminal on the last cell.

Connect row 1 of Module 1 with row 2 with a 4-inch length of red 18-gauge wire (fitted with #6 solderless ring connectors) from the positive terminal in row 1 to the negative end terminal of row 2. Make the same connection in Module 2. These two connections can also be completed with a metal strip cut to a C shape to go around the other terminals and drilled to fit the terminal posts, as shown in photo 19–2.

At this point, all 60 PV cells are fastened to the mounting board in two modules, wired in series of 30 cells each. There should be four unconnected terminal posts at this time. A negative post in row 1, a positive post in row 3, a negative in row 4, and a positive in row 6, as shown in illustration 19–5.

Now join the two modules in parallel with two 12-inch lengths of wire. Use red and black 18-gauge wire with #6 solderless ring connectors. As shown in photo 19–3, after cutting the wire the correct length, strip ⅜ inch of insulation from the ends, put one end into the connector, and crimp the connector firmly around the wire. Connect a 12-inch black wire from the free or unused negative terminal of the first cell in Module

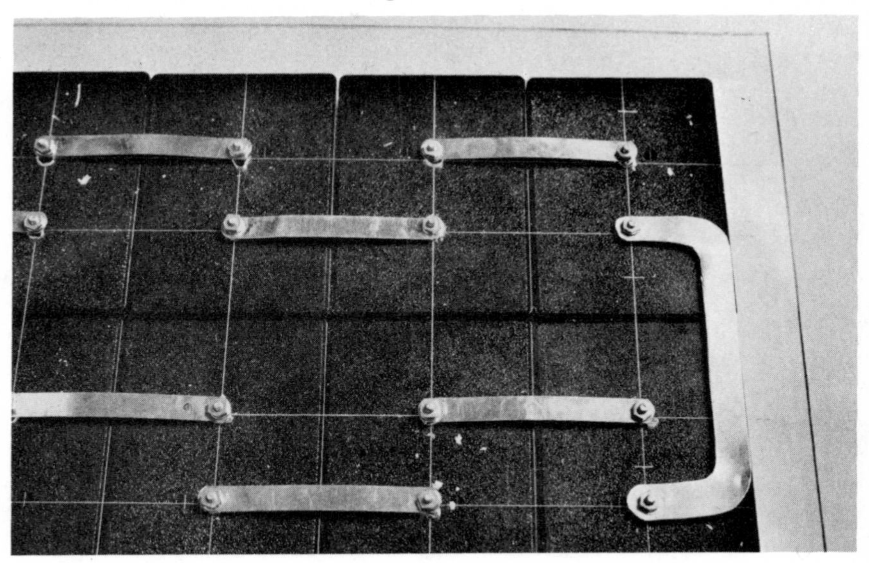

Photo 19–2—Make special connectors from metal strips in a C shape to connect the rows of PV cells.

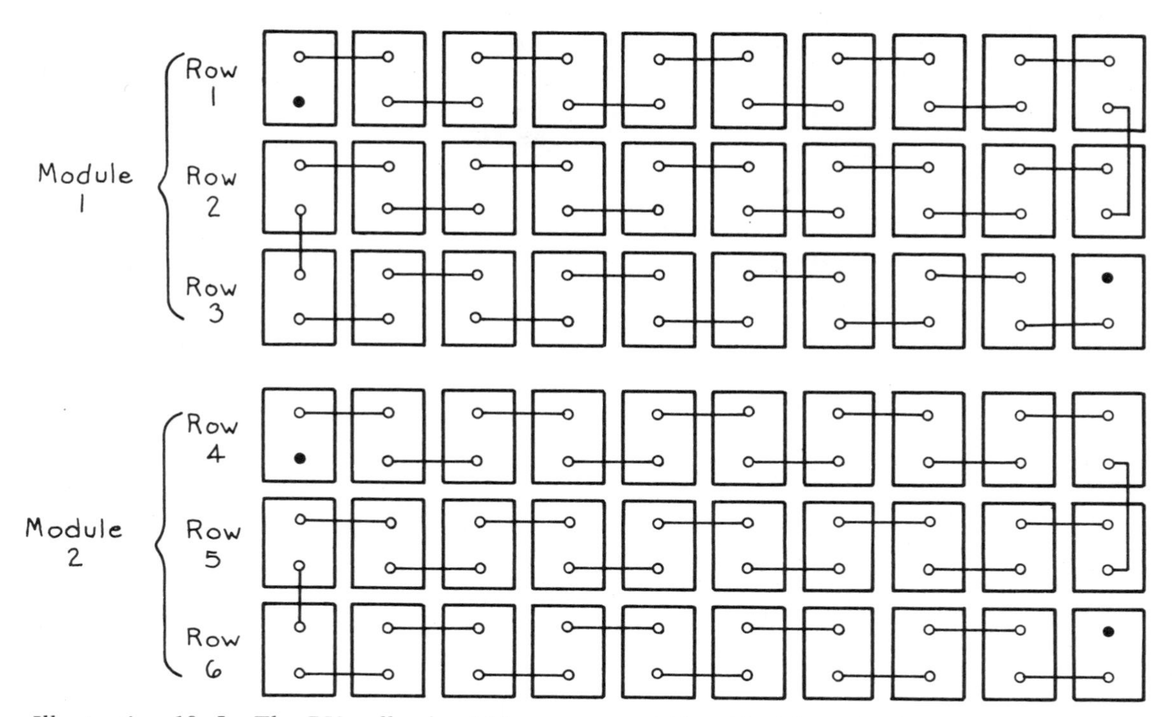

Illustration 19–5—The PV cells should be connected like this. Note the four empty terminals.

1, row 1, to the free negative terminal of the first cell in Module 2, row 4. Connect a 12-inch red wire from the last positive terminal in row 3 of Module 1 to the last positive terminal in row 6 of Module 2.

Now it is time to switch temporarily from electrical to carpentry work to build the array frame and stand. The frame is a pine border that holds the acrylic sheet and allows it to be fastened to the stand. Use 1 × 1¾-inch pieces of pine cut from the 5/4 × 4 pine. Cut 12 inches off the piece of 5/4 × 4 pine for the two side knobs. Use a rip guide on a circular saw or saber saw and cut the remaining length to the 1 × 1¾-inch width. The board can also be cut to size at your lumberyard. From the ripped board, cut two 25-inch pieces for the frame sides. Then cut two 28¼-inch pieces for the top and bottom of the frame. Arrange the four pieces, as shown in illustration 19–6, so that a 1-inch dimension forms the top edge. Miter the ends of the four pieces.

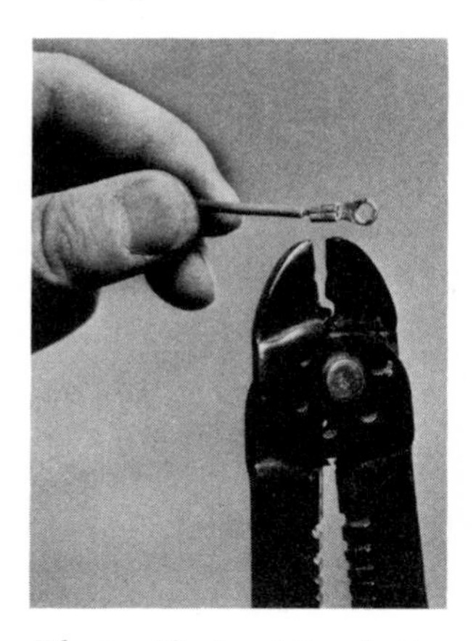

Photo 19–3—Attach the wire terminal ends with pliers or a crimping tool, as shown.

Cut a ⅛-inch-wide by ½-inch-deep dado into the inside face of all four frame members to hold the acrylic sheet. The dado is ⅜ inch from the edge. Use a router with a ⅛-inch bit, or a circular saw with a fine-tooth blade, to cut the dado. Clamp the frame piece to your bench, then clamp one or more boards of the same thickness alongside the board to support the router or saw. When you have cut the ½-inch depth, fit the pieces around the mounting board, cutting more deeply if necessary to have the corners fit together tightly. Also allow 1/16 inch of clearance depth on all sides to compensate for the acrylic sheet's expansion.

The frame is held to its stand by two tee nuts. In each sidepiece, drill a 3/8-inch-diameter hole 15 inches from one end. Center the hole in the 1¾-inch dimension of the sidepiece. Be sure to measure from the same end for both holes. Then, from the inside, counterbore a 3/8-inch-diameter hole ½ inch deep. In each hole, from the inside, insert a 5/16–18 tee nut and tap it in place.

In the back of the frame, again referring to illustration 19–6, cut a rabbet to accept the back. Use a router to cut the rabbet ½ inch wide (approximately half the width of the framing board) and ¼ inch deep (the thickness of the back). Use sandpaper, a rasp, or a router to cut a ¼-inch radius on the inside front edge of each frame piece, as shown in photo 19–4, which gives the frame a more finished look.

Assemble the frame around the acrylic mounting board and fasten it at the corners with one 1½-inch #8 brass flathead wood screw at each corner. Drill 3/32-inch-diameter pilot holes, and fasten the screws parallel with the frame piece. When installed, the back will also give the unit strength. The back of the frame is made of pegboard to provide ventilation, as PV cells work more efficiently at lower temperatures. Measure

Illustration 19–6—Frame details showing the rabbet and dado to hold the back and acrylic sheet and the mounting holes.

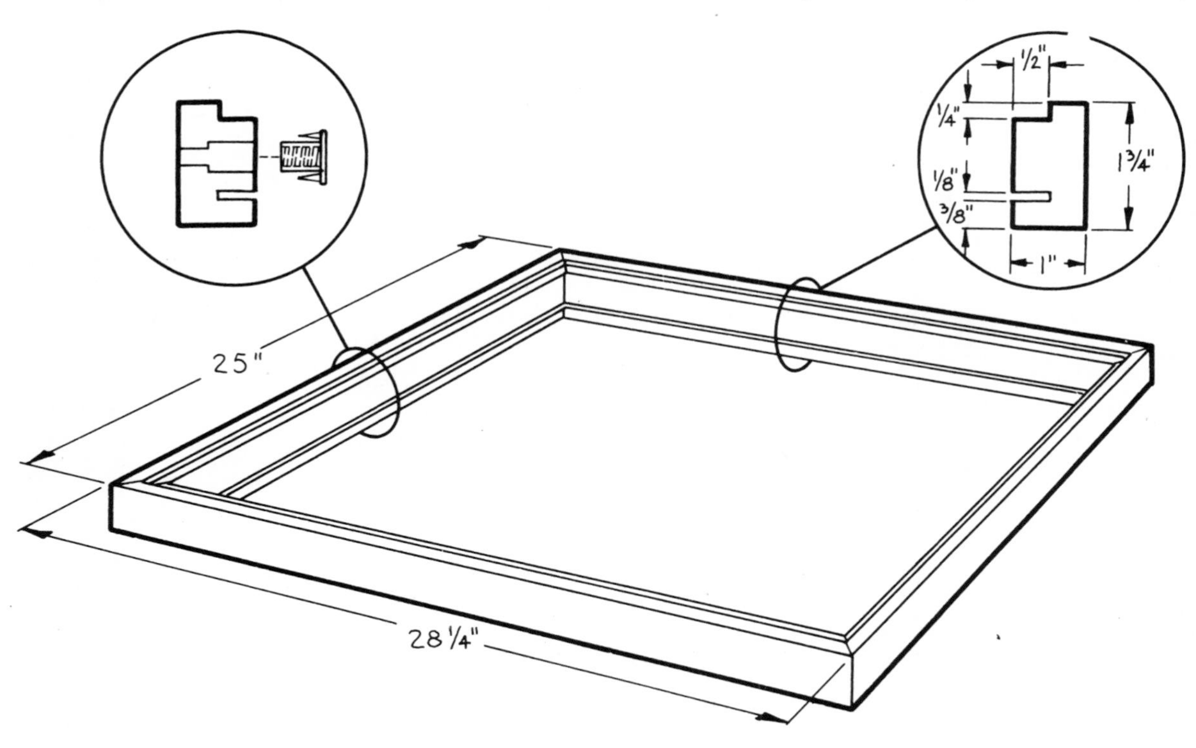

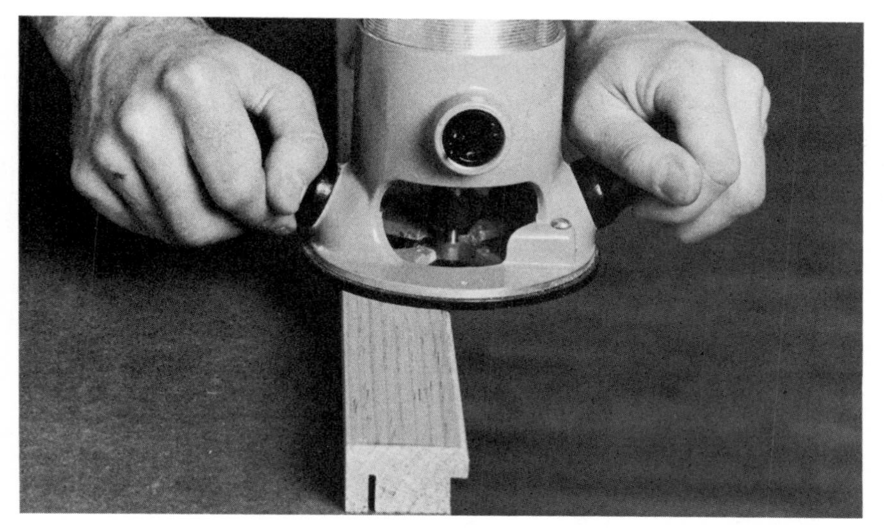

Photo 19–4—A router is the easiest tool to use for cutting the rabbet in the back of the frame pieces and to give the top edge a rounded shape.

the rabbeted opening in the frame, and cut a piece of ¼-inch pegboard approximately 24 × 27¼ inches.

The back holds six phone jacks, which provide the connections for different amounts of electricity from your panel. The phone jacks are held in counterbored holes, as shown in illustration 19–7. Lay out ¾-inch-diameter holes at regular intervals 2½ inches apart and approx-

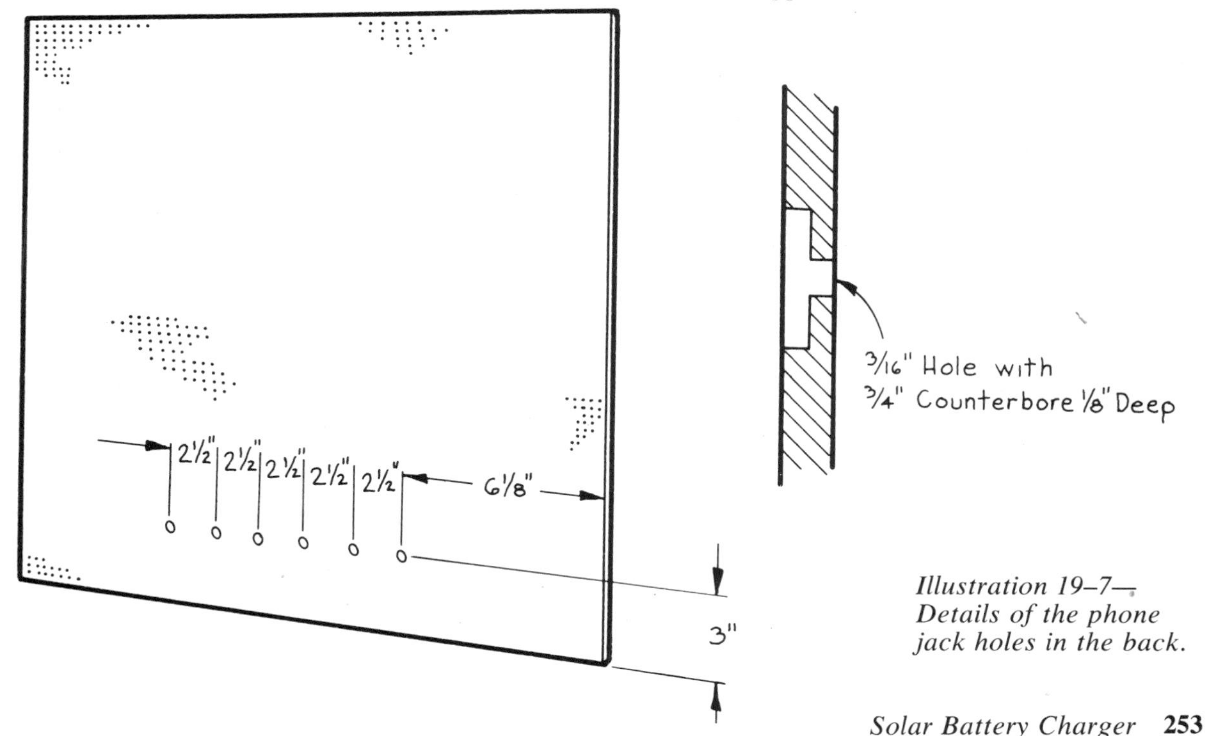

Illustration 19–7—Details of the phone jack holes in the back.

imately 3 inches from the lower edge of the back. Counterbore the holes ⅛ inch deep, then drill a 3/16-inch hole through the back in the center of each counterbored hole.

Complete the electrical wiring by referring to illustration 19–8. The wiring includes six different circuits to provide different amounts of power. Also included is a blocking diode to prevent the current from reversing and possibly damaging the PV cells. Cut eight 12-inch lengths of red wire, which will enable the phone jacks to reach the holes in the back board. Also cut one 12-inch and five 2-inch lengths of black wire.

Start by soldering the blocking diode into the negative, or left-hand, side of the circuit. The diode blocks current from flowing in the wrong direction, and it is necessary to wire the diode properly in order for it to work effectively in preventing damage to the PV cells. Diodes have a banded (anode) side and an unbanded (cathode) side. Current flows only from the unbanded side through the banded side of the diode.

Solder a 12-inch length of black wire to the banded side of the diode. Add a #6 solderless ring connector to the wire and fasten the connector to the first cell in the first row of Module 2, to the same terminal that holds the black wire to the negative end connection of Module 1.

Examine your phone jacks and identify the negative and positive terminals, according to illustration 19–9. Although your jack may look slightly different, it will work in the same way. Number the jacks 1

Illustration 19–8—Circuit connecting details. Note: *Metal connector bands are not shown.*

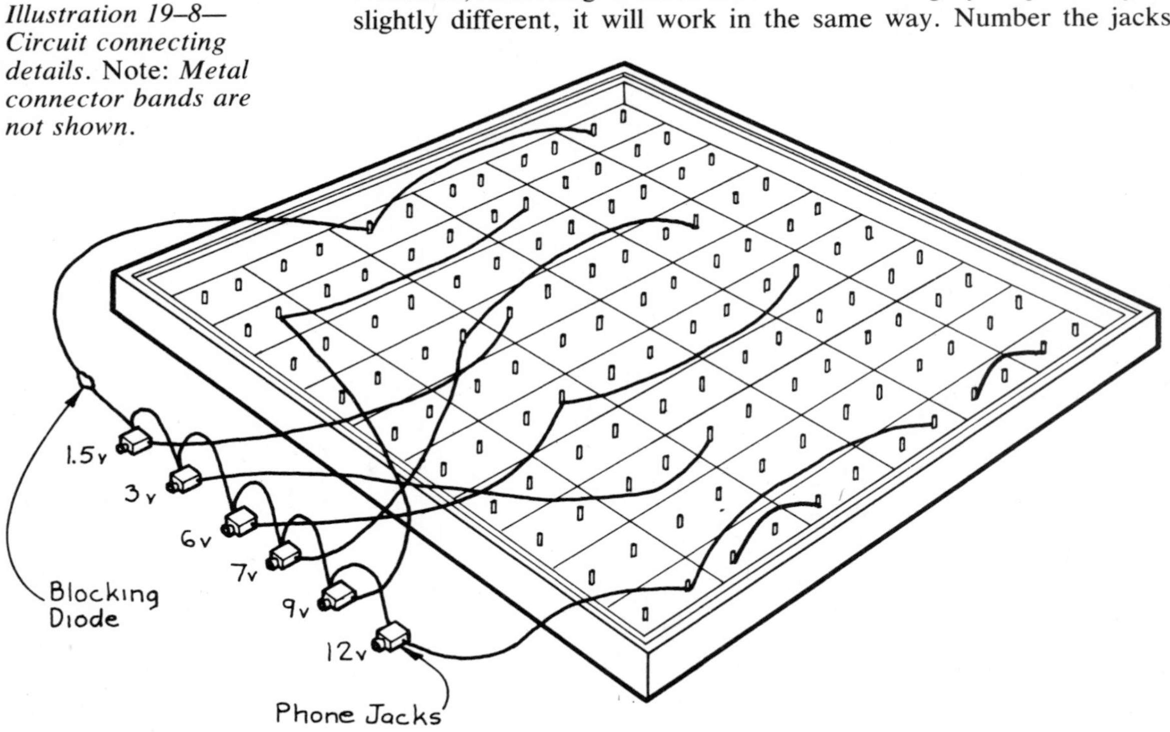

through 6, starting from the left. All connections to the phone jacks must be soldered.

Insert the end of the lead of the blocking diode into the first phone jack's negative connection, and crimp it over. Using solder and a soldering pencil with a fine tip, as shown in photo 19–5, melt a drop of solder onto the connection.

To complete the first circuit, which has 1.5 volts of power and a current of 0.7 amps, connect a red wire from the first phone jack's positive terminal to the positive terminal of the fourth cell in the first row of Module 2, as shown in illustration 19–8. In this way there is a complete circuit in series containing only four PV cells. Each cell delivers about 0.45 volts of electricity, so, since the blocking diode uses a small portion of the power, four cells deliver about 1.5 volts.

The second circuit is composed of four more PV cells from Module 2. Fasten a 2-inch black wire from the negative terminal of the first phone jack to the negative terminal of the second jack. Then connect a 12-inch red wire from the positive terminal of the jack to the positive terminal of cell 8 in Module 2. Since these cells are wired in series, the current stays the same at 0.7 amps, while the voltage is doubled to 3.0 volts.

The four remaining circuits use PV cells from both Module 1 and Module 2. In these cases, since the two modules are wired to each other in parallel, the current doubles to 1.4 amps and the voltage varies according to the number of cells in each circuit.

Circuit three gives you 6 volts, using a total of 30 PV cells. Connect a 2-inch black wire from the negative terminal of the second jack to the negative terminal of the third jack. Add a 12-inch red wire from the third

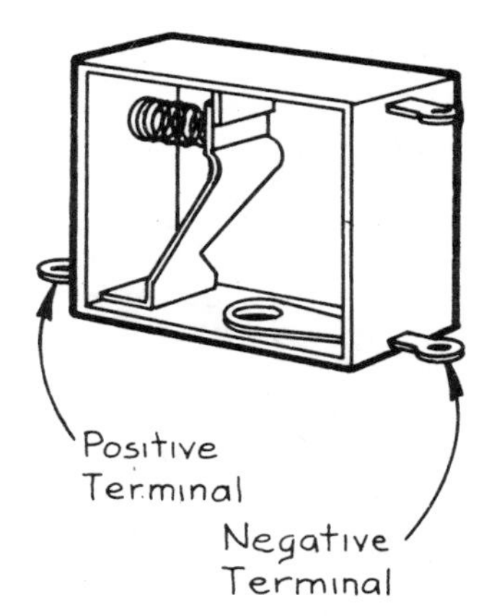

Illustration 19–9—Phone jack.

Photo 19–5—Solder the connections to the phone jacks with a fine-point soldering pencil.

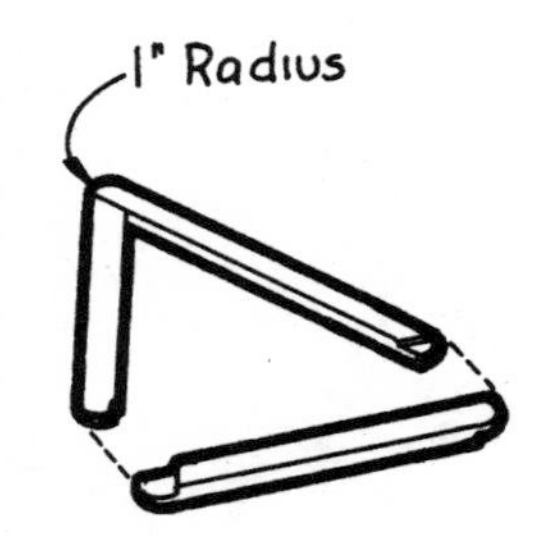

Illustration 19–10—Radius and half-lap details for the stand pieces.

jack's positive terminals to the positive terminal of PV cell 6 of row 2 of Module 2, and a second 12-inch red wire from the cell to the positive terminal of cell 6 in row 2 of Module 1. Refer again to the wiring diagram, illustration 19–8, if this is not clear.

Circuits four, five, and six, giving 7 volts, 9 volts, and 12 volts respectively, all carrying 1.4 amps, are wired in the same way. Circuit four uses 34 PV cells from Modules 1 and 2. Circuit five includes 40 cells from Modules 1 and 2. Circuit six is wired to include all 60 cells in the array. Complete the remaining circuits, using the wiring diagram in illustration 19–8 as a guide.

With the wiring complete, fasten the six phone jacks into the holes prepared in the array back. Then fasten the back to the frame with ¾-inch #6 brass flathead wood screws. Drill 1/16-inch-diameter pilot holes for the screws.

Use sandpaper to remove all sharp edges on the array frame, sand it, and finish it with two coats of polyurethane or enamel. Above each phone jack, mark the correct voltage rating of the outlet.

Construct the triangular array stand, if needed, from 1 × 2s. Cut six pieces, each 21¼ inches long. Lay out the pieces in position, as shown in illustration 19–10, to form 30-degree angled corners. Carefully mark where each board crosses the other. Using a handsaw, cut a half-lap just ⅜ inch deep, or half the thickness of the frame piece. Make several cuts in the waste area, then remove the excess wood with a chisel. Repeat these steps for all ends of the six stand pieces. Using glue and ⅝-inch #6 brass flathead wood screws, fasten the stand sections together. At each corner of the two sections, cut a 1-inch radius with a saber saw, again referring to illustration 19–10.

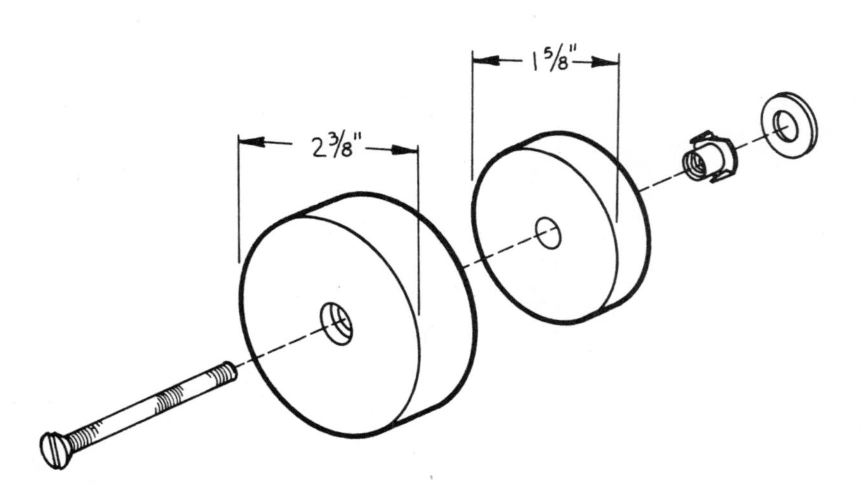

Illustration 19–11—Locking knob assembly details.

Cut the two crosspieces to 29¾ inches from 1 × 2, which will hold the stand together at the bottom. Lay the crosspieces out face down. Place each stand section over them, marking the location of the crosspieces, 2⅝ inches from the outer corners and 1½ inches wide. Cut two notches in the bottom piece of each stand section to fit the crosspieces.

In the center of the joint at the top of the stand section, drill a ⁵⁄₁₆-inch-diameter hole. This will hold the pivot bolt, which is attached to the array frame. Assemble the two triangular units with the crosspieces, using glue and one 1½-inch #8 brass flathead wood screw at each joint. Sand the stand and finish it by applying two coats of polyurethane or enamel.

Construct two knobs to lock the array in place. The knobs are shown in illustration 19–11. Cut two 2½-inch-diameter disks from 5/4 pine and two 1¾-inch-diameter disks from 1 × 2 pine. Drill a ⁵⁄₁₆-inch-diameter hole through the center of the two larger disks. Counterbore one side of the hole ⅝ inch deep and ⅝ inch in diameter to hold the stove bolt head, as shown in illustration 19–12. In the center of the smaller disks, drill a ⅜-inch-diameter hole. Tap a ⁵⁄₁₆–18 tee nut into each of the smaller disks. Join the knobs by gluing a large and a small disk together, with the counterbored hole and the tee nut on the outer sides. Pull them together with a 3-inch ⁵⁄₁₆–18 stove bolt. Sand the knobs, and finish them with polyurethane or enamel.

Using the assembled knobs and two 1¼-inch-inside-diameter by ⁵⁄₁₆-inch-outside-diameter washers, fasten the array to the stand. The stove bolts pass through the top corners of the stand and are secured in the tee nuts in the array frame.

A charging cord, shown in illustration 19–13, is needed to tap the electricity generated by the photovoltaic array. Make the cord from a red and a black alligator clip and a 6-foot piece of clear two-stranded speaker wire. Separate 6 inches of the speaker wire, strip away ½ inch of insulation from each wire, and fasten one alligator clip to each wire, designating the wire with the red clip as positive. Take apart a standard phone jack and put the wires through the casing. Solder the free end of the wire with the red clip to the positive (center) terminal of a phone jack plug and the wire with the black clip to the negative (side) terminal. Reassemble the phone jack.

Aim the solar battery charger directly at the sun for maximum power. Electricity is generated even on a cloudy day, but bright sun is more effective. Plug the phone jack plug into one of the six phone jacks in the back of the solar battery charger, and you can use the sun-generated electrical power for any project requiring direct current of as much as 12 volts. Let your imagination be your guide!

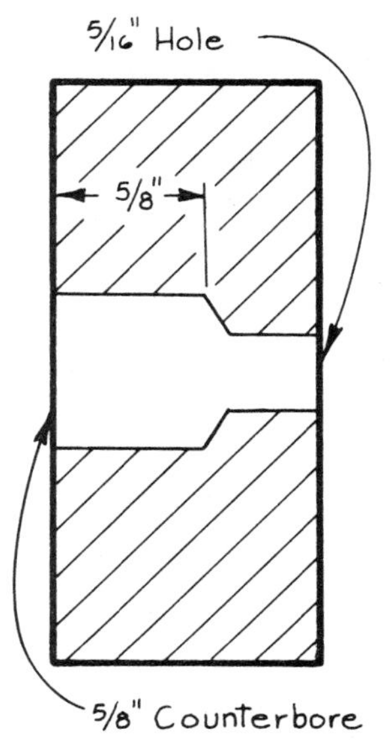

Illustration 19–12—Phone jack mounting hole.

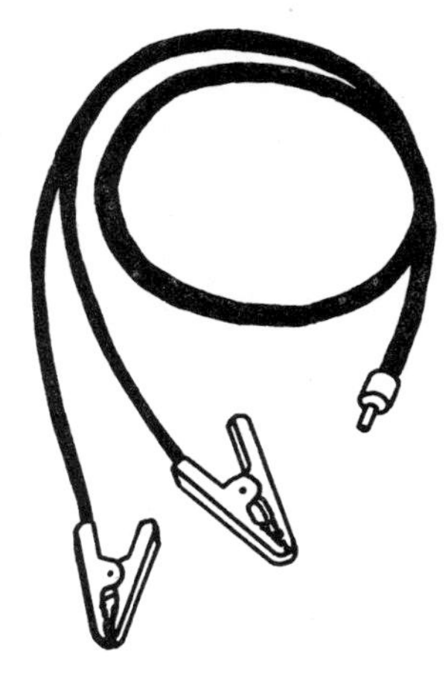

Illustration 19–13—Charger cords.

20 Solar Irrigation System

A small-scale solar-powered irrigation system can be extremely useful. Operating automatically when the sun shines, the small pump can put 8 to 10 gallons of water in your garden while you are away on vacation or busy with other chores. The water is delivered slowly so the plant roots do not become saturated.

The system uses photovoltaic (PV) power to run a low-volume pump. The water is supplied from a rain barrel that incorporates an optional

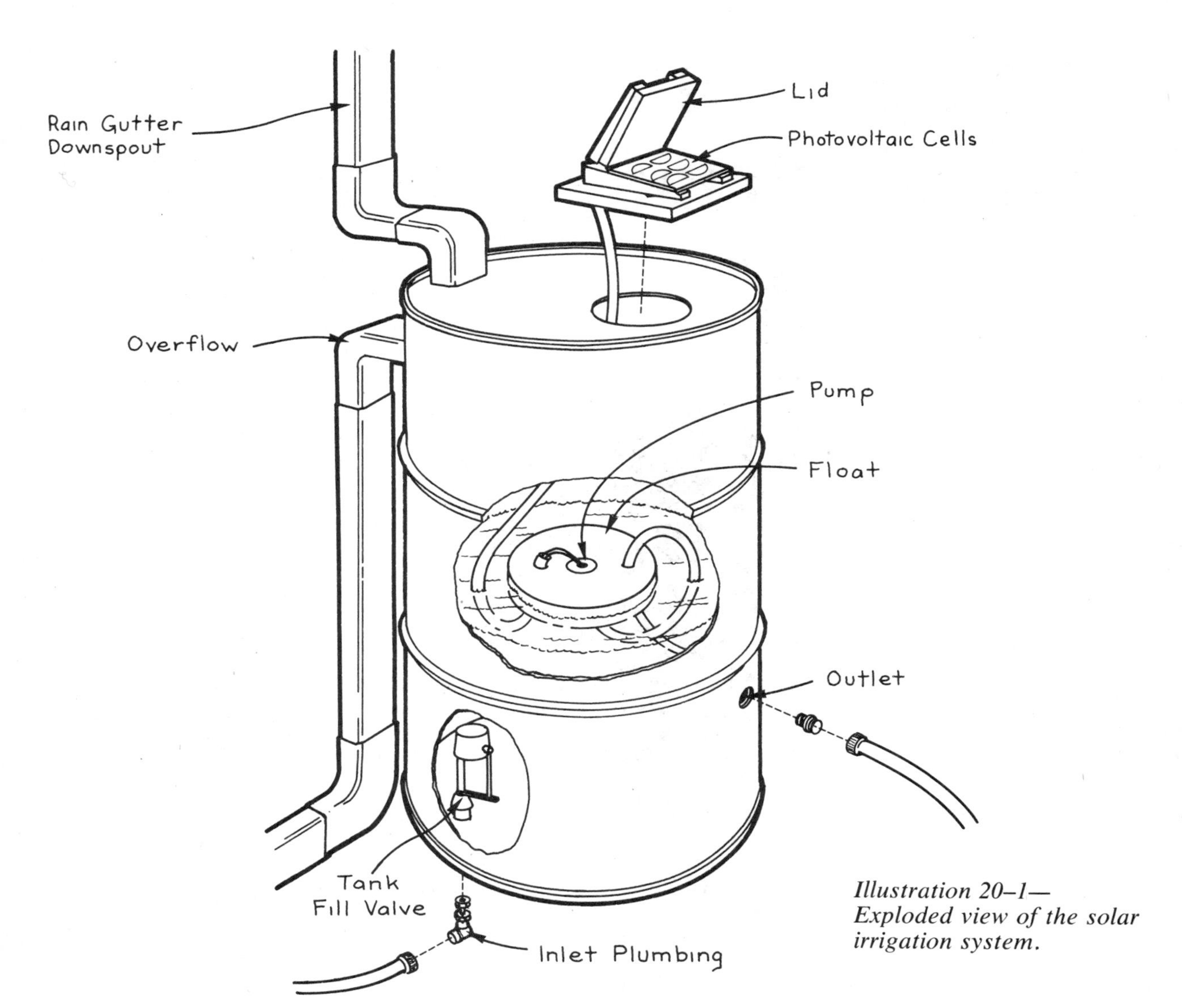

Illustration 20–1— Exploded view of the solar irrigation system.

domestic water backup system to keep the barrel from running dry. The volume of water delivered to the system is controlled directly by the strength of the sun, and this delivers water to the plants when they need it most. On a clear July day, the system will deliver about 10 gallons of water between 10:00 A.M. and 4:00 P.M. For best results, this water should be delivered to well-mulched plants no more than once a week. You can direct water to a different group of plants each day, or you can save water by closing the cover on the PV array. You will need to locate the

CHART 20-1—
Materials

DESCRIPTION	SIZE	AMOUNT
Lumber		
#2 Pine or Mahogany	1 × 10 × 15″	1
#2 Pine	1 × 2 × 26″	1
Hardware		
#10 Flathead Brass Wood Screws	2″	4
#6 Flathead Brass Wood Screws	1″	4
#6 Flathead Brass Wood Screws	¾″	4
Stove Bolts	⅛″ × ½″	2
Aluminum Rivets	⅛″ × ⅛″	6
Brass Hinges	¾″ × 1½″	2
Plumbing Supplies		
Inlet Valve (optional)	···	1
Toilet Tank Fill Valve	···	1
Valve Supply	9″ riser	1
Brass Adapter Elbow	⅜″comp. × ½″ MPT	1
Brass Adapter	½″ FPT × ¾″ FHT	1
Brass Adapter	½″FPT × ¾″MHT	1
Bushing	½″comp. × ⅛″ MPT	1
Teflon Tape	···	1 roll
Gutter Outlet	···	1
Downspout	···	4′
Downspout Elbows	···	3

CHART 20–1—*Continued*

DESCRIPTION	SIZE	AMOUNT
Plumbing Supplies—*continued*		
Flat Washers	¾″	2
Hose Washers	...	2
Plastic Tubing	3⁄16″ I.D. × 10′	1
Electrical Supplies		
Photovoltaic Array*	0.9 watt; 1.5 volts	1
Circulator Pump*	0.6 watt; #RSO,345	1
22-Gauge Insulated Wire	...	10′
Heat-Shrink Tubing	...	6″
Miscellaneous		
55-Gallon Drum with Two-Bung Top	...	1
Concrete Blocks (optional)	...	3
Reflective Mylar	5″ × 5″	1
Styrene Foam	1½″ × 8″	1
Waterproof Contact Cement	...	trace
Primer	...	½ pt.
Rust-Resistant Primer	...	1 qt.
Exterior Enamel	...	½ pt.
Rust-Resistant Paint	...	1 qt.
Silicone Caulk	...	1 tube
60/40 Tin-Lead Flux-Core Solder	...	trace

**The photovoltaic array and pump are available from Edmund Scientific Co., 101 East Gloucester Pike, Barrington, NJ 08007. Similar units could also be used in this system.*

CHART 20–2—
Tools

Tools
Saber Saw
Drill and Bits
Plane
Pop Rivet Tool
Compass
Soldering Pencil

system at a rain gutter downspout for filling the rain barrel, and on a south side of the building to receive maximum sun for the PV cells.

The components of the irrigation system are readily available, and assembly is not difficult. It will provide you with a fine application of solar energy to your gardening needs. Of course, the system can be enlarged by adding more PV cells, a larger pump, and a bigger holding drum.

Illustration 20–1 is an exploded view of the irrigation system. The materials you will need are listed in chart 20–1 and the tools in chart 20–2.

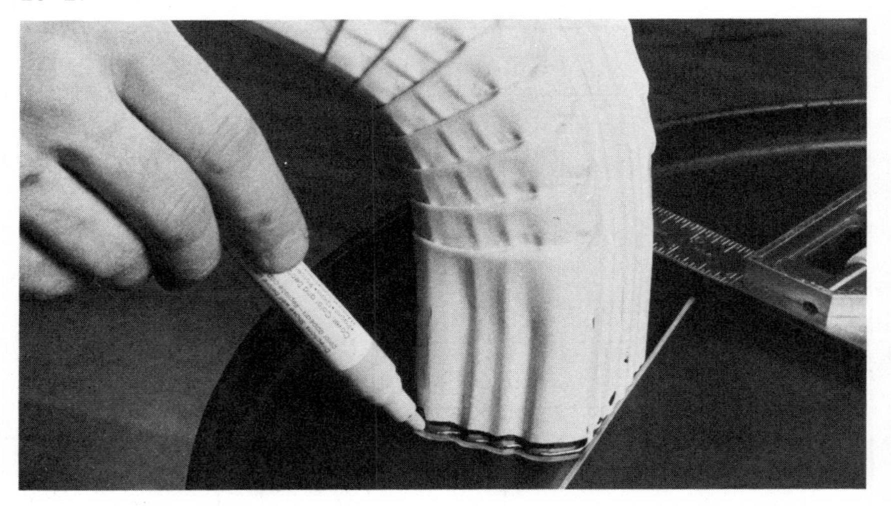

Photo 20–1—Measure 4 inches from the edge of the barrel, then trace around the downspout elbow to mark the intake hole.

Start construction of the irrigation project by obtaining a 55-gallon steel drum. It should be in good condition and watertight. The drum must have one solid end and one with two bungholes. It is necessary to have a stopper that fits the larger hole tightly. The barrel end with bungholes is the bottom, and the solid end of the barrel will be the top for this project.

Cut two openings in the top of the barrel: one to fit the downspout elbow and one as an access port. Measure 4 inches from the edge of the barrel and, as shown in photo 20–1, trace around the downspout elbow. On the opposite side of the top measure 14 inches from the rim, as shown in illustration 20–2. Use the 14-inch point as the center of a 8⅛-inch-diameter circle. Drill a hole within this area near the edge of the circle, and cut out the access port, using a saber saw fitted with a metal-cutting blade. Cut out the downspout opening the same way.

Cut an overflow port on the side of the drum, as shown in illustration 20–3. This port should be in line with the downspout hole so that the overflow pipe will fit between the barrel and the house wall. Cut the port 4 inches on center from the top of the barrel and to fit your downspout piece.

In a direction convenient for your garden, cut a ⅞-inch-diameter outlet hole in the side of the drum, about 9 inches from the bottom. Use a rat-tail file to smooth the barrel opening. Clean the barrel, using a solvent, such as paint thinner, and then soap and water. Paint the barrel with a rust-resistant primer and two coats of rust-resistant paint.

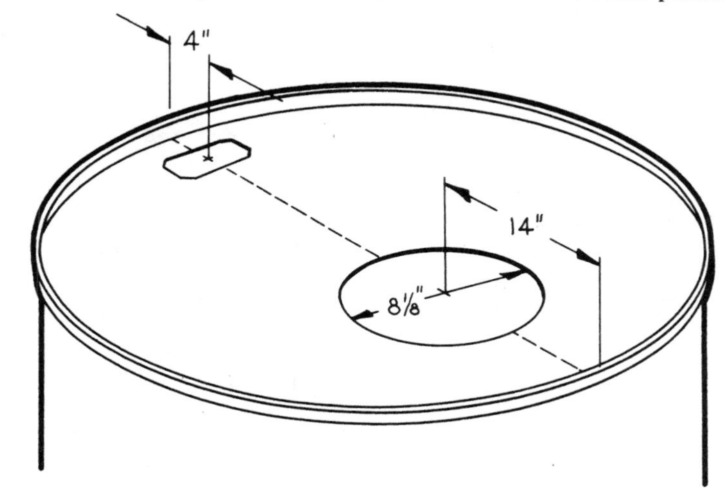

Illustration 20–2—Layout details for the inlet and access port on the top.

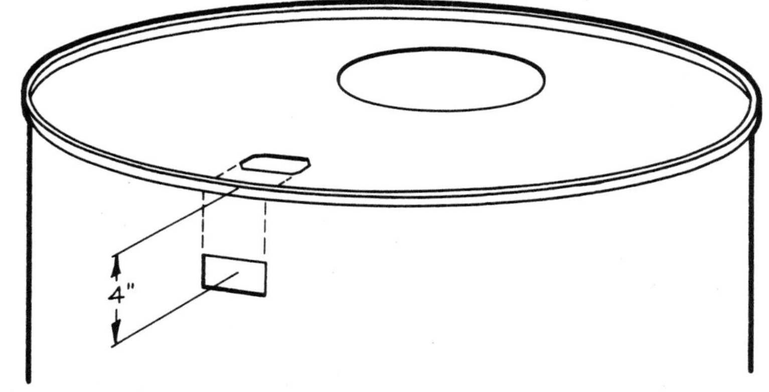

Illustration 20–3—Overflow drain positioning. Note alignment with the inlet.

Photo 20–2—Fasten the gutter outlet in place with bolts to make the overflow port.

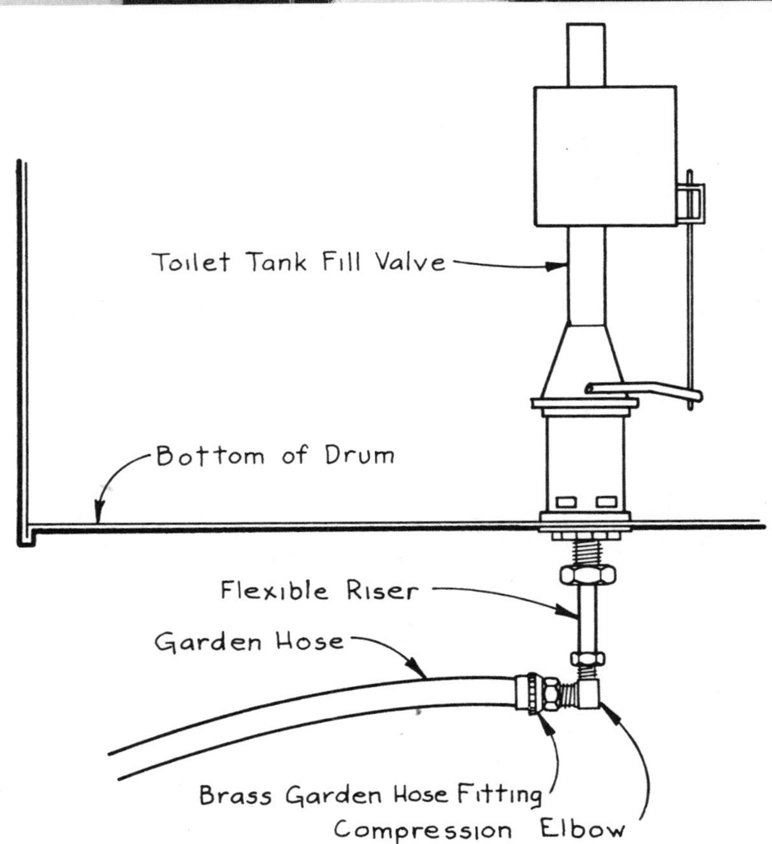

Illustration 20–4—The automatic fill valve assembly. This is an optional addition to the system.

When the paint is dry, fasten a gutter outlet fitting into the overflow port, as shown in photo 20–2. Drill two ⅛-inch-diameter holes, and use ⅛-inch stove bolts ½ inch long through the barrel and the outlet. Bend the lower edge of the outlet down slightly to form a drip edge. Seal the joint between the barrel and gutter outlet with silicone caulk.

It is not necessary to install an intake valve in the barrel if there will be sufficient water from rain, or if you can easily fill it by hand once a week during dry spells. If, however, you might need to leave the irrigation system untended for a period of time, it is useful to have an automatic intake valve. This works by attaching a garden hose to the coupling, turning on the water to the hose, and allowing the intake valve to keep a constant supply of water in the barrel. The valve will open when the water level drops below the float, and will close with about 6 inches of water in the drum.

Turn the barrel onto its side so that you can install the tank fill valve, shown in illustration 20–4. Use Teflon tape on all the pipe fittings to prevent leaks. Use the 1-inch bunghole in the bottom of the barrel and fit the toilet tank fill mechanism into it, following the manufacturer's instructions. Add a piece of ⅜-inch flexible pipe to the assembly, and cut it to a length suitable for the space you will have under the barrel. Then fit a ⅜-inch compression to ½-inch male pipe thread (MPT) adapter elbow to the flexible pipe. Add a ½-inch female pipe thread (FPT) to ¾-inch female hose thread (FHT) adapter to attach a garden hose from a faucet to the inlet valve.

Stand the barrel on three cement blocks, being careful to space the blocks around the inlet plumbing you have just installed. Adjust the barrel's position to fit the downspout elbows. First cut the existing downspout 5½ inches above the top of the barrel. Cut the spouting a second time to remove the 14 inches needed to fit the barrel, as shown in illustration 20–5. Then cut a new length of downspout, about 28 inches, to extend the length of the barrel from the downspout inlet to the bottom. Cut a semicircular notch at the lower end of the downspout to allow water to enter the tank at the bottom. Orient the notch to the rear of the tank. Following illustration 20–5, assemble two downspout elbows onto the house spouting. Adjust the lower end to fit into the top inlet hole of the barrel. Complete the overflow spouting with an elbow and enough spouting to reach the ground. Drill ⅛-inch holes and rivet all the connections with ⅛ × ⅛ inch aluminum rivets.

Next, assemble the outlet plumbing to carry irrigation water to the garden, as shown in illustration 20–6. Slip a ¾-inch flat washer on a ½-inch compression to ⅛-inch pipe bushing, and add a rubber hose washer. Cut a 36-inch length of $\frac{3}{16}$-inch-inside-diameter plastic tubing for the outlet tube. Press one end of the plastic tubing into the bushing, and

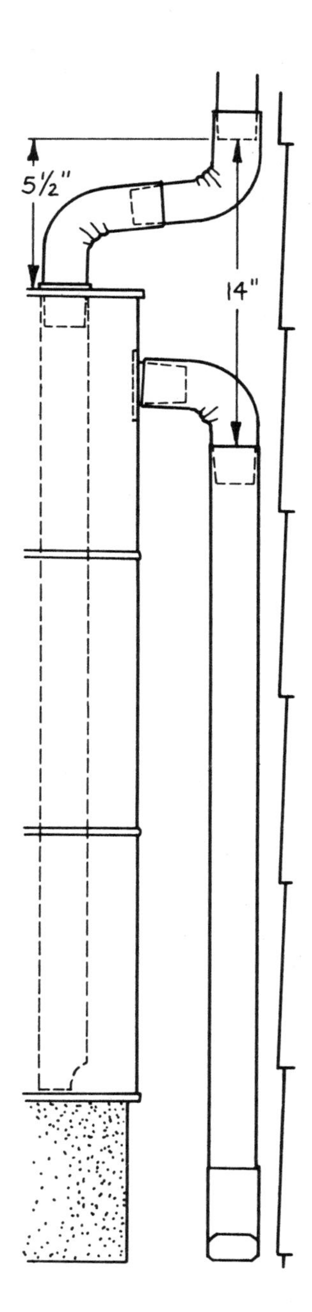

Illustration 20–5—Downspout assembly.

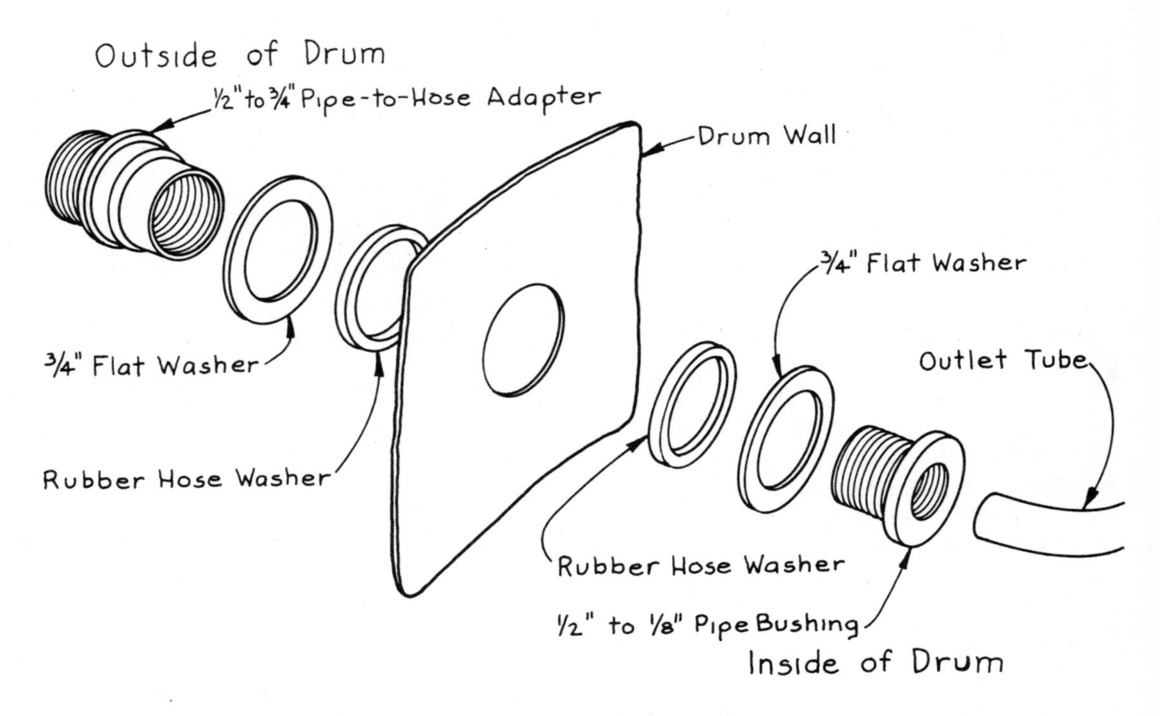

Illustration 20–6—The outlet tube assembly connects to a standard garden hose for water delivery.

silicone-caulk the joint. Install the bushing in the outlet hole from the inside wall of the barrel. On the outside, add another hose washer, a ¾-inch flat washer, and a ½-inch FPT to ¾-inch male hose thread (MHT) adapter. Tighten the entire assembly in place. This outlet may now be used to attach a garden hose leading directly to your thirsty plants.

Make a safety edge on the access port by cutting a piece of 3⁄16-inch-inside-diameter plastic tubing 25½ inches long. Slit the tubing from end to end. As shown in photo 20–3, slide the slit tubing onto the edge of the access port, and fasten it in place with silicone caulk.

Now you are ready to construct the access port lid and cover to hold the PV cells. Although the wood specified is pine or mahogany, pressure-treated wood or cedar are also excellent choices for outdoor use.

Cut the access port cover ¾ × 8¼ × 8¼ inches. The access port cover assembly is shown in illustration 20–7. Cut a piece 5 × 8 inches for a cover insert. From the center of the piece, draw an 8-inch-diameter circle with a compass; the circle should only catch the ends of the piece. Cut the curved ends with a saber saw.

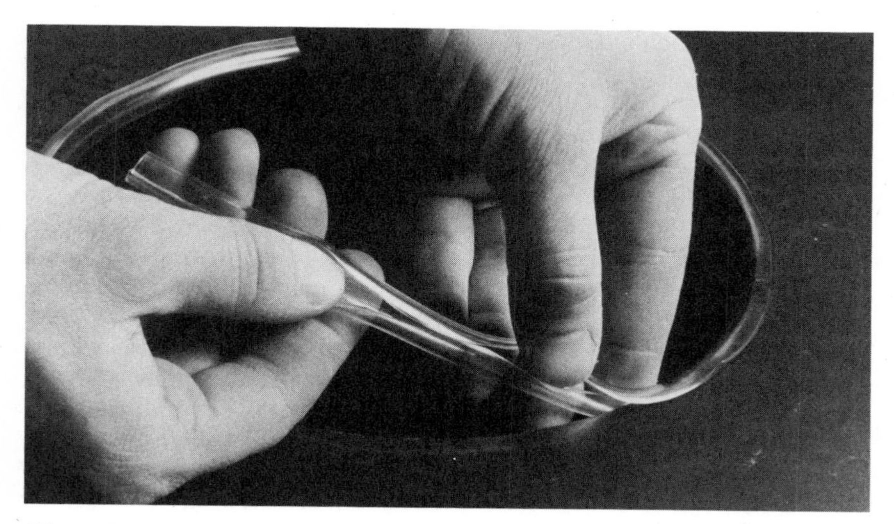

Photo 20–3—Slide slit tubing around the access porthole, and fasten it in place with silicone caulk to make a safety edge.

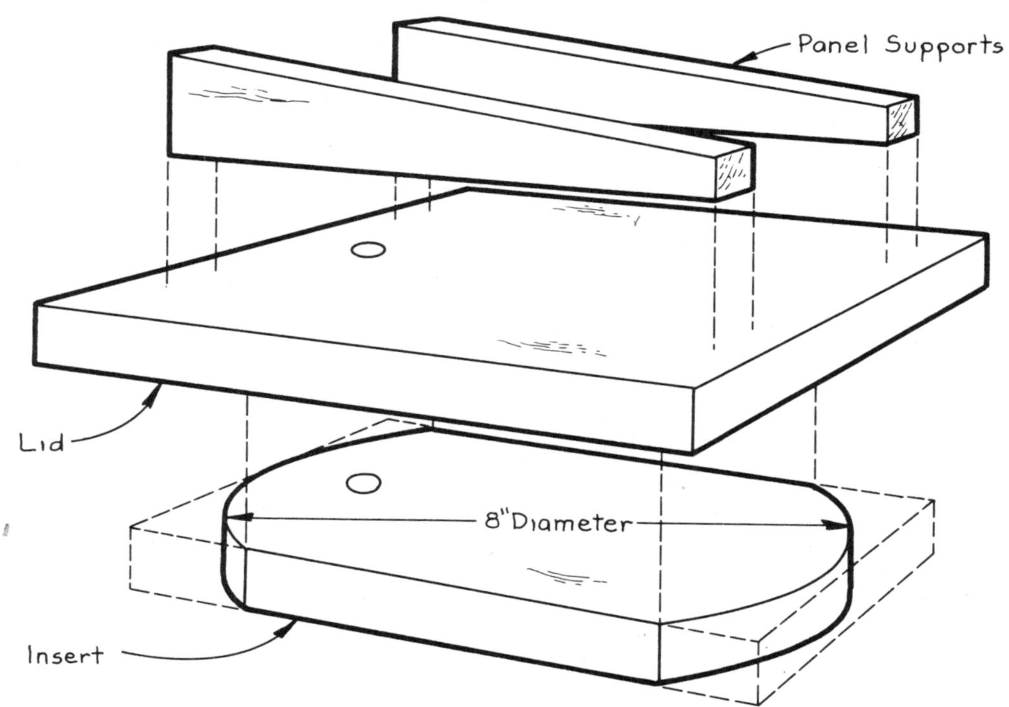

Illustration 20–7—Access port and cover assembly, with tapered panel supports.

Cut two 7-inch pieces of 1 × 2 for the panel supports. Cut the pieces to taper from ¾ inch to 1½ inches to mount the PV cells on a slant. Be sure to measure the 7-inch dimension along the grain of the wood.

Center the insert on the cover and clamp the two pieces together. Drill four ³⁄₁₆-inch shank holes through the insert and through the cover, ⅜ inch from the straightedge of the insert. Countersink the holes on the bottom of the insert. Place the panel supports against the cover, parallel with and directly above the outside edges of the insert, centered over the screw holes. Drill ⅛-inch pilot holes in the supports, and assemble the cover, insert, and panel supports with four 2-inch #10 brass flathead wood screws. Sand the cover unit and round the edges slightly. At one side of the cover, between the high ends of the supports, drill a ¼-inch hole for the wiring.

The next operation is to wire the photovoltaic (PV) array to the motor. Start by splicing 55 inches of 22-gauge insulated wire to each of the PV panel wires. Stagger the splices. Then, solder each splice. To solder, strip each end of the wire, twist them together, apply flux, and use a soldering iron to drop a bead of 60/40 tin-lead flux-core solder around the connection. Slide the heat-shrink tubing to the splice, and heat it with a match, as shown in photo 20–4. Make a waterproof cover for the wires with a piece of ³⁄₁₆-inch-inside-diameter plastic tubing cut to a length of 48 inches. Feed the wires from the panel through the tubing. Pull the splices 6 inches into the tubing. They will be pulled closer to the PV end when the motor splices are pulled into the opposite end. Feed the tubing through the hole in the cover, until only about ¼ inch extends above the cover.

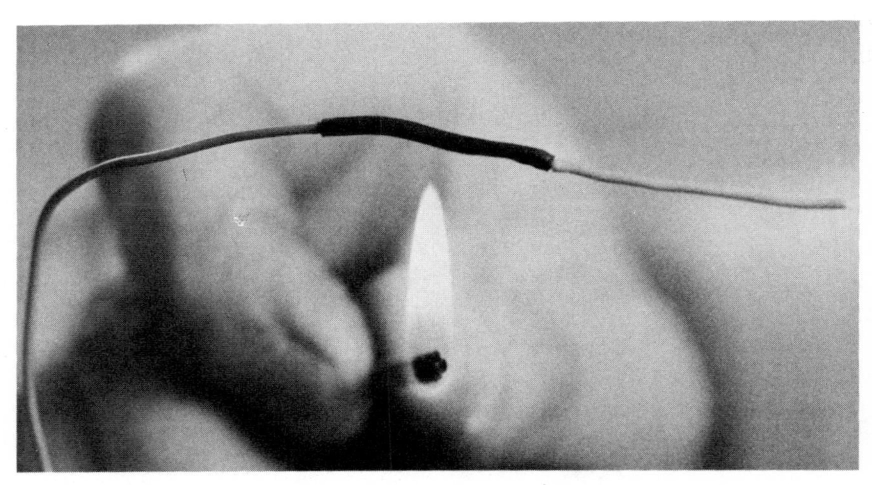

Photo 20–4—Heat the heat-shrink tubing gently with a match, and it will form around the soldered connection.

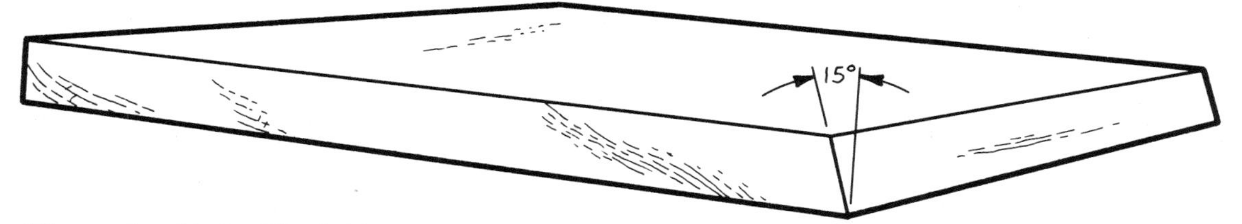

Illustration 20–8—The lid has a 15-degree bevel on one edge.

Set the PV panel on the panel supports, and mark the screw locations. Drill 1/16-inch-diameter pilot holes in the supports and fasten the panel to the supports with four 3/4-inch #6 brass flathead wood screws.

The lid over the PV panel both protects the cells and covers them so that you can switch the pump off even though the sun is shining. The underside of the lid will also be covered with reflective Mylar to intensify the light when your system is in operation. Cut a 5 × 5-inch piece of pine for the lid. On one edge, as shown in illustration 20–8, shape a 15-degree bevel, using a plane. Make the lid braces with two 5-inch pieces of 1 × 2 pine tapered from 3/4 inch to 5/16 inch. Also shape a 15-degree angle on the wide end of each brace. Mount the braces on the top of the lid, as shown in illustration 20–9. The braces are flush with the sides of the lid and flush at the front, which means they will extend past the back edge of the lid slightly. Drill and countersink 1/8-inch-diameter shank holes in the braces, and drill 1/16-inch-diameter pilot holes in the cover. Fasten the braces to the cover with four 1-inch # 6 brass flathead wood screws.

Sand the cover assembly and the lid unit. Finish them with a coat of primer and two coats of exterior enamel. Do not get any enamel on the PV cells.

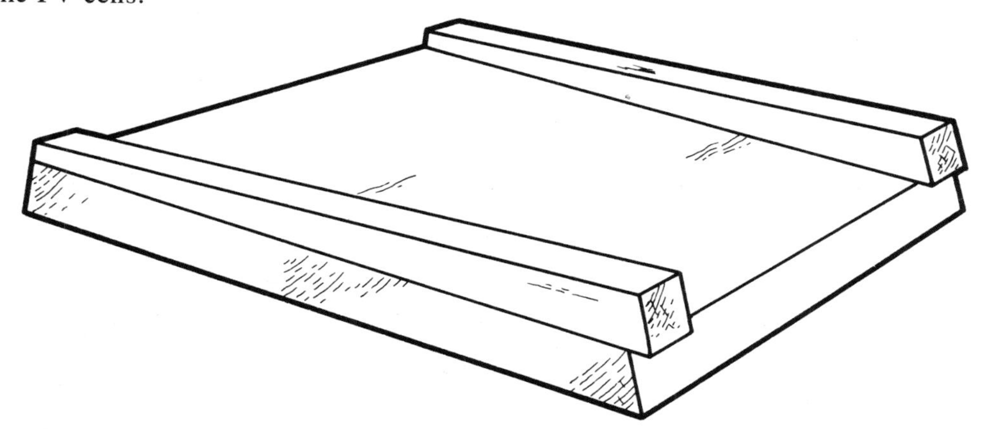

Illustration 20–9—The lid has two braces that overhang the beveled end slightly.

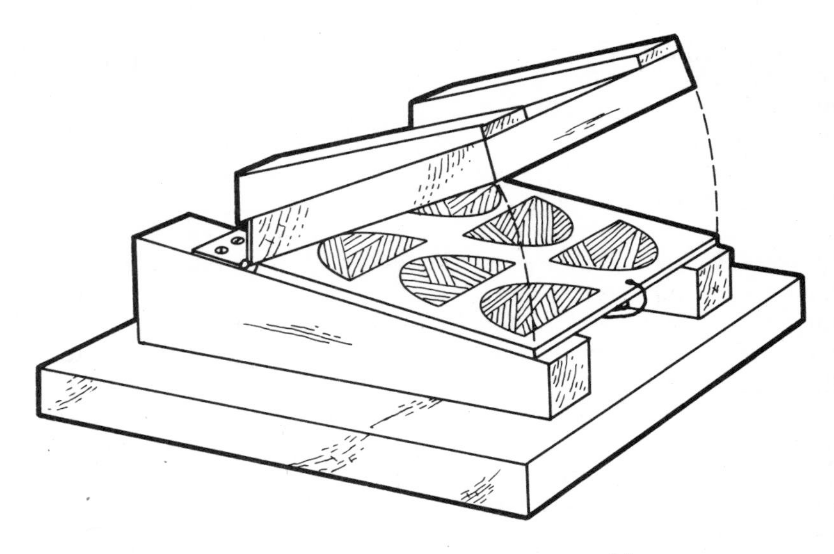

Illustration 20–10—The completed lid/cover assembly.

The reflective backing for the lid is a 5 × 5-inch piece of aluminized Mylar, or similar reflective material. Glue the Mylar to the underside of the lid with waterproof contact cement. When the lid is raised and supported by the back braces, additional sunlight is reflected onto the PV array. Mount the lid on the panel supports with two ¾ × 1½-inch brass hinges. The finished assembly is shown in illustration 20–10.

The pump floats in a piece of styrene foam, shown in illustration 20–11, so that it will follow the water level. Cut the float from 1½-inch-thick foam to a diameter of 8 inches. Through the center of the float, drill a hole the same diameter as the pump. Press the pump into the hole so that the intake opening is below the foam and the wires extend out of the hole on the upper side. Drill two ⅜-inch holes through the float on opposite sides of the pump. The holes will hold the plastic tubing in position and must be snug or caulked snugly around the tubing.

Pull the wires and tubing from the PV assembly through the underside of the float to the top. Connect the wires to the pump motor leads. Again stagger the splices so the bare wires will not touch, and solder the connections. Cover the splices with heat-shrink tubing. Pull the pump splice into the tube. Keep the PV array splices within the tubing. If both sets of splices are not within the tube, cut the wires shorter and resplice them. Seal both ends of the tube with silicone caulk so that the splices are protected from water. Caulk around the joint between the float and the tubing so the tube does not slip.

Refer again to illustration 20–11 and pull the outlet tubing up through the second hole in the float, around the side of the float and into the pump. At the highest point, cut a small nick in the tube to prevent the water from being siphoned out of the rain barrel. This way, the pump alone controls the water flow. Using the pump rather than siphoning regulates the amount of flow.

To operate the irrigation system, either fill your tank with a hose by hand, or wait until it is filled with rainwater. If you have installed the optional fill valve, connect a garden hose to the inlet and turn on the water. The float fill valve will allow about 6 inches of water into the tank. Set the foam float onto the water, and connect a garden hose to the outlet on the side of the drum. Set the PV array and lid unit into the access port on the top of the barrel. On a sunny day, set the cell cover up so that light hits the cells.

The PV system will operate most efficiently if the reflector faces south. You will find that the cells produce the most electricity during periods of bright sun. Be sure to maximize the usefulness of this irrigation system by thoroughly mulching your garden. You will find the pump delivers enough water to keep an area moist but not flooded—about 8 to 10 gallons per day. This is a real benefit, since the pump will operate even if you are not at home.

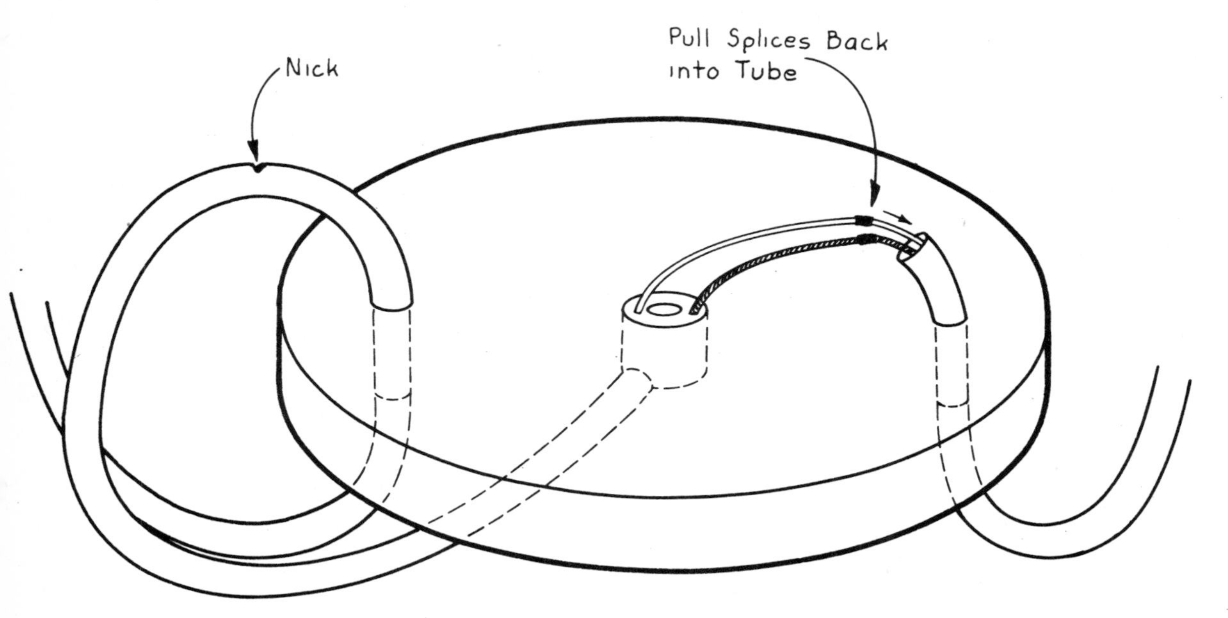

Illustration 20–11—Pump and float assembly details. Note the position of the nick in the tubing to prevent siphoning.